Irrigation Water Management

NIPA® GENX ELECTRONIC RESOURCES & SOLUTIONS P. LTD.
New Delhi-110 034

About the Author

Prof (Dr) A Zaman is an internationally renowned scientist specialization with agricultural water management having spectacular career of 46 years. His lifetime and original research contribution in the field of irrigation water management, rainfed agriculture, cropping system, watershed management, dry land agriculture, agricultural heritage and bio-diversity conservation, to name a few, globally appreciated leading to outstanding teachers in farm universities and agricultural extension services in India and abroad. A large number original concepts, strategies, techniques a tools those developed which have contributed enormously to agricultural science and benefitted the global society. He served as Irrigation Agronomist and formerly Dean, Faculty of Agriculture, Director of Research, Head, Department of Agronomy in Bidhan Chandra KrishiViswavidyalaya [SAU], West Bengal, India. He is a former Chief Scientist in AICRP on Water Management (ICAR) and former Associate Director of Research, Regional Research Station, Old Alluvial Zone (Presently UBKV, Pundibari). He has coordinated, associated and involved with post-graduate teaching programme at the University and has significant research experiences in field of water management and in academia and well as having considerable administrative experiences in different capacity at various time frame. Dr. Zaman worked as Head, Irrigation Expatriate Team, Government of Uganda for 2-year period (1994-1996) under ITEC Programme being deputed by the Government of India to design, lay out and installation of famers managed small scale irrigation for sustainable crop cultivation. He was also designated as IFFCO-Chair Professor. His scholarly publications include more than 200 research papers most of which published in peer-reviewed journals having national and international reputes, 9 book chapters, 7 books with global publishing houses. His academic contribution was profusely appreciated by the International Institute of Bengal and Himalayan Basin patronized by Nobel laureate Charles Townof California, USA where he became co-founder and organized International Conferences in India and Bangladesh bi-annually. He organized and chaired many prestigious international workshops, chaired many technical sessions in scientific meetings and platforms. He has delivered number of invited plenary lectures and key-note addresses and acting as member of Editorial Board members, Examiner and question setter in various agricultural universities in India and Bangladesh. He has 179 publications in the proceedings of national and international seminars and conferences held in the country and abroad in his credit. He guided and supervised 7 (seven) Ph D and at least 50 (fifty) students towards post graduate thesis work. He also worked as Programme Coordinator, West Bengal Accelerated Development of Minor Irrigation Project (World Bank) from 20 September 2014 to 2nd September 2016 after superannuation from the University Service. Prof (Dr) Zaman worked as Emeritus Professor, M S Swaminathan School of Agriculture, CUTM, Paralakhemundi, Odisha since 3rd September 2016 to 31 May 2018.

Prof Zaman was former Professor, In charge and founder of agricultural education in The Neotia University, Kolkata campus at Sarisa, West Bengal from May, 2018 to April, 2020, organized the School of Agricultural Science in the university and introduced the agricultural academic course curriculum in the university from the academic session of 2018-19 in the university. At present, working as the Dean, School of Agricultural Sciences, Sister Nivedita University, Kolkata being founder of agricultural education in the university since joining on and from 1 May 2020.

Irrigation Water Management

A Zaman

Presently Dean, School of Agricultural Sciences
Sister Nivedita University, Kolkata-700 156, India
Former Founder & Professor, School of Agriculture, TNU (2018-2020)
Former Emeritus Professor (CUTM), Odisha (2016-2018)
Former Programme Coordinator, Small Irrigation Project
World Bank (2014-2016), Former Dean & Director of Research
BCKV, Kalyani, Chief Scientist (Water Management)
Professor & Head, Agronomy, BCKV (1982-2014)
Head, Irrigation Expatriate Team, The Government of Uganda Deputed
by the Government of India (1994-1996)
Co-founder & President, International Institute of Bengal Basin
USA Eastern Region, Kolkata

NIPA® GENX ELECTRONIC RESOURCES & SOLUTIONS P. LTD.
New Delhi-110 034

NIPA® GENX ELECTRONIC
RESOURCES & SOLUTIONS P. LTD.
101,103, Vikas Surya Plaza, CU Block
L.S.C. Market, Pitam Pura, New Delhi-110 034
Ph : +91-11-43860225, Mob.: +91 9717133558, 9540816132
E-mail: newindiapublishingagency@gmail.com
Website: www.nipaersources.com

Print ISBN: 978-93-58878-74-5
ebook ISBN: 978-93-58879-44-5

Composed and Designed by NIPA®.

Preface

The sustainable management of water resources is crucial for food security, environmental health, and socio-economic development, particularly in agricultural regions. In line with the 6th Dean Committee of the Indian Council of Agricultural Research (ICAR), which stressed the need for comprehensive undergraduate education in irrigation water management for agricultural students, this book, "Irrigation Water Management," is designed as a foundational text.

This comprehensive treatise covers eighteen chapters, offering a holistic exploration of irrigation water management. It delves into essential aspects, from basic principles to advanced concepts, aiming to equip students with a thorough understanding of water resources, soil-water relationships, irrigation systems, water use efficiencies, and modern approaches like participatory and virtual water management. The book progresses logically from introductory concepts to practical applications, emphasizing sustainable practices, water conservation, and the judicious use of water to meet diverse agricultural demands.

The book begins by establishing the global context and significance of irrigation water management in Chapter 1, followed by an overview of declining water supplies and the history of irrigation in Chapter 2. Crucial for effective strategies, Chapter 3 explores soil properties, while Chapter 4 details the intricate soil-plant-water relationships. Practical water assessment and management are covered in Chapter 5. Furthermore, the book addresses broader land and water management approaches in Chapter 6 and specific insights for maximizing productivity in rainfed lowlands in Chapter 7.

In pivotal chapters focus on practical irrigation management: Chapter 8 guides on optimal irrigation scheduling, Chapter 9 describes various irrigation systems and methods, and Chapter 10 analyzes the economic and environmental performance of these systems, with Chapter 11 detailing the measurement of irrigation water. To address water scarcity challenges, Chapter 12 focuses on soil moisture stress and conservation, while Chapter 13 offers solutions for using problematic water sources and addressing water quality. Chapter 14 explores strategies for managing excess water and preventing issues like waterlogging and salinization.

The book's scope extends to practical applications and contemporary concepts, with Chapter 15 providing tailored strategies for different crops. Chapter 16 integrates the human element through participatory water management, advocating for community involvement. Modern global perspectives are introduced in Chapter

17 with the concept of virtual water, and Chapter 18 showcases advances in irrigation, including smart systems and emerging trends.

"Irrigation Water Management" would serve as a valuable resource for students, educators, and practitioners committed to sustainable irrigation and responsible water resource management. It is believed that well-informed future agricultural professionals will be pivotal in promoting water-efficient and environmentally sustainable systems, ultimately contributing to a water-secure future for agriculture.

This textbook, "Irrigation Water Management," addresses the critical need for efficient and sustainable water use in agriculture, especially in the face of climate change, population growth, and increasing food demands. Designed for undergraduate Agricultural Sciences students, it aligns with the Indian Council of Agricultural Research (ICAR)'s 6th Dean Committee recommendations.

The book features 18 chapters structured for a progressive learning experience, starting with foundational concepts like water resources, soil properties, and soil-plant-water relationships. It then moves to practical strategies such as dryland and watershed management, efficient rainfed lowland utilization, and irrigation scheduling.

Key aspects of irrigation are meticulously covered, including application systems, methods, efficiency analysis, and water measurement. The book also delves into water conservation, managing soil moisture stress, irrigation water quality, the use of problematic water, and agricultural/bio-drainage.

Beyond core topics, "Irrigation Water Management" explores crop-specific water needs, participatory approaches, the concept of virtual water, and the latest advancements in irrigation technology. Notably, it includes chapters like "Effective Utilization of Rainfed Lowland," "Agricultural and Bio-drainage," "Participatory Water Management," and "Virtual Water in Agriculture" for a more comprehensive understanding, even if these extended beyond the strict syllabus.

Each chapter builds logically on the previous, with clear explanations, practical examples, and case studies to bridge theory and application. The book aims to be a holistic guide, equipping students with the knowledge and skills to manage agricultural water effectively, foster resilience, and ensure a water-secure future. It aspires to inspire a new generation of agricultural scientists to be responsible stewards of water resources, contributing to national food security and sustainability. The guidelines prescribed in the syllabus of the ICAR for the agricultural students in UG level, the author is duly acknowledged.

Authors

Contents

1

Introduction

Irrigation: definition and objectives; Importance: Function of water for plant growth, water resources and irrigation development for different crops in India.

Introduction

The effect of supplemental water application as irrigation for crop production at its needs during rainless period is well established throughout the world. The plant has life, so it requires water for growth and development to give us proper quality and good yield. Water is required for metabolic activities including solubilization, absorption and translocation on the way of biochemical processes in photosynthesis, transpiration and respiration. India with 2.4% of the world's total area has 16% of the world's population; but has only 4% of the total available fresh water. This clearly indicates the need for water resource development, conservation, and optimum use.

Importance of irrigation water management

The importance of irrigation in the world is well stated "Irrigation in many countries including India is an age-old art–as old as civilization – but for the whole world it is a modern science–science of survival". The pressure of survival and the need for additional food supplies are necessitating a rapid expansion of irrigation throughout the world including India. Production and productivity of agricultural lands is primarily dependent upon the availability of irrigation water to crops. Expansion of irrigation has paid rich dividends in India. fter independence, production registered nearly 4.52 times increase from around 50.8 million tons in 1950-51 to about 230 million tons in 2007-08. This has been mainly attributed to three factors viz., high yielding cereals, expansion of area under irrigation and fertilizer use. Irrigation may be considered as the major factor of modern intensive agriculture as it aids in efficient use of costly fertilizers and other inputs besides meeting crop water needs.

On the other hand, improper irrigation causes wastage of large amounts of water, leach mineral nutrients from the root zone into the deeper layers contaminating groundwater, and impairs the productivity of soil, or yield losses may occur if insufficient water applied. Excessive application of water causes waterlogging and increase high water tables or seepage spots to develop which may be corrected only by the construction of expensive drainage systems. In addition, salts accumulate and an alkali soil may develop. Water for irrigation and other uses is becoming more and more valuable due to increasing cost of irrigation projects and a limited supply of good quality water. Therefore, we must learn how to make optimal use of water to prevent an excessive wastage to preclude the degradation of the land and bring about its improvement for maximizing water productivity.

Objectives

The objective of irrigation management is different where availability of water is abundant that of limited or restricted water supply. The former is associated with yield increase per unit of land whereas later is to increase crop yield per unit of water use. Water requirement of crops response to water use vary with the different growth stages of the crop. The certain physiological growth stages are more sensitive to water shortage. So particularly where the water supply is limited, it is necessary to take into account the physiological critical growth stages with respect to moisture requirement to get more water use efficiency. If the crop suffers from water shortage to such growth stages, either quality or yield of the crop reduces drastically. Such as in case of maize: vegetative, tasselling and silking at about 20-40 and 50-60 days, respectively, after sowing have been observed as critical in the demand for water. In case of pulses including beans profit comes from irrigation during flowering and podding; as in case of oilseeds, the flowering and fruiting are considered as critical growth stages. If wheat crop suffers from water shortage at crown root initiation stage (21 days after sowing), it will hamper the tillers production, which would give poor quality and yield of crop. So the optimum moisture conditions in the soil profiles are very essential to maintain qualitative and quantitative standard of the crop.

Water Requirement of Crop

The demands for water to the crop usually comes from evaporation and transpiration which is commonly known as consumptive use associated with unavoidable conveyance losses during application and some special needs during land preparation and transplantation.

The supply of water to meet the crop needs usually comes from effective rainfall; irrigation water application and contribution from storage moisture at the soil profiles.

Equation ally, WR= IR+ER+S where, WR=Water requirement; IR= WR-(ER+S)

IR= Irrigation requirement; ER= Effective rainfall & S= Storage moisture in the soil profile.

So to calculate the water requirement of the crop, it is essential to know the existing moisture of the soil on which the crop is growing. It needs the knowledge of rooting habits of a particular crop and procedure to determine the moisture present in the different soil profiles up-to root zone depth. The crops usually extract moisture for essential activities through root system. Water comes up to the tip of the leaves to complete the food production processes and goes down up to roots again.

- The crop which the farmer wants to grow including its rooting habits;
- Physiological critical growth stages of the crop; and
- The existing moisture status soil profiles up-to root zone depth.

Water Balance

It is very important to harvest excess rainwater as run-off during medium to heavy rainfall period for its utilization in crop production during rainless period and development of rainfed agriculture through scientific management. The gains of water would come from total rainfall over the period and application of water as irrigation. The losses constitute a total of run-off (R), downward draining (D), evaporation (E) and transpiration (T). thus the balance sheet shows as 'surplus' and 'deficit' over a particular period and it used to give changed situation of the stored moisture in the soil profiles as well as in the plants.

Equation ally

Gains=P(Rainfall)+I(Irrigation)

Losses=Runoff(R)+Downward drains(D)+ Evaporation(E)+Transpiration(T)

Gains-Losses=S+V where,

S and V are changed moisture status in the soil profiles and plants, respectively.

Water used for crop production

To get proper understanding on water use, it needs experimentations, observations and scientific information on these aspects mentioned above. Only the research findings on water requirement of crops and irrigation water use for crop production would lead to formulate an guidelines on effective utilization of irrigation potential whatsoever it has been created.

The economic and judicious water application was also ensured by these guidelines. So the entire principles of irrigation, its theories and practices, is depended on:

i) How much water to apply? (ie: measurement of irrigation water);

ii) When to apply the supplemental water as irrigation? (ie:proper timings or scheduling of irrigation application; and

iii) How best to apply irrigation? ie: methods of water application to the crop for its sustainable production.

Irrigation Development and Management of Field Crops

Water is one of the most important inputs essential crops. It profoundly influences photosynthesis, respiration, absorption, translocation and utilization of universal nutrients and cell division besides some other processes. Both it's shortage and excess affects the growth and development of the plants, yields and quality of produce. Rainfall is the cheapest and major source of natural water supply. Its distribution is very uneven and uncertain. Artificial water supply through irrigation on one occasion and removal of excess water through drainage on the other becomes imperative. Water management, thus comprises of irrigation and drainage or both. With proper combination of water and soil nutrients the crop yields can be boosted manifold under irrigated agriculture. Water is a costly input when canals supply it misuse of water leads to water logging, salt imbalance etc. rendering agricultural lands unproductive. Proper appreciation of the relationships among soils, crops, climate and water is essential for an efficient and economic use of water resources for maximum crop production.

Sources of Water

There are two sources of water:

1. Surface or canal water; and
2. Groundwater or dug wells and bore wells

Canal Water

This water becomes available from rivers after construction of dams on them and storing it in those dams. In India, rains are received during specific season from June to September called rainy season and in rest of the period there are no rains. This causes the problem of shortage or non-availability of water not only for irrigation but also for drinking. Therefore, artificial storages are created by constructing dams across rivers. The rivers are small to big and accordingly dams of various sizes and capacity are built.

These are mainly classified in three categories viz. 1. Minor 2. Medium and 3. Major or large

1. Minor- Command less than 5000 ha
2. Medium- Command 5000 to 10,000ha
3. Major- Command more than 10,000 ha

Although the command of minor dams is small, their benefits are accrued in short period. The designing, planning and construction of large dams take long time and hence there benefit is available after a very long period. There are controversies over small Vs large dams. Water from these dams is made available for drinking and other human purposes, for industries and for irrigation. Huge capital operating and administrative costs are involved in making dam water available for above-mentioned purposes. Therefore, canal water is a costly input. Few dams were constructed during British Period before Independence. But looking to the pressing needs for water large scale programmers of development of river dams were undertaken on priority basis. By now thousands of dams have been constructed, thousands of Crores of rupees have been spent and several lakhs of hectares of land has been brought under irrigation, largest ever project on interlinking of rivers for diverting water from surplus areas to deficit ones was taken under consideration besides providing water for drinking in cities and towns and for industrial other uses.

Well Water

Well is traditional source of water. Wells are mostly constructed by individual farmers from their own funds or bank loans as it does not require much technical know-how. Wells can be completed in a short period and their benefits are available immediately. The farmers have full control on the use of waste for crops. They decide what crops to grow, when to irrigate and how much water to apply. There is generally no misuse or excess use of well water and hence no spoilage of lands. The command of each well is very small (5 to 10 ha) but the number of wells is very large. Therefore total area irrigated by wells is substantial. Groundwater recharging and its availability in wells are uncertain due to erratic and uneven distribution of annual rainfall. But since farmers have control on use of water, delicate crops or crops requiring frequent watering but less water at a time, are grown on well water. Fruit, vegetables' and differences in cropping patterns under canal and flower crops are commonly grown on well water. So, there are differences in cropping patterns under canal and well irrigations.

Water Requirements of Crops

Although water is essential for crop growth, its requirement for each crop is definite. It may change according to the season (*Kharif, Rabi* or Summer) or according to soil type (light, medium and heavy) but net requirement may not change. The total water requirements, critical stages of water application and number of irrigation turns through which this water is to be given have been decided by experimentation.

Soil and water

Soil water is of interest because of its influences on plant growth and crop production. Water is needed for many growth functions of plants/crops starting from germination, sometime even from land preparation, to ripening/maturity in crop cultivation. Soils serve as storage reservoir for water for plant use. Water is stored in pore space, which constitutes about 1/3rd to more than half of the soil volume.

All the field crop needs soil, water, air and sunlight to grow. The soil provide anchorage for plant roots and also serves as storage for water and nutrients which plants absorb through their roots. The soil voids-atmosphere continuum also provides the air the plants to breath. Crops have to absorb water from the soil in order to live and grow. Part of the water that absorb is used for cooling effect. That converted into vapour subsequently and escapes into atmosphere through plant leaves in the process called Transpiration. Water can also escape directly from the soil or open water surface into atmosphere, called Evaporation. Evaporation and transpiration are influenced by the same process of dissipation the heat load on plants and their surroundings. Hence, they are jointly referred as Evapo-Transpiration, is index of crop water needs. The demand of water for growing plants neither solely depended on soil nor crop that is depended as well atmospheric continuum. The state of demand for water is influenced by the climatic factors also that are solar radiation, relative humidity, wind speed and temperature where the supply of water is mainly governed by the soil moisture regime. So the water requirement of crops is based on Soil-Plant-Atmospheric Continuum (SPAC).

Soil Moisture

The proportion of the soil volume occupied by the liquid is called moisture content. Some other terms like soil moisture conditions or soil wetness are also used to describe the situation. This indicates the amount of water present in the soil. This can be expressed by weight basis or volume basis or depth of water per unit depth of soil.

It is ratio of mass of water to mass of soil solid and referred as gravimetric water content. The ratio of volume of water to the total soil volume is volumetric water content. Gravimetric water content can be converted to volumetric multiplying the former by bulk density or apparent specific gravity.

Example: When an amount of 150 mm of water is present in depth of one metre of soil, soil moisture content is 100 mm/m. In volumetric method, this is 15% by volume.

Role of Water in Crops

1. Helps in nutrient availability to the plants;
2. Helps to absorb through roots used for cooling;
3. Helps in converted in to vapour and escapes into atmosphere through plant leaves, the process is called transpiration;
4. Water can also escapes directly either from soil surface or open water surface into atmosphere, process is called evaporation;

Evapo-transpiration: measure of crop water needs; evaporation & transpiration are influenced by the process of dissipation the heat load of the plants and the surroundings.

The state of demand for water is influenced by climatic factors (i) Solar radiation (ii) Relative Humidity (iii) Wind Speed and (iv) air temperature;

The state of water supply is governed by soil moisture regimes. Low evaporative demand along with unlimited soil moisture supply, the control to atmosphere and ET at its potential rate;

High evaporative demand and low moisture supply, the control to the potential rate will be less. Under intermediate conditions, the control may be either from soil, the plant or the atmosphere;

Water Used

To get proper understanding on water use, it needs experimentations, observations and scientific information on these aspects mentioned above. Only the research findings on water requirement of crops and irrigation water use for crop production would lead to formulate an guidelines on effective utilization of irrigation potential whatsoever it has been created.

The economic and judicious water application was also ensured by these guidelines. So the entire principles of irrigation, its theories and practices, is depended on:

iv) How much water to apply? (ie: measurement of irrigation water);

v) When to apply the supplemental water as irrigation? (ie:proper timings or scheduling of irrigation application; and

vi) How best to apply irrigation? ie: methods of water application to the crop for its sustainable production.

References

Gupta, V. K., & Kumar, S. (2018). Water Resources Development in India: Challenges and Opportunities. Journal of Agricultural Science, 10(2), 456-472.

Hoekstra, A. Y., & Mekonnen, M. M. (2012). A Global Assessment of the Water Footprint of Farm Animal Products. Ecosystems, 15(3), 401-415.

Jha, A. K., & Bhattacharya, P. (2021). Integrated Water Resource Management for Sustainable Agriculture in India. Agriculture and Water Management, 248, 106704.

Kumar, S., & Khanna, P. (2019). The Role of Irrigation in Indian Agriculture: A Comprehensive Review. Indian Journal of Agricultural Economics, 74(3), 181-203.

Mishra, R. K., & Singh, A. (2020). Irrigation Management: Concepts, Tools and Applications. New Delhi: Wiley India.

National Commission on Agriculture. (1976). Report on Irrigation, Water Management and Drainage. New Delhi: Government of India.

Ramesh, V. (2014). Economic Analysis of Irrigation and Agricultural Development in India. Indian Journal of Economics and Development, 10(1), 68-76.

Rao, K. S. (2015). Challenges and Innovations in Indian Irrigation: A Policy Framework. Environmental Science & Policy, 55, 197-208.

Saleth, R. M., & Dinar, A. (2005). Water Institutional Reforms: Theory and Practice. Water Policy, 7(1), 1-14.

Singh, B., & Sharma, R. (2017). Function of Water for Plant Growth and Crop Production. Journal of Plant Nutrition, 40(18), 2128-2145.

Multiple Choice Questions

1. What are the main components of water gains in crop production?
   a) Runoff and Evaporation
   b) Rainfall and Irrigation
   c) Transpiration and Downward drains
   d) Soil moisture and Air
2. What is the formula for calculating Irrigation Requirement (IR)?
   a) IR = WR + (ER + S) b) IR = WR - (ER + S)
   c) IR = WR * (ER + S) d) IR = WR / (ER + S)
3. What is the primary factor that influences the production and productivity of agricultural lands?
   a) Soil type b) Irrigation water availability
   c) Fertilizer use d) Crop rotation
4. What is the traditional source of water that farmers have full control over?
   a) Canal water b) Rainwater
   c) Well water d) River water

5. What is the main purpose of harvesting excess rainwater during heavy rainfall periods?
   a) To increase soil erosion
   b) To utilize it in crop production during rainless periods
   c) To reduce groundwater levels
   d) To improve irrigation efficiency
6. What is the formula for calculating the water requirement (WR) of crops?
   a) WR = IR + ER + S b) WR = IR - (ER + S)
   c) WR = ER + S d) WR = IR + S
7. What are the main components of water losses in crop production?
   a) Runoff, Downward drains, Evaporation, Transpiration
   b) Rainfall, Irrigation, Evaporation
   c) Soil moisture, Air, Sunlight
   d) Irrigation, Runoff, Downward drains
8. What is the primary benefit of well water for farmers?
   a) Immediate availability of benefits
   b) High cost of operation
   c) Requires technical know-how
   d) Limited control over water use
9. What is the main purpose of irrigation in agriculture?
   a) To reduce soil moisture
   b) To provide water for crop production
   c) To increase evaporation
   d) To eliminate the need for rainfall
10. What is the classification of minor dams based on command area?
   a) Less than 5000 hectares b) 5000 to 10,000 hectares
   c) More than 10,000 hectares d) No specific classification

True or False Questions

1. Irrigation is considered a major factor of modern intensive agriculture.
   - True - False
2. Well water is a costly input compared to canal water.
   - True - False

3. Evaporation and transpiration are jointly referred to as Evapo-Transpiration.
   - True
   - False
4. The command of minor dams is more than 10,000 hectares.
   - True
   - False
5. Water management includes both irrigation and drainage.
   - True
   - False
6. Irrigation is considered a major factor of modern intensive agriculture.
   - True
   - False
7. Well water is a costly input compared to canal water.
   - True
   - False
8. Evaporation & transpiration are jointly referred to as Evapo-Transpiration.
   - True
   - False
9. The command of minor dams is more than 10,000 hectares.
   - True
   - False
10. Water management includes both irrigation and drainage.
    - True
    - False

Short Questions

1. What are the three main factors attributed to the increase in agricultural production in India after independence?
2. Explain the role of water in crop production and its importance for plant growth.
3. What are the three categories of dams based on their command area, and how do they differ?
4. What is the significance of proper irrigation scheduling and methods in crop production?
5. Discuss the impact of groundwater recharging on well water availability for irrigation.

Answer Key

1	b	2	b	3	b	4	c	5	b	6	a	7	a
8	a	9		10	a								

1	Ture	2	False	3	Ture	4	False	5	Ture	6	Ture	7	False
8	Ture	9	False	10	Ture								

2

Water Resources and Irrigation Development

Water Resources: Classifications of water: Groundwater, surface water and Virtual Water; Conjunctive use of water in agriculture; Water resources in different states and Stakeholders of water use.

Introduction

Despite fact that more than two-third of the earth's surface is covered with water, currently, near about 450 million people in 29 countries are facing severe water shortage and at least 20 percent more water would be required to feed additional 3 billion populations by 2025. About 97.5 percent of ocean-sea water is not available due to its salinity and 2.49 percent is locked up in ice and only 0.01 percent is technologically available and economically accessible water either from surface or ground water sources for human uses. So water saving technologies developed so far is very pertinent to propagate amongst the users. Water is an essential natural resource for sustaining the environment and supporting life where the agriculture is the major user of water resource. Water is an indispensable for sustainable agricultural growth and development. Previously, rainfed agriculture was associated primarily with conservational land treatments to check soil erosion and land degradation.But in the areas of medium to heavy rainfall, there is ample scope of tapping excess rain water through suitable water harvesting structures constructed for this purpose for its subsequent uses as irrigation or to apply lifesaving irrigation to the crops which should also come under the purview of water management for crop production. Although there is a difference in the objectives of water management where there is abundant availability of water from continuous flow connected with perennial sources or lakes and rivers and that of from restricted or limited supply from collection of excess rain through water harvesting structures.The former is associated with to obtain as much as yield of crop per unit area under water application and later is associated with increasing productivityof crop per unit of water application even under rainwater recycling and its effective management for the purpose of irrigation. Hence, both the rainwater and irrigation water are considered to be equally important to context of agricultural development of the country to get higher water use efficiency and proper utilization of water resources (Zaman, et.al., 2017).

Water Resources

Water is an essential requirement for sustaining the environment and supporting life and agricultural sector is the major user of the water resource. Despite fact that more than two-third of the earth's surface is covered with water, currently, near about 450 million people in 29 countries are facing water shortage and at least 20 percent more water would be required to meet the total water requirement for additional 3 billion populations by 2025. Amongst the total available water about 97.5 percent of is not available due to its salinity of ocean-sea and 2.49 percent is locked up in ice and only 0.01 percent is technologically available and economically accessible water either in form of surface or ground water sources for human uses including water in river, lake and water bodies (Table 2.1). As the water is running out due to indiscriminate uses, water resource stands one of the major crises and a threat of dire consequences resulting from poverty, hunger, eco-system destruction, climatic changes tying together other severe world problems are inevitable. As much as two-thirds of the world population could be water stressed by 2025. In this context, conservation of water for agricultural and industrial sectors has got paramount importance. It is stated that agricultural water management must be integrated with an adequate address to water conservation and transfer of water to uses with higher economic return as well as environmental benefits.

Table 2.1: World water availability pattern

Forms of water	Per cent availability
Sea and Ocean	97.208
Ice and Iceland	2.150
Atmospheric water	0.006
Ground water (Under the ground)	0.625
Surface water (Lake, River, Pond)	0.001

Sources of Water

Precipitation is the main source of water that varied differently within agro-ecological zones. The amount of annual rainfall exhibits certain typical characteristics of distribution. India received about 80% of the total annual rainfall within three-month time in 80% of the area and average annual rainfall comes about 1170 mm in comparison to 800 mm world average. The total rainfall hours as estimated about 200 hours wherein more than 50% occurred within 20-30 hours' time. The wastage of water mainly through run-off became unavoidable. The infiltrated part of rainfall usually converted in to soil water that used by the crops/vegetation, part of which evaporated and rest gone down as ground water or appeared as springs and streams wherein the major part flown as run-off on the land surface. Only about 20 percent of run-off water could be possible to utilize effectively. As a consequence, more that one-third of total area became drought prone in our country.

Status of Groundwater

It is reported that nature's way of storing water is under ground, where about 98% of the entire world's liquid and fresh water occurs. And mostly 2% occurs in streams and lakes, which often are fed by groundwater as a dependable source, less affected by the vagaries of climate. Indiscriminate use of groundwater particularly with introduction of high irrigation responsive photo-insensitive short duration rice varieties throughout the year caused a drastic depletion of the ground water table and posed a complicated problem. Groundwater reservoir was primarily replenished by the annual precipitation received in a particular area. The rate of entering water to ground water reservoir was depended on pattern of rainfall, run-off, stream flow, permeability of the soils and earth materials present prior to reach the water up to the zone of saturation. The total water resource of the country comes about 400 million ham and total amount of surface flow throughout the year in our country is estimated at about 180 Mham, out of which about 105 Mham comes from annual rainfall (Table 2 and 3). The estimated ground water recharge per year is about 67 Mham, out of which about 50 Mham comes from rainfall. Hence there is enough scope to increase the amount of rechargeable ground water adopting ways and means. Introduction high yielding irrigation-responsive varieties of rice and intensive cropping system three to four crops in a year in the same piece of land required huge amount of water. The water was supplied from ground water resources. The deep and shallow tube wells were dug indiscriminately. As results, the depletion of water table occurred to be severe.

In this context, it was very much important to have the access as much as the excess water could be added to underground reservoir during the period high to medium rainfall; the possibilities of occurrence of flood would be drastically reduced. And as the water table would come to rise, the re-cycling of water during the period of drought would become easier and safer. There was clear indication that several soil conservation and water-harvesting measures could increase the ground water recharge up to a certain limit. In India, groundwater recharge from open wells and pits are very common practices in Orissa, Gujarat and Rajasthan. That's why greater exploitation of ground water prior to rainy season provided scope of greater quantity to infiltrate during high rainfall period. The use of surface water would become also to be more effective and helpful to mitigate the arsenic like problems. As the possibility of contamination of arsenic are very less in surface water; only the flowing surface water is the safest source that is free from arsenic. In this context, it is essential to add ground water as much as possible taking all possible measures for its purification (avoiding all sorts of contamination following de-germs chambers to be constructed for the purpose) and taking off water at need without hampering the water table depletion.

Table 2.2: Demand and supply of water resources in India

Demand and supply	**Km^3**
Amount of surface water resources	1880
Amount of usable surface water	690
Amount of usable ground water	418
Total usable water	1108
Water demand in 2002	780
Water demand in 2025 (Estimated)	1200

Table 2.3: Water resources scenario in India

Annual Average Precipitation	**1140 mm**
Total Available water	400 million ha
Net area sown	145 million ha
Gross cropped area	175 million ha
Irrigated area	70 million ha
Water demand for irrigation	46 million ha

India is one of the few countries in the world having abundant water resources as evident from demand and supply of water resources in the country. Water resources demand is lesser than supplies. But water scarcity became evident from frequent occurrence of drought at different locations of the country mainly due to lack of proper conservation and maintenance as well as uneven distribution of annual rainfall. As it was not possible to harvest excess rainwater during heavy to medium peak rainfall period, either that caused the disasters in form of flood or flown to the sea and ocean or followed by drought in the period when there is no possibility of occurring rainfall. The erratic and uneven distribution of annual rainfall coupled with lack of proper water harvesting technologies along with improper management of water resources at different locations of the country is responsible for such situations.

Irrigation Potential

The irrigation potential of the country was reportedly created about 22.6 mha with a food grain production of 50 million metric tons in the year 1951 and that reached 96.9 mha having food grain production of 200 million tons resulting per capita food grain production is about 200 kg/year/person. The gap between irrigation potential whatsoever created at the huge investment of national exchequer and its utilization started from the year 1960, is also becoming wider and that was serious concern of the researchers, policy makers and planners.

Irrigated Agriculture

The effect of supplemental water application as irrigation for crop production as it needs during rainless period is well established throughout the World. The plant has life, so it requires water for growth and development to give us proper quality and good yield. Water is required for metabolic activities including solublization, absorption and translocation on the way of bio-chemical processes in photosynthesis, respiration and transpiration. The objective of irrigation management is different where availability of water is abundant that of limited or restricted water supply. The former is associated with yield increase per unit of land, irrespective of amount of water used for the purpose, where, later is to increase crop yield per unit of water use. Water requirement of crops and crop response to water use vary with different physiological growth stages. The certain growth stages are more sensitive to water stress situation. So particularly, where water supply is scarce or limited, it is necessary to take account of application of water to the critical growth stages with respect to moisture requirement to get higher water use efficiency. If the crop suffers from water shortage during this period, either quality or yield of the crop reduces drastically. The optimum moisture conditions in the soil profiles are very much essential to maintain for getting the qualitative and quantitative standard of the crop output.

Crop Water Requirement

The demand for water to crop usually comes from evaporation and transpiration, which is commonly known as consumptive use, associated with unavoidable conveyance losses during application and some special needs during land preparation and transplanting. The supply of water to meet the crop needs usually come from effective rainfall, irrigation water application and contribution from storage moisture at the soil profile.

Equationally, $WR=IR+ER+S$ where WR= Water Requirement, IR= Irrigation Application, ER= Effective Rainfall and S= Storage moisture in soil profile

So, $IR=WR-(ER+S)$

To calculate the water requirement of the crop, it is essential to know the existing moisture status of the soil at sowing and harvesting on which the crop is growing. It needs the knowledge of root system and rooting habit of the particular crop and procedure to determine the moisture content in the different soil profile up to root zone depth of the crop. The crop usually extracts moisture for essential activities through root system. Water comes up to the tip of the leaves to complete the food production processes and goes down to the root again.

So before going to apply irrigation, it is very important to know

- The crop which the farmer wants to grow with its rooting habits;
- The physiological critical growth stages; and
- The existing moisture status of the soil profiles up to root zone depth.

Water Balance

It is very important to harvest excess rain water as runoff during medium to heavy rainfall period for its utilization in crop production during rainless period and development of rainfed agriculture through scientific management of rain water, particularly in the East African countries. The gains of water would come from total rainfall over period of application of water as irrigation. The losses constitute a total runoff (R), downward draining (D), Evaporation (E) and Transpiration (T). Thus the balance sheet shows as Surplus and Deficit over a particular period and it uses to give a changed situation of stored moisture in the soil profiles as well as in the plants.

Equationally,

Gains = P (rainfall) + I (Irrigation); Losses = Runoff (R)+ Down ward Drain (D) + Evaporation (E) + Transpiration (T); Gains-Losses = S + V where,

S and V are changed moisture status in soil profiles and plants, respectively.

Water Use for Crop Production

To get proper understanding on water use, it needs experimentation and observation on the basis of scientific information on these aspects mentioned above. Only the research findings on water requirement of crops and irrigation water use for crop production would lead to formulate a guide lines on effective utilization of irrigation potential whatsoever it has it has been created. The economic and judicious water application was also ensured by these guidelines. So the entire principles of irrigation, its theories and practices, depends on:

1. How much water to apply? i.e. Measurement of irrigation water;
2. When to apply the supplemental water as irrigation? i.e. proper timings and scheduling of irrigation application; and
3. How best to apply irrigation? i.e. methods of water application to the crop for sustainable production.

Classifications of Water

Rainwater: Surface Water and Groundwater

Primary source of water is rainfall / precipitation. Rainwater is nothing but it is precipitation, harvested water in liquid form. Various methods largely used to collect and store rain such as a roof, land surface or rock catchment. The water is mostly stored in a rain water tank or directed to recharge groundwater by means of channel. Rainwater infiltration is another aspect of rainwater harvesting drawing an important role in storm water management and in the replenishment of the groundwater levels with specific feasible places. Rainwater harvesting practiced for over 4,000 years ago throughout the world, traditionally in arid and semi-arid areas, and has provided drinking water, domestic water and water for livestock and irrigation purposes. Now a day, rainwater harvesting has attained much on significance as a modern, water saving and simple technology around the world.The average annual rainfall in the country is about 1170 mm. This is considerable variation in rain both temporarily and spatially. Most rain falls in the monsoon season (June-September), necessitating the creation of large storages for maximum utilization of the surface run-off. Within any given year, it is possible to have both situations of drought and of floods in the same region. Regional varieties are also extreme, ranging from a low value of 100 mm in Western Rajasthan to over 11,000 mm in Meghalaya in North-Eastern India. Possible changes in rainfall patterns in the coming decade, global warming and climate change and other predicted or observed long-term trends on water availability could affect India's water resources.

Storm Water

Storm water is rainwater that flows across the surfaces, roofs and paved areas like roads, driveways, footpaths or yards, and flows to the storm water system. The storm water system includes street gutters, drains, underground pipes and channels that transport rain water to water ways and ground water. Storm water is not treated to remove some pollutants which irrigate the field.

River Water or Lake Water

Rain water running down according to the slopes ends up merging to form small streams which then channel into a river. Due to the kinetic energy of the moving water, the river develops various land forms through channel making processes. Rivers refer to any natural stream or water ways of water that flows in a channel with defined banks. Lakes refer to any large body of water surrounded by huge area. The basis of assuming rivers and lakes is the hydrological cycle part of rain forming media. The hydrological cycle refers to the movement of water from the atmosphere to Earth's surface and ultimately to the oceans. Specific variations of the relative significance of the various components of the hydrological cycle determine the amount of water held for rivers and lakes. The land surface area that contributes water to a river is known as the drainage basin. Drainage basin lithology and structural fabric determines the type of drainage pattern that a river displays. The ability of a river to transport material also has an impact on river channel pattern. Sediment transport ability is strongly linked to relief; Bloom has calculated that the total energy available for sediment transport is the equivalent of a 7.5 kW engine removing material from 1 km 2 for 24 hours a day, 365 days a year. Rivers are subject to changes in climate and base-level. Floods are shown to be an important part of the normal hydrological, geo-morphological, and ecological functioning of a river it is argued that rivers should be seen as dynamic systems, and that any impact in one part of the system will be experienced in other parts of the system. Lakes are linked to rivers through the hydrological cycle. Lakes have many origins; major lakes are usually formed by glacial erosion or through tectonic activity. Lakes are dynamic, and evolve through time; young lakes tend to be deep, with limited biological activity, while older lakes tend to be shallow with high levels of biodiversity. Since the middle of the twentieth century, dam building has accelerated to the point where over 60% of the world's rivers are now regulated. The impacts of this regulation are widespread, and include detrimental hydrological, ecological, geo-morphological, and socio-economic effects. It is resulted that humans need to develop a reverence for rivers and lakes, and to recognize the inter dependence of all biodiversity.

Ocean Water

The ocean water is dynamic in nature. Its weather parameters like temperature, salinity, density and the external forces like of the sun, moon and the winds influence the current of ocean water. The horizontal and vertical current is common in ocean water bodies. The horizontal motion is known as ocean currents and waves. The vertical motion refers to tides and its result of motion of earth. Ocean currents are the continuous flow of large amount of water in a specific direction while the waves are the horizontal motion of water. Movement of water of ocean is one place to another through ocean currents while the water in the waves does not move. The vertical motion is to the rise and fall of water level in different height at the oceans and seas. The result of the attraction of the sun and the moon according to the specific direction and the ocean water is raised up and falls down twice a day.

Domestic Water

Domestic water is supplied by the municipality, rural and urban corporation for household uses. The good quality source, treated and well infrastructure that provides water to households. The source of supplied water may be a stream, a spring in a bigger way like a piped water supply with tap stand or house connection, or water vendors. When it is private initiation it may be a hand dug well, a bore hole with hand pump, a rainwater collection system. Households use water for many purposes i.e. drinking, cooking, washing hands and body, washing clothes, cleaning cooking utensils, cleaning the house, watering animals and irrigating the garden. The supplied water may be different sources of water for different activities, and the water sources available may change with the seasons and climate change or aberrant weather conditions. Some place it is very hard job to browse the water for dirking purposes and the source may be inadequate, however, it may be far away, difficult to reach, unsafe, or give little quantity of water, making it inaccessible or unavailable. The problems play an important role in people's health and well-being the availability of water is often more important than quality.

Industrial Water

Industrial water is effluent unsafe water and is one of the important sources in the pollution of the water environment. A huge amount of polluted water was discharged into the river, lakes, and coastal areas resulting serious water pollution problems in all types of water environment and caused bad effects to the all the eco system. Industrial water is of no use and it may recycle with after treatment and waste material may cause the serious problem in different levels.

Potable / Drinking Water

Drinking wateris defined as it does not represent any significant risk to health over a lifetime of intake, including much sensitivity that occurs between life stages. The greatest risk of water borne diseases is infants and young children, people who are debilitated or living under unsanitary conditions. Safe drinking water is very much needed for all usual domestic purposes, including personal hygiene. The packaged water and ice intended for human consumption is maintained as per drinking water guidelines. High quality water may be required for some special purposes, such as renal dialysis and cleaning of contact lenses, or for certain purposes in food production and pharmaceutical use etc. Severely immune compromised may need to take additional awareness, such as boiling drinking water, due to their susceptibility to organisms that would not normally be of concern through drinking water.

Agricultural Water

Agricultural water is defined as the edible portion of a crop during growing, harvesting, processing and packing. Water used for farming or packing operations, such as the water used for irrigation of crops, pesticide or fertilizer applications, preventing frost damage or crop dehydration or grain drying and washing or cooling of produce. Agricultural water is a potential way of contamination and by which pathogens can be spread to agricultural produce. Out breaks of *E. coli* and Salmonella point to irrigation water because of contamination in the production environment of fresh fruits, perishable products and vegetables. Water quality should have pure otherwise it dealing with fresh produce that is consumed raw. While the product arrival, it is difficult to separate pathogens from fresh produce and it is depending on the water contamination where from products comes. Many perishable products, fruits and vegetables have rough surfaces that make it difficult to dislodge microorganisms from the surface during washing and cleaning. The enough washing is needed; even though enough of a pathogen left on the contaminated produce which results serious health hazards. Agricultural water is water that is used to grow fresh produce, food grains, fruits, flower, and vegetables, spices and livestock production. The use of agricultural water gives assurance to produce fruits and vegetables and raise livestock. Agricultural water management includes the management of water used in crop production, livestock production and inland fisheries in efficient manner. Efficient and improve agricultural water management; stable and enrich the production system for sustainable water use and food security. Current food production may be doubled in order to meet the food needs of the world by 2050 by using with the latest technology and judicious use of water.

Irrigation Water

Irrigation is the artificial application of water to land for providing the water to the plant. The effective irrigation has an influence in the entire growth process from seedbed preparation, germination, root growth, nutrient utilization, plant growth and re-growth, yield and quality. The prime role to maximizing irrigation efforts is uniformity in irrigation system. The farmer has a sense of control over how much water to supply and when to apply it but the irrigation system determines uniformity. Decision taken up by grower in irrigation systems is best for your operation requires knowledge of equipment, system design, plant species, growth stage, root structure, soil composition and land capability class. Irrigation systems should allow to the plant growth once minimizing salt imbalances, leaf burns, soil erosion, and water loss may happen in the production process. The efficiency of water may hamper due to evaporation, wind drift, run off and water nutrients, leaching and seepage loss.

Sewage Water

Sewage systems are underground pipes and main lines which transport sewage from homes and waste water from industry, to waste water treatment plants. Once the sewage water is treated and the solids removed, the liquid or effluent is discharged to the ocean. Several system of treatment stage have to pass for getting quality water for irrigation purposes and other purposes which is very much needed to combat water crises in near future. Not only that, a large quantity of waste water or sewage water are of no use; by approaching modern water treatment process this kind of huge water may have the use in agriculture production system.

Waste Water

Waste water is the water which has been affected in quality by human use and agricultural use. Waste water after treatment it may use in agricultural activities, urban water use, and sewer inflow and storm water. Waste water from a municipality is also called sewage and precipitation of this is called sludge. Water demands increasing day by day, this will become a much more common occurrence worldwide for many purposes. Waste water is any water used or contaminated in conducting your plan and activities. Waste water from industry may contain pollutants such as litter, sediment, chemicals, oil and petrol and many more. Industry must have appropriate approvals before discharging waste water. It is important not to allow waste water without treating it and before use of it.

Grey Water

Grey water also not in big scale but it has significant contribution for each and every house hold in many times. Waste water are output of clothes washing machines, showers, bath tubs, hand washing, lavatories and sinks are called as grey water. This water generally is not in use except few emergency or option not available to fulfilling the need of that good quality of water.

Virtual Water

The fresh water availability per person having shrunk by two thirds over the past 40 years, a new technology needs to be developed. Some eminent are of the opinion that a number of countries may be required to export water in the form of agricultural goods with virtual water embedded. In some cases, actual water may be exported from one country to another due to water scarcity critical issues. The numbers will rest on those 31 countries which were come down from 62 that are projected to still have water in abundance in 2050.

The ideas of virtual water were given in the year of 1990 and were first defined by Prof. Allan in the year of 1993 as the water is embedded in commodities. Basically, water used to produce agricultural and industrial product is called the virtual water of the product. The third World Water Forum held in Kyoto, Japan, and was marked by the popularity of a phrase that had emerged in the 1990s. The emphasis was given to discussions on global food trade during the decade and added a new dimension to the debate on world water management. The highlighted phrase, "Virtual Water", can be traced to Israeli economists. The term "virtual water" was finally coined at a seminar at the School of Oriental and African Studies (SOAS), University of London in the year of 1993. The idea had been described even earlier by Professor J.A. Allan as embedded water. Water is essential for the production of food such as cereals, vegetables, and meat and dairy products. The amount of water consumed or used in the production process of a product is called the "virtual water" contained in the product. The 'virtual' named, because it is not contained anymore in the product but it is transform only. For easy understand, to produce a kilogram of wheat we need about 1,000 liters of water and for one kilo meat production we need about 5,000 liter of water. Now a day rising global population and increasing water scarcity and human requirements for drinking and sanitation are pushing demand to scarcity levels. Daily requirement of drinking water approximately three liters, people consume "virtual water" in the form of food and fiber that require huge quantity of water for their production. As per United Nations World Water Development (UNWWD) Report, people in developed nations have on average 3,000 liter virtual water consumption per day. The virtual water content of the products is given in Table. This "virtual water trade" is nothing but the virtual water content of the product times the trade volume of that product as mention here for easy understands. The way for the importing country is that it does not have to consume that amount of water in domestically producing the product.

Conjunctive Use of Surface and Groundwater

Groundwater is an important source of irrigation and caters to more than 45% of the total irrigation in India. The contribution of groundwater irrigation to achieve self-sufficiency in food grains production in past three decades is phenomenal. In the coming years, the groundwater utilization is likely to increase manifold for expansion of irrigated agriculture and to achieve national targets of food production. Although the groundwater is annual replenish-able resource, its availability is non-uniform with space and time. Hence, precise estimation and utilization in agriculture sector is a prerequisite for planning its development. In eastern states of India, the ground water development estimated is only 18 % in Orissa, 21 % in Jharkhand, 39 % in Bihar, 20 % in Chhattisgarh and 42 % in West Bengal as compared to the national average of 54%. In these states, rice is dominated and water exhaustive. However the productivity is quite low, though the amount of rainfall received is high as compared to other region of the country.

Groundwater being one of the most valuable resources of earth needs to be utilized properly for healthy and economic growth of any region. Present study on ground water and surface water use was carried out in Eastern part of India, where it is blessed with plenty of water resources but its utilization is very poor as compared to other part of India.

Evaporative Demands of the Field Crops

However the annual crop evaporative demand is 1657.8 mm. This total annual crop evaporative demand (ETo) is more by only 11% over the rain fall received, but water supply through rain is highly erratic during different seasons. So to fulfill crop water demand during off season, it is highly essential to exploit ground water and improve cropping intensity and water productivity.

In coastal area of Orissa, ground water balance model to estimate usable quantity of ground water gas been studied in area of 1066 sq. km (north latitude 21 0 27′ to 21 0 45′ 45″ and south longitude 860 56 ′ 15″ to 87 0 20′ 30″). They considered assessing optimal crop planning with linear model under various soil types, non –saline and saline ground water, irrigated and rainfed situation to make best use of available ground and surface water.

The water balance model developed by them showed that additional water resources available is 400.84 M m3(after withdrawing 255.03 M m3) for further use, is more than present demand due to more recharge from rainfall and base flow from river. In coastal area, though ground water is exploited during off season, it is again recharged with rainwater and not depleted to that extent, as it occurs in other region of the country.

This type of conjunctive use study of multi-sources and multi-quality water is being practiced in India and elsewhere to use the bad quality saline / alkali water for crop production and improve water use efficiency.

Particularly in salt affected soils as well as the sodic water, the crop suffers due to excess salt, osmotic effect and specific ion effect in crop root zone.

So application of such water by blending, alternate use with permissible limit is essential. Such type of irrigation water is available in northern region of India and it is followed in Punjab, Haryana and Rajasthan. In eastern region, only saline water exists but it is not being used intensively as sufficient quantity of good quality water is available for crop production.

Conjunctive Use of Surface and Groundwater

In water management practices conjunctive use is considered important as surface and groundwater are the related systems. Uncertainty in the parameters involved makes it difficult for the decision maker to derive the water use policy and estimate the returns from the system. Such uncertainties are efficiently tackled through fuzzy logic. Scientists have developed a mathematical model for achieving an optimal conjunctive use policy incorporating spatially and temporally distributed groundwater withdrawals and spatially distributed cropping patterns. The groundwater withdrawals were constrained to keep the water table elevations within an appropriate range. Murthy illustrated with a hypothetical example the case of conjunctive use of surface and groundwater to replace surface water. Scientists have developed a stochastic dynamic programming model to derive the optimal operating policy and also a lumped simulation model is used to evaluate the alternative policies for each alternative and a multiple criteria decision making is used to select the most satisfactory alternative plan. A farmers level programme model to derive optimal crop plans for an irrigation system for conjunctive use of surface and groundwater has also been formulated.

Water Conveyance and Irrigation Efficiency

Water from the sources, when it reaches to the crop field as irrigation, it requires any distributory system like watercourses or channels. It involves a loss of water through the distribution system associated with percolation, evaporation and losses occurred by runoff at borders and furrows. So the ratio of actual water that has been utilized as irrigation out of total water delivered from the source is known as irrigation efficiency. The ratio between the water stored in the root zone depth and the total water reached to the crop field is known as water application efficiency. The ratio between water stored in the root zone depth and the actual requirement of water prior to irrigation water application is known as water storage efficiency. These could be transferred and termed as operational or economic efficiency of the system.

The annual global flow of water ranges from 4000 to 4700 million hectare meters (m ha m), of which 1400 m ha is utilizable for irrigation and other purpose. The present level of utilization of available potential is just around 25%, of which irrigation constitutes about three-forth (Harswarup Singh, 1990). As per estimates of the national commission on Agriculture (1976), the average annual rainfall over the total geographical area of India is around 370 m ha m, added another 30 m ha m, which is brought in from catchment area lying outside the country that makes 400 m ha m. The amount is estimated that about 85 m ha m of water soaks in to the soil (contributing to the profile storage/

ground water) and 130 m ha m evaporates in to the atmosphere. The remaining 180 m ha m constitutes the average annual flow in different rivers, of which only about 70 m ha m is utilizable owing to the topographical, hydrological, climatic, soil and other limitations (Harswarup Singh, 1990). If policy decision is taken on groundwater development in command areas, an additional area of about 1 (one) million hectares in Andhra Pradesh, Tamil Nadu, West Bengal, Orissa, Bihar, and Assam could be brought under irrigated agriculture. Wherein *boro*/summer, even crops like groundnut, maize, potato, and linseed could successfully be grown under rainfed lowland ecology after the rainy season rice on the availability groundwater as well as residual moisture. The conjunctive use of surface and groundwater as well as residual moisture including water productivity assessment under rainfed ecology should come under purview of integrated water resource management for sustainable agriculture.

Agriculture accounted for at last 90 % of annual withdrawal of renewable water resource. The factors involving exploitation of natural resource, irrigation is by and far the most important. Out of total 253 m ha irrigated land is the world, China and India alone have over 100 m ha (Fredricksen *et.al.* 1993). In pursued of water demand the competitive supply of irrigation water is declined rapidly due to lack of sufficient budget allocation for infrastructure development and degradation along with lack of proper maintenance of existing infrastructure for irrigation. The over exploitation of groundwater resource are also playing a crucial role as the indiscriminate use of irrigation water from groundwater sources, resulted continuous dropping of water table with possible resultant health and environment hazards effect. The per capita withdrawal of water resource of about 612 m^3, is too high as compared only 298 m^3 in South Korea and 462 m^3 in China, where India, South Korea, China is having renewable water resource availability was 2464, 1452 and 2427 m^3 $year^1$ person respectively (Fredricksen *et. al.*,1993). The per capita availability of water resource declined by 40-60% in most of the Asian countries over 1955-1990 periods (Feder and Keck, 1996).

Out of 400 m ha m annual rainfall of our country, only 50 % is exploitable for uses of various purposes in agriculture, industry, navigation, forestry and power generation wherein about half of water resources are used for irrigation. This leads to an ultimate irrigation potential projected op to 114 m ha, of which only 92 m ha has already been achieved. In Eastern India, many multipurpose irrigation projects were developed due to presence of a number of rivers and due to high rainfall intensity. The groundwater resources are also rich still now that has significantly contributed towards development of irrigated agriculture.

Table 2.4: Source-wise and state-wise net irrigated area in Eastern India (in '000 ha)

State	Source			
	Canals	Tanks	Tube wells	Other source
Assam	362	NA	NA	210
Bihar	1072	117	1794	641
Orissa	949	305	836	NA
West Bengal	717	263	712	219
NEH	28	5	4	240

In Eastern India, only 36 % of the total net cropped area (7% in Mizoram and 49 % in Bihar) has been brought under irrigation exploiting all possible sources. In West Bengal, there are three major irrigation projects operating. These are Damodar Valley (Bankura, Hooghly, Burdwan, and Howrah), Mayurakshi (Birbhum, Mursidabad, Burdwan), and Kangsabati (Bankura, Midnapore, and Purulia). A part form these, Teesta-Torsa and Subarnarekha Barrage projects are also started and under construction. Work on eighteen medium irrigation schemes are continuing in the district of Birbhum, Bankura, Purulia, and Burdwan. In addition, water-harvesting structures, de-siltation basins, run off retention tanks and other measures are being taken into serious consideration. National Watershed Development Project in Rainfed Area (NWDPRA) involved the common farmers in conservation of land and water wherein at least 95 thousand ha of land have been brought under the program. Thus the irrigation potential created up to 45 %or arable land to cover in the state. Through the potential created at the cost of huge exchequers are being underutilized. Indiscriminate use of groundwater particularly with introduction high irrigation responsive photo-insensitive short duration rice varieties growing throughout the year caused a drastic depletion of the ground water table which results to have complicated problems. Groundwater reservoir was primarily replenished by the annual precipitation received in a particular area. The rate of entering water to ground water reservoir was depended on pattern of rainfall, run-off, stream flow, permeability of the soil and earth materials present prior to reach the water up to the zone of saturation. Total amount of surface flow through out of the year in our country is estimate as about 180 m ha m, out of which about 105 m ha m, comes from annual rainfall. An estimate of ground water recharge per year is about 67 m ha m, out of which about 50 m ha m, comes from rainfall. So there was enough scope to increase the amount rechargeable groundwater adopting ways and means. Introduction high yielding irrigation-responsive varieties of rice and intensive cropping system three rate to four crops in a year in the same piece of land required huge amount of water. The water was supplied from groundwater sources. The deep and shallow tube wells were dug indiscriminately. As result, the not only depletion of water table occurred to be severed but also per capita availability of water would come down from 2100 m^3 at present to 1700 m^3 in 2030.

Water (H_2O)

Water is indispensable for human, animals and plant life. It is a part of all organisms, some of which contain more than 90 percent. Water is essential part of protoplasm. It is an important ingredient in photosynthesis. About 400 to 500 liters of water is necessary for production of a one kilogram of plant dry matter. Water is also required for translocation of nutrient and dissipation of heat.

Properties of Water

Water molecule contains two hydrogen ions and one oxygen ions. The space occupied by each water molecule is mainly due to oxygen ions while two hydrogen Ions do not occupy practically any space. The shape of the water molecule is sphere and the position of two hydrogen ions is at the corners of a tetrahedron that exists within a sphere.

The positive valences of hydrogen ions are partially neutralized by negative valency of oxygenion. Thus, one, end of water molecule has positive charge and another end has negative charge. This makes water molecules a dipole. Water molecules do not exist in individually. Hydrogen in water serves as connecting link from one molecule to the other and it is known as hydrogen bonding. Water sticks to itself with great energy & this property is called cohesion, whereas water attaches itself to surface of many substances and this property is known as adhesion. By adhesion, water is held tightly at the soil water interface and water is retained in the soil by adhesion and cohesion. The water molecules hold other water molecules by cohesion forces. Because of these forces, water fills small pores in the soil and is in fairly thick film in large pores.

Hydrological Cycle

The earth's outer solid crust is called Lithosphere. Most of the earth's total water is contained in Oceans (96%), small portion (2%) as snow and ice and rest (2%) in the water bodies of the continents. Oceans, lakes, rivers and other water bodies of the Earth are called Hydrosphere.

Continuous circulation of water between hydrosphere, atmosphere and lithosphere is known as hydrological cycle. This has neither a beginning nor an end.

The physical and biological processes in the environment are sustained by the Hydrological cycle. Water from various water bodies evaporates due to energy provided by solar radiation and enters the atmosphere as water vapour. Oceans account for 85% of worldwide evaporation. Evaporation is the chief source of water vapour in atmosphere. Water vapour in the atmosphere constitutes only 0.001 percent of the total global water. Even if all water vapour present in the atmosphere at any amount could be precipitated, an average depth of only about 2.5 cm of water is added to the oceans. Though the quantity of water vapour present in the atmosphere is small, it provides vital hydrological link between oceans land.

Clouds are formed as the water vapour rises above in to the atmosphere. When condensation takes place in the atmosphere, water precipitates mainly as rain or to some extent as snow. Thus, water is constantly added to the atmosphere through evaporation and lost through precipitation.

The annual average world precipitation is 1000 mm. As Oceans, occupy 2/3 of the total surface of earth, most of the precipitation that falls over oceans. Of the precipitation that falls over continents, about 65% is returned to Atmosphere through evapo-transpiration and the rest goes as a surface run off into the rivers and finally into the oceans. Thus, water occurs on earth in three forms viz solid, liquid and gaseous.

Water Resources of the World

About 97 percent of worlds in the oceans and this are not useful for irrigation. Of The total quantity of water, only 2.6 percent is fresh water, which is in the form of ice caps, icebergs and glaciers and only small fraction of water is present in the ground, rivers and atmosphere that can be harvested for irrigation of crops.

Water Resources of India

The average rainfall of India is 1194 mm. When considered over geographical area of 328 million hectares, this rainfall amounts to 392 million hectare meters (m. ha. m). This may round off to 400 (m. ha. m) by including the contribution of snowfall which is not yet fully determined. Out of 400 (m. ha. m) of rainfall, 75% is received during South-West Monsoon period (June to September) and rest in remaining months as shown below. A Major portion of water (215 m. ha. m) soaks into the soil, while 70 (m. ha. m) is lost as an evapo-transpiration.

Flow chart showing the water resources of India

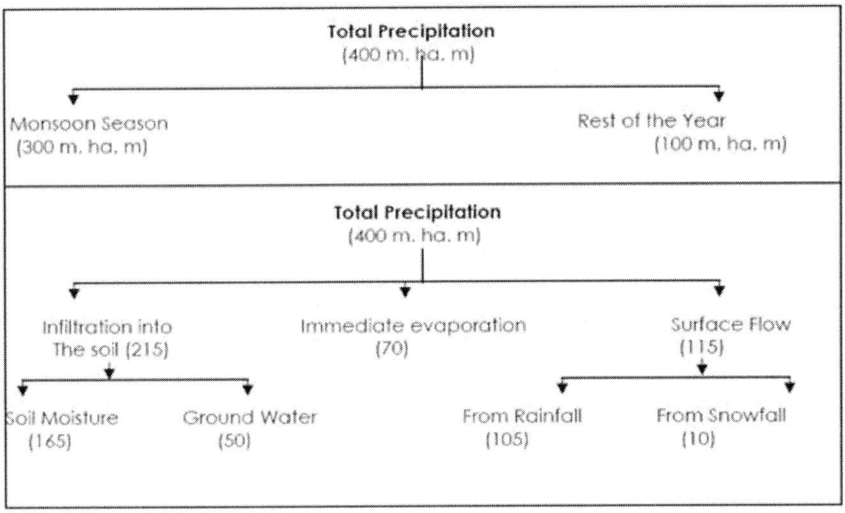

Sources of Water and Functions of Water

Sources of Water

The major sources of water available either for agriculture or for human consumption is obtained from the precipitation in the form of rainfall or snowfall. Run-off from precipitation drains through streams and rivers or collects in surface depressions forming tanks or ponds. Water of streams stored in reservoirs or is diverted directly through canal system for irrigation. Run-off water stored in tanks or ponds is also regulated for irrigation through suitable conveyance system. Part of rainfall is stored as a Ground water. Of the annual rainfall of 400 (m. ha. m) about 215 (m. ha. m) infiltrates into the soil. A major part of it amounting to about 165 (m. ha. m) is retained as a soil moisture, which is essential for growth of vegetation. It is only after the soil has absorbed water to field capacity that water starts percolating down to water table and adds to ground water reservoir.

Functions of Water

Ecological importance

The distribution of vegetation over the surface of earth is controlled by the availability of water than any other single factor. In heavy rainfall area, flush vegetation (forest) is observed.

Physiological importance

The ecological importance of water is result of physiological importance.

- **It is a constituent of protoplasm:** Water is as important quantitatively as qualitatively Constituting 80 to 90 Percent of fresh weight of most herbaceous plant parts and 50 percent of the fresh weight of woody plant.
- **It is a very good solvent:** Water acts as a solvent in which gases, minerals (plant Nutrients) and other solutes are dissolved. The dissolved plant nutrients are absorbed by Plant through soil solution. It acts as a carrier of food nutrients.
- **It is a reagent:** Water acts as a reagent in many important processes, such as photosynthesis and hydrolysis of starch and sugar.
- **It maintains turgidity of plant:** Maintenance of turgidity is essential for cell Enlargement and growth. Turgidity is also important in opening of the stomata, movement of leaves, flower, petals etc.
- **It controls the temperature of plant and soil:**
- **It is a major part of plant body:**

Table 2.5: Water availability and water use indicators (Asian Countries)

Country	Renewable water resource ($m^3yr^{-1}preson^{-1}$)	Per capita withdrawal(m^3)	Withdrawal as % of availability	% share of agriculture
China	2427	462	19.1	87
India	2464	612	24.8	93
Indonesia	14020	465	3.3	94
Bangladesh	11740	211	1.8	96
Thailand	3274	599	1803	90
Vietnam	5638	81	1.4	78
Myanmar	25960	103	4.0	90
Philippines	5180	693	13.4	61
South Korea	1452	298	20.5	75
Pakistan	3962	1250	31.5	98
Nepal	8686	155	1.8	95
Malaysia	26300	765	2.9	47
Sri Lanka	2498	503	20.1	96
Cambodia	10680	69	0.6	94

Source: Fredericksen et. al. (1993)

Water Productivity

Water productivity concepts provide information on the amount food can can be produce with amount of available water per unit volume of water (Viets, 1962; Molden, 1997). Evaporation, transpiration, gross inflow, net inflow, percolation and seepage, the important components of water productivity can be included increasing yield per unit of land area. Use of better varieties improved agronomic practices and water-efficient m management for growing crop identifying water saving opportunities as well as to develop water saving technologies in crop cultivation. The concept that holds good where water supply for the purpose of irrigation.

Strategies to Improve Water Productivity

Groundwater Exploitation and Recharge: In this context, it was very much important to have the access as much as the excess water could be added to underground reservoir during the period of drought would become easier and safer. There was clear indication that several soil conservation and water-harvesting measures could increase the ground water recharge up to a certain limit. Groundwater recharge from open wells and pits are very common practice in Orissa, Gujarat and Rajasthan. That's why greater exploitation of ground water prior to rainy season provided scope of greater quantity to infiltrate during high rainfall period. The use of surface water would become also to be more effective and helpful to mitigate the arsenic like problems. As the possibility of contamination of arsenic are very less in surface water, only the flowing surface water is the safest source that is free from arsenic. So all that an endeavour be undertaken by the planners and policy-makers to find out ways and means to add the groundwater as much as possible (avoiding all sorts of contamination following de-germs chambers to be constructed for the purpose) and taking off water at need without hampering the water table depletion with integrated and judicious management of water resources right from beginning to avoid the global disasters.

On-farm Irrigation Water Management

Adoption of proven and efficient water management technologies are essential to maximize crop production and minimizing gap between irrigation potential created and utilized. Application losses due to conveyance, methods and scheduling of irrigation also are effective in this direction. Selection of crop(s) and crop sequence(s) also played an important role for enhancement of water use efficiency through crop cultivation and related agronomic practices. Conservation, distribution and utilization of irrigation water are the basic parameters of on-farm water management. Optimum scheduling of irrigation, suitable method adoption, conjunctive use of rain, surface and ground water for crop cultivation having improved agro-technology adoption and provision of drainage. Application of proper amount of water at proper time increased the water use efficiency and crop yield maximization with given amount of water reducing evaporation and deep percolation. Scheduling of irrigation with limited water availability is a big challenge to the irrigation experts that needs rigorous research. It is considered the irrigation system as a complex socio-technical phenomenon that involved interaction between a physical, environmental and technological in the process of adoption. However, the participatory action research appeared to be more effective and communicative measures in the developmental processes. It is described that the participatory action research as a method of merging both developmental intervention and research activities as an effort to involve the users in systemic process of change to make the socio-technical system more effective and efficient. Majority of the farmers are reluctant to apply irrigation scheduling technologies as they do not expect a net benefit, out of its application. Some workers elaborated their work on-farm participation and encouraged the farmers to adopt irrigation scheduling as cost-benefit approach. They formulated user-friendly format for a wide variety of crops irrigated in South Africa. The on-farm irrigation management and irrigation service is improved considerably by involving farmers in water management practices and introducing an acceptable water-saving irrigation schedule for growing crops and increasing agricultural production. On-farm water management with emphasis on active participation of the farmers at each and every stage is the corner stone of success.

A series of on-farm water management research activities were conducted as a collaborative project of Indian Council of Agricultural Research, New Delhi and Ministry of Water Resources, Government of India in some of the selected command area either by package intervention or conducting component technology field trials in the irrigation command. On-farm participatory field trials brought positive improvements towards better water use efficiency following the recommended methods of water management practices in the farmers' field in DVC command. Several research workers also reported success story of efficient soil and water management for enhancing agricultural productivity in diverse soil groups of India.

Research Prioritization

1. Generation of suitable technologies on rain water harvesting and recycling of excess water for its effective use for crop production;
2. Evaluation of suitable rice and other crop cultivars along with improved agro-techniques for Coastal saline zone;
3. Study on reclamation of salinity, use of saline water for irrigation, breeding of salt tolerant varieties of crops and better agronomic practices suitable for Sundarban area for higher yield and effective utilization of saline soil;
4. Effective utilization of low land eco-system through integrated farming with rice-vegetable and fish components;

5. To mitigate the arsenic problem in water with suitable technology available to the farmers;
6. Studies on system management for scheduling irrigation to crop(s) and crop sequence(s) with judicious and conjunctive use of rain/canal and ground water keeping parity with optimum level of ground water;
7. Studies on ground water fluctuations, its adverse effect on soil health as well as to evolve the methodology to raise ground water table;
8. To studies on the crops like date palm, swamp taro, colocasia, lettuce, water chestnut and other in view to meet the ever increasing food demand;
9. Management of horticultural crops in orchards, flowers and vegetable to find out suitable alternative of rice crop and to introduce less budget high value crops;
10. Evaluation of traditional rice varieties for its selection for suitable bye-products as well as co-products employing indigenous technologies for alternate use of rice;
11. To develop suitable agro-technologies for rain fed farming based on farmers participation and introduction of crops with less water requirement;
12. To replace the summer paddy by other substitute crop(s) without hampering the economic benefits of the farmers;
13. To utilize the plant protection measures more efficiently introducing IPM;
14. To introducepaira/mixed crop in risk prone areas to ascertain the farmers benefits more effectively;
15. To increase the area under adaptive and innovative trials on several aspects of agricultural production enhancement activities based on participatory approach.
16. Development of submerged and mercy land (*jheels and beels*) having hydromorphic characteristics for its for utilization to increase agricultural productivity.

Water resources management is an important consideration to the researchers to address the issues like global climate changes, natural disasters, pollution, environmental degradation, soil health and food security to fit the system in mitigating occurrence of such situations. In addition to the procedures to be adopted for rainwater harvesting and it's effective utilization, it is essential to explore the possibilities of inter basin transfer of water from surplus to deficit basin to increase maximum water resources to get full potential of irrigation development.

1. It is also essential to find out the ways and means to use the contaminated as well as reclaimed water for its effective utilization in crop cultivation;
2. Efficient utilization of water saving irrigation technologies for growing crops;
3. Introduction of supplemental and protective irrigation that could save at least 30% of irrigation water minimizing conveyance and other losses; and
4. Adoption of micro-irrigation could save 30-60% of water with 30-50% increased crop yield. Save water and increase the area under crop cultivation as well as more crop yield with less water application is the basic purpose of adapting integrated water resources management. .

References

Cullen, H., 2010. The Weather of the Future.Heat Waves, Extreme Storms, and Other Scenes from a Climate-Changed Planet. HarperCollins.

Dinar, A., 2004.Cooperation in Managing Transboundary Water Resources: Evaluation Approaches and Experiences. Paper presented at the Fourth Biennial Rosenberg International Forum on Water Policy, Ankara, Turkey, September 3-9, 2004. Available at: http://rosenberg.ucanr.org/documents/TransboundaryWaterResources.doc.

Fishman, C., 2011. The Big Thirst.The Secret Life and Turbulent Future of Water. New York: Free Press.

Gleick, P.H., 2004. "Environment and Security: Water Conflict Chronology, Version 2004-2005" in The World's Water. The Biennial Report on Freshwater Resources, 2004-2005. Washington, D.C.: Island Press, pp. 234-255.

Gleick, P.H., 2009. "Water Conflict Chronology, Version 2004-2009" in The World's Water. The Biennial Report on Freshwater Resources, 2008-2009. Washington, D.C.: Island Press, pp. 151-193.

Gosselin, P. et al., 2010. The Royal Society of Canada Expert Panel: Environmental and Health Impacts of Canada's Oil Sands Industry. Ottawa, Canada: The Royal Society of Canada.

IPCC, 2012. "Summary for Policymakers" in Field, C.B., V. Barros, T.F. Stocker, D. Qin, D.J. Dokken, K.L. Ebi, M.D. Mastrandrea, K.J. Mach, G.-K.Plattner, S.K. Allen, M. Tignor, and P.M. Midgley (Eds.), Managing the Risks of Extreme Events and Disasters to Advance Climate Change Adaptation. A Special Report of Working Groups I and II of the Intergovernmental Panel on Climate Change. Cambridge, UK and New York, USA: Cambridge University Press, pp. 1-19.

Jury, W.A. and H.J. Vaux Jr., 2007."The Emerging Global Water Crisis: Managing Scarcity and Conflict between Water Users", Adv. Agron.95: 1-76.

MacGregor, A., 2011. Insurers Worried About Recent Water Damage. Colonial Times [online] 31 August, 2011. Available at: http://www2.canada.com/victoriatimescolonist/news/business/story.html?id=310d146a-5ef6-4a84-bb0a-4dc3911802a7.

Milly, P.C.D., J. Betancourt, M. Falkenmark, R.M. Hirsch, Z.W. Kundzewicz, D.P. Lettenmaier and R.J. Stouffer, 2008. "Stationarity Is Dead: Whither Water Management?",Science 319: 573-574.

Prud'homme, A., 2011. The Ripple Effect. The Fate of Freshwater in the 21st Century. Scribner.
Safriel, U., 2011, "Balancing Water for People and Nature" in Garrido, A. and H. Ingram (Eds.), Water for Food in a Changing World: Contributions from the Rosenberg International Forum on Water Policy. New York and London: Routledge, pp. 135-170.
Solomon, S., 2010.Water.The Epic Struggle for Wealth, Power, and Civilization.HarperCollins. United Nations Secretary-General's High-Level Panel on Global Sustainability, 2012. Resilient People, Resilient Planet: A Future Worth Choosing. New York: United Nations. 99 pp.
Vaux, H. J., 2012. "Water for Agriculture and the Environment: The Ultimate Trade-off", Water Policy [in press].
Vidaurre, R. et al., 2011. Climate Change, Hydro-conflicts and Human Security: Achievement of and Gaps in Current Policies. 1st Policy Brief. EU: Ecological Institute.
Wolf, A., 1999. "Water Wars and Water Reality: Conflict and Cooperation along International Waterways" in Lonergan, S. (Ed.) Environmental Change, Adaptation, and Security. Doredect, The Netherlands: Academic Press.
Zou'bi, M., 2011.Water Issues in the Jordan River Basin. Paper presented by the Secretary-General of the Islamic World Academy to the Interaction Council's High-Level Expert Group Meeting "The Global Water Crisis: Addressing an Urgent Security Issue", Toronto, Canada, March, 2011.

Multiple Choice Questions

1. What percentage of the total precipitation in India reaches the groundwater table?

   a) 12.5% b) 25%

   c) 50% d) 75%

2. What is the annual exploitable groundwater potential estimated by the Central Groundwater Board?

   a) 43.3 million ha meters b) 50 million ha meters

   c) 165 million ha meters d) 215 million ha meters

3. What is the primary use of the majority of groundwater utilized in India?

   a) Drinking water b) Industrial use

   c) Irrigation d) Domestic use

4. What is the basic purpose of adapting integrated water resources management?

   a) Save water and increase crop yield

   b) Increase industrial water usage

   c) Reduce drinking water quality

   d) Minimize agricultural water usage

5. What is the term used for the water consumed or used in the production process of a product?
   a) Virtual water b) Groundwater
   c) Surface water d) Rainwater
6. What is the main source of water that varies within agro-ecological zones in India?
   a) Groundwater b) Surface water
   c) Precipitation d) Desalination
7. What percentage of run-off water can be effectively utilized according to the document?
   a) 10% b) 20%
   c) 30% d) 50%
8. What is the estimated annual rainfall in India compared to the world average?
   a) 800 mm b) 1000 mm
   c) 1170 mm d) 1500 mm
9. What is the greatest risk of waterborne diseases according to the document?
   a) Elderly people b) Infants and young children
   c) Healthy adults d) Teenagers
10. What is the term used for the water used in the production of food products?
    a) Virtual water b) Surface water
    c) Groundwater d) Irrigation water
11. What percentage of the total precipitation in India reaches the groundwater table?
    a) 12.5% b) 25%
    c) 50% d) 75%
12. What is the annual exploitable groundwater potential estimated by the Central Groundwater Board?
    a) 43.3 million ha meters b) 50 million ha meters
    c) 165 million ha meters d) 215 million ha meters
13. What is the primary use of the majority of groundwater utilized in India?
    a) Drinking water b) Industrial use
    c) Irrigation d) Domestic use

14. What is the basic purpose of adapting integrated water resources management?
    a) Save water and increase crop yield
    b) Increase industrial water usage
    c) Reduce drinking water quality
    d) Minimize agricultural water usage
15. What is the term used for the water consumed or used in the production process of a product?
    a) Virtual water b) Groundwater
    c) Surface water d) Rainwater
16. What is the main source of water that varies within agro-ecological zones in India?
    a) Groundwater b) Surface water
    c) Precipitation d) Desalination
17. What percentage of run-off water can be effectively utilized according to the document?
    a) 10% b) 20%
    c) 30% d) 50%
18. What is the estimated annual rainfall in India compared to the world average?
    a) 800 mm b) 1000 mm
    c) 1170 mm d) 1500 mm
19. What is the greatest risk of waterborne diseases according to the document?
    a) Elderly people b) Infants and young children
    c) Healthy adults d) Teenagers
20. What is the term used for the water used in the production of food products?
    a) Virtual water b) Surface water
    c) Groundwater d) Irrigation water

True or False Questions

1. Water is considered indispensable for human, animal, and plant life.
    - True
    - False

2. The majority of India's annual rainfall occurs over a period of six months.
   - True
   - False
3. Industrial water is a significant source of pollution in the water environment.
   - True
   - False
4. The term 'virtual water' refers to the water contained in a product after it has been produced.
   - True
   - False
5. Efficient utilization of water-saving irrigation technologies can save 30-60% of water.
   - True
   - False
6. Water is considered indispensable for human, animal, and plant life.
   - True
   - False
7. The majority of India's annual rainfall occurs over a period of six months.
   - True
   - False
8. Industrial water is a significant source of pollution in the water environment.
   - True
   - False
9. The term 'virtual water' refers to the water contained in a product after it has been produced.
   - True
   - False
10. Efficient utilization of water-saving irrigation technologies can save 30-60% of water.
    - True
    - False
11. Is the ocean water dynamic in nature and influenced by weather parameters like temperature and salinity?
    - True
    - False
12. Is the total rainfall in India approximately 800 mm, which is equal to the world average?
    - True
    - False
13. Agricultural water is defined as the edible portion of a crop during growing, harvesting, processing, and packing.
    - True
    - False

14. The document states that establishing a basic constitutional right to water is a guaranteed solution to the water crisis.
    - True
    - False
15. The document mentions that only about 20% of run-off water can be utilized effectively.
    - True
    - False

Short Questions

1. What are the main challenges mentioned in the document regarding water management and availability?
2. How does the document describe the relationship between rivers and lakes?
3. What is the significance of ethical considerations in water management?
4. What is the estimated annual rainfall in India and how does it compare to the world average?
5. What are some methods suggested in the document for improving water resource management in agriculture?

Answer Key

1	a	2	a	3	c	4	a	5	a	6	c	7	b
8	c	9	b	10	a	11	a	12	a	13	c	14	a
15	a	16	c	17	b	18	c	19	b	20	a		

1	Ture	2	False	3	False	4	False	5	Ture	6	Ture	7	False
8	Ture	9	False	10	Ture	11	Ture	12	False	13	Ture	14	False
15	Ture												

3

Soil Properties and Soil Characteristics

Soil physical properties

The physical properties of soil, particularly its mechanical behavior, significantly impact its suitability for plant growth. Factors like plant support, root penetration, drainage, aeration, moisture retention, and nutrient availability are all influenced by the soil's physical condition. These properties also affect the soil's chemical and biological activities.

A soil's physical properties are determined by the quantity, size, shape, arrangement, and mineral composition of its particles, as well as its organic matter content and pore spaces.

Important Physical Properties of Soils

1. Soil Texture
2. Soil Structure
3. Surface Area
4. Soil Density
5. Soil Porosity
6. Soil Color
7. Soil Consistence

Soil Texture

Definition

Soil texture refers to the relative proportions of different-sized particles, specifically the percentage by weight of sand, silt, and clay. It's a fundamental soil property that is difficult to alter. Soil textural determinations typically exclude particles larger than 2 mm in diameter (stones and gravels). While larger particles can affect land management and tillage, they contribute minimally to water-holding capacity (WHC) and nutrient storage/supply.

Particle size ranges

- **Gravels:** 2 – 4 mm
- **Pebbles:** 4 – 64 mm

- **Cobbles:** 64 – 256 mm
- **Boulders:** > 256 mm

Particles smaller than 2 mm are considered "fine earth" and are typically analyzed in chemical and mechanical assessments; Fine earth comprises sand, silt, and clay, which are referred to as "soil separates." Various organizations have established size limits for these fractions, leading to different classification systems:

- USDA (United States Department of Agriculture)
- BSI (British Standards Institution) / English system
- ISSS (International Society of Soil Science)
- European system

i) USDA System

Soil Separate	**Diameter (mm)**
Clay	< 0.002
Silt	0.002 – 0.05
Very Fine Sand	0.05 – 0.10
Fine Sand	0.10 – 0.25
Medium Sand	0.25 - 0.50
Coarse Sand	0.50 - 1.00
Very Coarse Sand	1.00 – 2.00

ii) BSI System

Soil Separates	**Diameter (mm)**
Clay	< 0.002
Fine Silt	0.002 – 0.01
Medium Silt	0.01 – 0.04
Coarse Silt	0.04 – 0.06
Fine Sand	0.06 - 0.20
Medium Sand	0.20 - 1.00
Coarse Sand	1.00 – 2.00

iii) ISSS System

Soil Separates	**Diameter (mm)**
Clay	< 0.002
Silt	0.002 – 0.02
Fine Sand	0.02 – 0.2
Coarse Sand	0.2 – 2.0

iv) European System

S.No	Soil Separates	Diameter (mm)
1	Fine clay	< 0.0002
2	Medium clay	0.0002 – 0.0006
3	Coarse clay	0.0006 – 0.002
4	Fine silt	0.002 - 0.006
5	Medium silt	0.006 - 0.02
6	Coarse silt	0.02 - 0.06
7	Fine sand	0.06 - 0.20
8	Medium sand	0.20 - 0.60
9	Coarse sand	0. 60 - 2.00

Characteristics of Soil Separates

- **Sand:**
  1. Primarily composed of quartz, but may include feldspar, mica, and heavy minerals (zircon, tourmaline, hornblende).
  2. Relatively uniform in size.
  3. Often represented as spherical particles.
  4. Surface is not always smooth, often jagged
- **Silt:**
  1. Intermediate in size between sand and clay.
  2. Larger surface area compared to sand.
  3. Often coated with clay.
  4. Exhibits some physico-chemical properties similar to clay, but to a lesser extent.
  5. Sand and silt together form the "skeleton" of the soil.
- **Clay:**
  1. Particle size less than 0.002 mm.
  2. Plate-like or needle-like in shape.
  3. Belongs to the alumino-silicate group of minerals.
  4. May contain significant concentrations of non-alumino-silicate fine particles (e.g., iron oxides, $CaCO3$).
  5. Secondary minerals derived from primary minerals in rocks.
  6. Considered the "flesh" of the soil.

Understanding soil texture is crucial for determining land value, land use capability, and appropriate soil management practices.

Methods of Textural Determination

Numerous methods exist for both laboratory and field use:

i) Elutriation method (Water & Air)
ii) Pipette method
iii) Decantation/beaker method
iv) Test tube shaking method
v) Feel method (Field applicable, quick method by feeling soil between thumb and fingers)

Feel Method

This method involves moistening the soil and attempting to squeeze it into a thin ribbon by pressing and rolling it between the thumb and forefinger, or alternatively, rolling it into a thin wire. Assess:

i) Feel by fingers
ii) Ball formation
iii) Stickiness
iv) Ribbon formation

Soil Textural Classes

Soil textural class names are used to provide an idea of the textural composition of soils and their physical properties. They are based on the proportion of the three main fractions: sand, silt, and clay. This name provides information about the soil's composition and its general properties. Soils are classified into various textural classes like sands, clays, silts, loams, etc.

Sands

The sand group includes soils with at least 70% sand and 15% or less clay. Their properties are characteristic of sand, contrasting with the stickier nature of clays. Two textural classes are recognized: sandy and loamy sand. Loamy fine sand and loamy very fine sand are also used in practice.

Silt

The silt group includes soils with at least 80% silt and 12% or less clay. The properties of this group are dominated by those of silt. Only one textural class, "Silt," is included in this group.

Clays

For a soil to be classified as clay, it must contain at least 35%, and generally not less than 40%, clay particles. In these soils, the properties of the clay fraction are distinctly dominant. These soil classes are named clay, sandy clay, and silty clay. It's important to note that sandy clays can contain more sand than clay, and silty clays typically have a higher silt content than clay content.

Loams

The loam group is a more complex soil textural class with numerous subdivisions. An ideal loam is a mixture of sand, silt, and clay particles that exhibits the beneficial properties of each separate in roughly equal proportions. Loam soils do not display the dominant physical characteristics of sand, silt, or clay individually. While they don't contain equal percentages of each component, they demonstrate approximately equal contributions from each.

Determination of Textural Class

The United States Department of Agriculture (USDA) system proposes twelve textural classes.

The Textural Triangle

Because the percentages of sand, silt, and clay must add up to 100%, knowing the amount of any two fractions automatically determines the percentage of the third. To use the textural triangle diagram:

1. Locate the percentage of clay on the clay axis and project a line inward, parallel to the sand line.
2. Locate the percentage of silt on the silt axis and project a line inward, parallel to the clay line.
3. (Optional, but often helpful) Locate the percentage of sand on the sand axis and project a line inward, parallel to the silt line.

The point where the projections intersect identifies the soil's textural class. Sometimes, the intersecting point falls directly on the line between textural classes. In this case, it's customary to use the name of the finer fraction. For example, a soil containing 40% clay and 30% sand would be classified as clay.

Importance of Soil Texture

1. Each type of soil particle contributes to the overall nature and properties of the soil.
2. Soil texture significantly impacts soil management and productivity.
3. Sandy soils are typically open, loose, and friable.
4. This texture makes them easy to work with during tillage.
5. Sand promotes good drainage & aeration, allowing for rapid evaporation and percolation.

6. However, sandy soils have a low water-holding capacity.
7. This means they are prone to drought and are generally unsuitable for dryland farming.
8. Sandy soils are also poor at storing plant nutrients.
9. They tend to have low organic matter content.
10. Leaching of applied nutrients is a significant problem in sandy soils.
11. Suitable crops for sandy soils include potatoes, groundnuts, and cucumbers.
12. Clay particles are crucial for soil fertility.
13. Clayey soils are difficult to till, requiring considerable skill.
14. When moist, clayey soils are very sticky, and when dry, they become hard and difficult to break apart.
15. They have fine pores, leading to poor drainage and aeration.
16. Clayey soils have a high water-holding capacity but poor percolation, which can result in waterlogging.
17. They are generally fertile, with a good content of plant nutrients.
18. Rice, jute, and sugarcane can be grown successfully in clayey soils.
19. Loam and silt loam soils are highly desirable for cultivation.
20. The best agricultural soils typically contain 10-20% clay, 5-10% organic matter, and a balanced proportion of silt and sand. A soil with 30% silt is generally classified as clay rather than clay loam.

Grades of Soil Structure

Grade refers to the distinctness of individual soil aggregates (peds) and is determined by their stability. Soil moisture content, organic matter, and texture influence the grade of soil structure. The four commonly used terms to describe soil structure grade are:

1. **Structure less:** No noticeable aggregation is present, as seen in loose sand.
2. **Weak Structure:** Poorly formed, indistinct peds that are not durable, with much unaggregated material.
3. **Moderate Structure:** Moderately well-developed peds that are fairly durable and distinct.
4. **Strong Structure:** Very well-formed peds those are quite durable and distinct.

Structure Naming

When naming a soil structure, the standard sequence is grade, class, and type. Examples include: "strong coarse angular blocky," "moderate thin platy," and "weak fine prismatic."

Factors Affecting Soil Structure

The development of structure in arable soil depends on several factors:

1. **Climate:** Climate significantly influences the degree of aggregation and the type of structure. Arid regions exhibit very little aggregation of primary particles, while semi-arid regions show a greater degree of aggregation.
2. **Organic Matter:** Organic matter improves the structure of both sandy and clay soils. In sandy soils, the sticky and slimy materials produced by decomposing organic matter and associated microorganisms cement the sand particles together to form aggregates. In clayey soils, organic matter modifies the properties of the clay by reducing its cohesiveness, making the clay more crumbly.
3. **Tillage:** Cultivation implements break down large clods into smaller fragments and aggregates. Optimum soil moisture content is necessary for obtaining good granular and crumbly structure. If the moisture content is too high, large clods will form upon drying. If it is too low, some of the existing aggregates will be broken down.
4. **Plants, Roots, and Residues:**
   - The excretion of gelatinous organic compounds and exudates from roots acts as a binding agent.
   - Root hairs cause soil particles to cling together (grass and cereal roots are particularly effective).
   - Pressure exerted by the roots also helps hold the particles together.
   - Dehydration of soil due to shrinkage strains the soil, resulting in cracks that lead to aggregation.
   - Plant tops and residues shade the soil, preventing extreme and sudden temperature and moisture changes, as well as raindrop impact.
   - Plant residues serve as food for microbes, which are crucial for aggregate building.
5. **Soil Fauna:** Small animals within the soil, such as earthworms, moles, and insects, play a significant role in aggregating finer soil particles through their burrowing activities.

6. **Microbes:** Algae, fungi, actinomycetes, and bacteria contribute to soil particle cohesion. Fungi and actinomycetes bind soil particles mechanically through their mycelia. Furthermore, cementation occurs via decomposition products and materials synthesized by bacteria.
7. **Fertilizers:** Some fertilizers, like sodium nitrate, can disrupt soil granulation by decreasing the stability of aggregates. However, other fertilizers, such as calcium ammonium nitrate (CAN), can promote the development of favorable soil structures.
8. **Wetting and Drying:** When dry soil becomes wet, soil colloids swell as they absorb water. Conversely, as the soil dries, shrinkage creates tension within the soil mass, leading to cracks and the formation of clods and granules of varying sizes.

Effect of Soil Structure on Other Physical Properties

Porosity: Soil porosity is easily influenced by soil structure. Plate-like structures typically exhibit lower porosity, while crumby structures have higher porosity.

Temperature: Crumby soil structures promote good aeration and water percolation, contributing to a more consistent and optimal soil temperature compared to plate-like structures.

Density: Bulk density is inversely related to total pore space. Soil structure significantly impacts pore space. Platy structures, characterized by limited pore space, generally have high bulk density, while crumby structures with abundant pore space have low bulk density.

Consistence: Soil consistence is also influenced by structure. Plate-like structures often exhibit strong plasticity.

Color: Bluish and greenish soil colors are commonly associated with poor drainage. Platy structures often impede proper drainage.

Importance of Structure: Soil structure indirectly affects soil properties by influencing the formation of pores in various shapes and sizes. These pores act as critical regulators of water, air, and temperature within the soil.

The Role of Soil Structure in Plant Growth

1. Soil structure determines the amount and characteristics of soil porosity.
2. Structure regulates the air and water content in the soil. In addition to the amount of water and air, their movement and circulation are also influenced by structure.
3. Soil structure affects tillage practices.
4. Structure influences runoff and erosion.

5. Platy structures tend to hinder drainage, while sphere-like structures (granular and crumby) facilitate it.
6. Crumby and granular structures promote optimal infiltration, water-holding capacity, aeration, and drainage. They also create a favorable habitat for microorganisms and nutrient availability.

Retention of Water by Soil: Soils retain water due to their colloidal properties and aggregation capabilities. Water is held on the surface of colloids and other particles and within the pore spaces. After drainage ceases, water retention is primarily attributed to surface tension and surface attraction, collectively known as surface moisture tension. This concept relates to the energy required for moisture retention. The force with which water is held is also termed suction.

Mechanisms of Water Retention in Soil

1. **Cohesion and Adhesion:** These two fundamental forces govern water retention. Cohesion refers to the attraction between water molecules themselves, while adhesion describes the attraction between water molecules and solid soil surfaces. Adhesion allows soil solids to tightly bind water molecules at the soil-water interface. These water molecules, in turn, hold other water molecules together through cohesion. These combined forces enable soil solids to retain water.
2. **Surface Tension:** This phenomenon occurs at water-air interfaces. Water behaves as if its surface is covered by a stretched elastic membrane. At the surface, the attraction between air and water molecules is significantly weaker than the attraction between water molecules themselves. This results in a net downward force on the surface molecules, creating a compressed film or membrane at the surface, which is called surface tension.
3. **Polarity (Dipole Character):** The retention of water molecules on the surface of clay micelles is based on the dipole character of water molecules. Water molecules are held by electrostatic forces present on the surface of colloidal particles. Due to their dipole nature and under the influence of electrostatic forces, water molecules become oriented (arranged) on the surface of clay particles in a specific manner.

Each water molecule possesses both negative and positive charges. The clay particle carries a negative charge. The positive end of a water molecule attaches to the negatively charged clay surface, leaving its negative end exposed. Water molecules attached to the clay surface in this way create a layer of negative charges, which then attracts another layer of oriented water molecules. This process continues, forming successive molecular layers. As the molecular layer thickens, the orientation weakens. At a certain distance from the particle surface, the water molecules cease to orient and capillary water (free liquid water) begins to appear. Due to the attractive forces exerted by the surface of soil particles (adsorption), water adheres to the soil surface. The force of gravity also acts simultaneously, attempting to pull the water downwards. The surface force is considerably stronger than gravity, allowing the water to remain attached to the soil particle. Whether the water remains attached or moves downward into lower layers depends on the magnitude of the resultant force.

Soil Water Potentials

The movement and retention of water in soil, its uptake by plants, and its loss to the atmosphere are all driven by energy relationships. The strength with which water is held in the soil is directly proportional to the energy required to remove it. This relationship is described by the soil water energy concept, where water moves from areas of high free energy (e.g., a water table) to areas of low free energy (e.g., dry soil). Essentially, work must be done to overcome the forces holding water in the soil.

The free energy of soil water is affected by

- **Matric Forces:** These are attractive forces between soil solids and water (adsorption). They significantly reduce the free energy (mobility) of water molecules. Matric potential is always negative.
- **Osmotic Forces:** These are attractive forces between ions/solutes and water in the soil solution, also reducing free energy. Osmotic potential is also negative.

  Matric and Osmotic potentials are negative and reduce the free energy level of the soil water. These negative potentials are referred as suction or tension.
- **Gravity:** Gravity pulls water downwards towards the earth's center. This force contributes a positive potential.

The *soil water potential* is defined as the difference in energy state between soil water and pure, free water.

Soil Moisture Constants

An older classification system categorized soil water as gravitational, capillary, and hygroscopic. Hygroscopic and capillary water are in equilibrium with the soil under certain conditions. Soil moisture constants represent specific soil moisture relationships and the soil's ability to retain moisture.

While maximum capillary capacity represents the maximum amount of capillary water a soil can hold, it is not all available for plant use. The same is true of hygroscopic water. Therefore, the concepts of *field capacity* and *wilting coefficient* were developed to better describe soil-plant-water relationships under field conditions.

Here's a breakdown of key soil moisture constants

1. **Field Capacity:** After a soil is saturated and excess (gravitational) water has drained away, the remaining water held in the micropores represents the field capacity. It's the soil's ability to retain water against gravity. At this stage only micropores or capillary pores are filled with water and plants absorb water for their use. At field capacity water is held with a force of 1/3 atmosphere. Water at field capacity is readily available to plants and microorganism.
2. **Wilting Coefficient (Permanent Wilting Point):** This is the moisture content at which plants can no longer extract water from the soil to maintain turgor. Even if placed in a saturated atmosphere, the plant will remain wilted unless water is added to the soil. Water at wilting coefficient is held with a force of 15 atmosphere.
3. **Hygroscopic Coefficient:** The maximum amount of hygroscopic water absorbed by 100 g of dry soil under standard conditions of humidity (50% relative humidity) and temperature (15°C). This tension is equal to a force of 31 atmospheres. Water at this tension is not available to plant but may be available to certain bacteria.
4. **Available Water Capacity:** The amount of water held in the soil that plants can readily use. It's calculated as the difference between the water content at field capacity (-0.3 bar) and the water content at the permanent wilting point (-15 bars).
5. **Maximum Water Holding Capacity (Maximum Retentive Capacity):** The total amount of water a soil can hold when all pore spaces, both micro and macro, are filled. It is a rough measure of total pore space of soil. Soil moisture tension is very low between 1/100th to 1/1000th of an atmosphere or pF 1 to 0.
6. **Sticky Point Moisture:** The moisture content at which the soil no longer adheres to other objects. It represents the upper limit of friability for the soil equivalent of soils.
7. **Soil Moisture Constants:** A table outlines moisture classes (chemically combined, hygroscopic, capillary, gravitational), their corresponding tension (atm), and pF values. Terms like moisture equivalent, field

capacity, sticky point, maximum water holding capacity, and wilting point are defined. Moisture equivalent is the percentage of water held by a 1 cm thick soil layer after being subjected to a centrifugal force of 1000 times gravity for 30 minutes.

8. **Soil Water Movement:** Explores saturated flow (all pores filled, driven by gravity – infiltration and percolation), unsaturated flow (water held at potentials lower than -1/3 bar, moving from wetter to drier areas via adhesion and capillarity), and water vapor movement (internal and external, driven by vapor pressure gradients).
9. **Factors Affecting Water Movement:** Texture, structure, organic matter, soil depth, water content, temperature, and pressure influence both saturated and unsaturated flow.
10. **Darcy's Law:** Defines vertical water flow rate based on hydraulic gradient.
11. **Entry of Water (Infiltration):** Downward movement of water into the soil surface. Influenced by surface conditions, vegetative cover, soil temperature, texture, structure, organic matter, and initial moisture content; Specific factors include clay minerals, soil texture/structure, moisture content, vegetative cover and topography.

Soil Water Dynamics

Percolation: This is the process of water moving downward through a soil column. It's vital for two primary reasons:

- **Groundwater Recharge:** Percolation is the main source of replenishing groundwater, which can be accessed through wells for irrigation.
- **Nutrient Leaching:** As water percolates, it can carry plant nutrients downwards, potentially moving them beyond the reach of plant roots, a process known as leaching.
    - Percolation rates vary; they are negligible in dry regions and high in areas with substantial rainfall.
    - Sandy soils exhibit greater percolation than clayey soils.
    - Vegetation cover and a high water table can reduce percolation losses.

Permeability: This describes the ease with which water moves within the soil. It refers to the characteristics of the soil that determine how quickly air and water can flow through it. Hydraulic conductivity is a related term indicating the soil's ability to transmit fluids.

Hysteresis

Hysteresis refers to the phenomenon where the moisture content of a soil at a given tension during wetting differs from the moisture content at the same tension during drying. This discrepancy arises due to the presence of both capillary and non-capillary pores within the soil.

- Moisture content is typically lower during sorption (water uptake) and higher during desorption (water loss).
- Hysteresis in soil minerals is linked to shrinking and swelling, which affect pore size at both microscopic and bulk density levels.
- Factors influencing hysteresis include the shape and size of soil pores, their interconnection, pore configuration, the nature of soil colloids, soil bulk density, and entrapped air.
- Air entrapment during rewetting is a significant factor, as it can clog pores and hinder effective contact between others.

Methods for Determining Soil Moisture

Two main types of measurements are used to assess soil water:

1. **Moisture Content Measurement:** Methods that directly or indirectly quantify the amount of water in the soil.
2. **Soil Moisture Potential (Tension/Suction) Measurement:** Techniques that determine the energy status of water in the soil.

Laboratory Measurement of Soil Moisture Content

1. **Gravimetric Method:**
   - A moist soil sample is collected and dried in an oven at 105°C until a constant weight is achieved.
   - The percentage of moisture is calculated based on the weight loss.
   - *Drawback:* Time-consuming and labor-intensive, involving sampling, weighing, and drying.
2. **Electrical Conductivity Method:**
   - This method relies on the relationship between soil moisture and electrical conductivity.
   - Gypsum blocks containing two electrodes at a fixed distance are used.
   - The blocks are buried in the soil at the desired depth, and conductivity across the electrodes is measured using a modified Wheatstone bridge.
   - *Drawbacks:* Requires calibration of blocks and is affected by salt concentration, making it unsuitable for highly saline soils.

In-situ (Field) Measurement of Soil Moisture Potential

- **Suction/Equilibrium Tension Method (Tensiometers):**
  - Tensiometers measure the tension (suction) with which water is held in the soil.
  - Used to determine when irrigation is needed.
  - A tensiometer consists of a porous cup connected to a water-filled glass tube, which is connected to a mercury manometer.
  - The cup is inserted into the soil, and water flows through it until equilibrium is reached.
  - The tension readings on the manometer (expressed in cm or atmospheres) indicate the soil's suction.
  - In dry soil, water moves out of the cup, creating a negative

Soil Temperature and Related Factors

- **Soil Structure, Texture, and Moisture:**
  - Compact soils have higher thermal conductivity than loose soils.
  - Natural soil structures exhibit higher conductivity than disturbed structures.
  - Mineral soils have higher conductivity than organic soils.
  - Moist soils tend to have more uniform temperatures with depth due to better conductivity compared to dry soils.
- **Soluble Salts:** Soluble salts indirectly affect soil temperature by influencing biological activity, evaporation rates, and other processes.
- **Soil Air:** Soil air is a continuation of atmospheric air within the soil pores. It is in a constant state of motion, flowing between the soil and the atmosphere. Soil aeration is the continuous movement or circulation of air in the soil mass, resulting in the renewal of its component gases.

Exchange of Gases Between Soil and Atmosphere

The exchange of gases between the soil and the atmosphere occurs through two primary mechanisms:

1. **Mass Flow:**
   - During rainfall or irrigation, soil air is displaced and moves into the atmosphere.
   - As moisture is lost through evaporation and transpiration, atmospheric air enters the soil pores.

- Temperature variations in the soil cause the soil air to expand and contract, leading to air movement in and out of the soil.

2. **Diffusion:**
   - Most gaseous interchange in soils occurs through diffusion.
   - Gases like nitrogen, oxygen, and carbon dioxide move according to their partial pressure gradients.

Importance of Soil Aeration

1. **Plant and Root Growth:**
   - Adequate soil aeration is essential for healthy plant growth.
   - Sufficient oxygen supply to roots and removal of carbon dioxide (CO2) are critical.
   - Insufficient oxygen can retard or halt plant growth due to CO2 accumulation.
   - Poor aeration can lead to abnormally shaped roots, particularly in root crops.
   - Underdeveloped root systems resulting from poor aeration cannot absorb sufficient moisture and nutrients from the soil.
   - Microorganisms and Oxygen: Microorganisms present in the soil require oxygen for respiration and metabolism. Many vital microbial activities such as the decomposition of organic matter, nitrification, and sulfur oxidation depend on the presence of oxygen in the soil air. The absence or deficiency of oxygen in soil reduces the rate of microbial activity. For example, the decomposition of organic matter is slowed down, and nitrification is arrested. Poor aeration also negatively affects the microorganism population in the soil.
   - Oxygen Deficiency and Nutrient Absorption: A lack of oxygen can hinder the absorption of nutrients and water by plants. The energy generated from respiration is used in the absorption of water and nutrients. Under poor aeration conditions, such as waterlogging, plants show signs of water and nutrient deficiency.
   - Soil Aeration and Plant Diseases: Insufficient aeration of the soil can lead to the development of plant diseases. For example, wilt of gram and dieback of citrus and peach.

Soil Organic Matter

Soil organic matter is made up of carbon-containing substances, including decomposing plant and animal residues, as well as other organic substances of animal or plant origin, either living or dead. Soil organic matter plays a crucial role in maintaining soil physical conditions and influences soil chemical properties, particularly cation exchange capacity. Organic matter also serves as an energy source for soil microorganisms and influences soil development.

Factors affecting soil organic matter

1. **Climate:** Temperature and rainfall significantly impact the amount of organic matter and nitrogen found in soils. Organic matter and nitrogen content in soils tend to increase as one moves from warmer to cooler areas. The decomposition of organic matter is accelerated in warm climates compared to cooler climates. For every 10°C decline in mean annual temperature, the total organic matter and nitrogen increase by two to three times. Rainfall also affects organic matter, with an increase in organic matter found under higher rainfall conditions.
2. **Natural Vegetation:** Soils developed under grasslands have higher total organic matter compared to those under forests.
3. **Texture:** Fine-textured soils generally have higher organic matter content than coarse-textured soils.
4. **Drainage:** Poorly drained soils, due to their high moisture content and relatively poor aeration, tend to have higher organic matter and nitrogen content compared to well-drained soils.
5. **Cropping and Tillage:** Cropped lands have lower nitrogen and organic matter content compared to virgin soils. Modern conservation tillage practices can help maintain high organic matter levels compared to conventional tillage.
6. **Crop rotations, residues, and plant nutrients:** Crop rotations of cereals with legumes result in higher soil organic matter. Higher organic matter levels are preferable, especially when following a crop rotation.

Composition of organic residues: Plant residues contain 75% moisture and 25% dry matter. This 25% is made up of Carbon (10-12%), Oxygen (9-10%), Hydrogen (1.5-2.5%), Nitrogen (1-2%), and mineral matter (1-3%).

Composition of plant tissues: Carbohydrates

- **Celluloses:** 20-50%
- **Hemicellulose:** 10-30%
- **Starch, Sugar:** 1-5%
- **Proteins:** 1-15%
- **Fats, waxes, tannins:** 1-10%
- **Lignins:** 10-30%

Inorganic residues (mineral matter)

- **Water insolubles:** Proteins, Peptides, Nitrogenous Peptones and S containing materials
- **Water solubles:** (NO_3, NH_4 compounds)

Soil organic residues

- Non Nitrogenous Carbohydrates (celluloses, hemicellulose, starch, sugar, etc.)
- Ether solubles (Fats, oils, waxes, resins, etc.)
- Lignins

The organic matter is also classified based on their rate of decomposition:

1. Rapidly decomposed: Sugars, starches, simple proteins, etc.
2. Less rapidly decomposed: Hemicelluloses, celluloses, etc.
3. Very slowly decomposed: Fats, waxes, resins, lignins, etc.

Decomposition of soil organic matter: Different organic residues contain different organic compounds, and the rate of decomposition varies. Sugars, starches, and simple proteins are rapidly decomposed, while fats, waxes, and lignin are very slowly decomposed. Hemicellulose, celluloses, and proteins are intermediate. However, the end products are more or less the same. The general reactions taking place during decomposition are:

1. Enzymatic oxidation of the bulk with the release of CO_2, water, energy, and heat.
2. Essential elements are released (N, P, S, etc.) and immobilized by a series of reactions.
3. Formation of compounds resistant to microbial action.

Ammonification: The transformation of organic nitrogenous compounds (amino acids, amides, ammonium compounds, nitrates, etc.) into ammonia is called ammonification. This process occurs as a result of hydrolytic and oxidative enzymatic reactions under aerobic conditions by heterotrophic microbes.

Nitrification: The process of converting ammonia to nitrites (NO_2) and then to nitrate (NO_3^-) is known as nitrification. It is an aerobic process by autotrophic bacteria.

$NH4 + O_2 \rightarrow NO_2 + 2H+ + H_2O$ + energy $NO_2 + O_2 \rightarrow NO_3^- +$ energy

Role of Organic Matter in Soil

Organic matter plays a crucial role in maintaining healthy soil and supporting plant life. Its influence extends to soil structure, water management, nutrient availability, and overall biological activity.

Key Benefits of Organic Matter

1. **Improved Soil Structure:** Organic matter promotes the formation of soil aggregates (granules), leading to better aeration and permeability. This improves drainage and root penetration.
2. **Enhanced Water Retention:** Increased organic matter boosts the soil's water-holding capacity. This reduces surface runoff and erosion by facilitating infiltration.
3. **Temperature Regulation and Erosion Control:** Surface mulching with coarse organic materials minimizes wind erosion and moderates soil temperature fluctuations, keeping soil cooler in summer and warmer in winter.
4. **Nutrient Source and Microbial Energy:** Organic matter serves as a food source for beneficial soil microbes and acts as a reservoir of essential plant nutrients, hormones, and even antibiotics.
5. **Increased Biological Activity:** Fresh organic matter provides sustenance for earthworms, ants, and rodents, fostering a healthy soil ecosystem. It also enhances phosphorus (P) availability in acidic soils.
6. **pH Buffering and Mineral Solubilization:** Decomposing organic matter releases organic acids that can help neutralize alkaline soils. These acids, along with released CO2, dissolve minerals, making them more accessible to plants.
7. **Cation Exchange Capacity:** Humus, the highly decomposed form of organic matter, acts as a storehouse for exchangeable cations, enhancing the soil's ability to retain essential nutrients.
8. **Buffering Capacity:** Organic matter buffers the soil against rapid pH changes and extreme soil reactions, maintaining a more stable environment for plant growth.

Carbon Cycle and Organic Matter Decomposition

The carbon cycle is fundamentally linked to organic matter decomposition. Plants fix atmospheric carbon dioxide (CO_2) into organic compounds through photosynthesis. When plants and animals die, their remains become part of the soil organic matter. Decomposition by micro-organisms releases the carbon back into the atmosphere as CO2, thus closing the carbon cycle. This process is essential for making carbon and other nutrients available to new plant life.

Decomposition transforms plant and animal residues, recirculating carbon as CO_2 and releasing nutrients like nitrogen (N), phosphorus (P), sulfur (S), and micronutrients in plant-available forms (e.g., ammonium NH_4^+ and nitrate NO_3^-). Microorganisms assimilate some of these nutrients, incorporating them into their own biomass.

- **Mineralization:** The conversion of organic forms of C, N, P, and S into inorganic or mineral forms.
- **Immobilization:** The conversion of inorganic forms of those elements into organic forms.

The carbon cycle is powered by CO_2 fixation by plants. This organic carbon feeds the animal world. Dead plants and animals are then decomposed by microbes into microbial cells and humus. Final decomposition of humus releases CO_2, completing the cycle.

Carbon to Nitrogen Ratio (C:N Ratio)

The C:N ratio is the proportion of carbon to nitrogen in organic materials and soil. It is a critical indicator of decomposition rate and nutrient availability in the soil.

- **Wider Ratio (e.g., >30:1):** Fresh plant residues typically have a high C:N ratio (e.g., 40:1). This leads to nitrogen immobilization as microbes consume available N to decompose the carbon-rich material. A temporary nitrogen deficiency for plants may occur.
- **Narrower Ratio (e.g., <20:1):** As decomposition progresses, the C:N ratio narrows (e.g., 10:1 for humus). When the C:N ratio reaches approximately 20:1, nitrogen is released (mineralized) and becomes available to plants.
- **Typical Soil C:N Ratio:** Cultivated soils generally have a C:N ratio ranging from 8:1 to 15:1, with an average of 10:1 to 12:1.
- **Examples:**
    - **Legumes and Farm Manure:** 20:1 - 30:1
    - **Straw:** 100:1
    - **Saw Dust:** 400:1
    - **Microorganisms:** 4:1 to 9:1

The C:N ratio reflects the balance between organic matter and N. The C:N ratio in soils is fairly constant and is lower in soils of arid regions, and is smaller in subsoils.

- The C:N ratio plays a key role in regulation of available nitrogen, total organic matter and the rate of decomposition.

Environmental Factors Affecting C:N Ratio

Climate influences the C:N ratio. Warmer, drier (arid) soil conditions tend to have lower C:N ratios compared to cooler, wetter (humid) conditions. Subsoils also tend to have narrower C:N ratios than surface soils.

"The carbon to nitrogen (C:N) ratio differs significantly between plants and microorganisms. In plant material, this ratio is quite variable, ranging from 20:1 to 30:1 in legumes and farmyard manure, and reaching as high as 100:1 in certain straw residues. In contrast, the C:N ratio in microbial biomass is more consistent and falls within a much narrower range of 4:1 to 9:1. Consequently, when fresh organic materials are added to soils, they typically contain a large proportion of carbon relative to nitrogen. This results in soil C:N ratios that fall between those of higher plants and microorganisms."

References

Borken, W., & Matzner, E. (2009). Soil and Ecosystem Properties as Influenced by Land Use. In: Soil Ecology and Ecosystem Services. Edited by S. E. H. W. A. C. Buckland. Oxford University Press.

Brady, N.C., & Weil, R.R. (2010). The Nature and Properties of Soils. 14th Edition. Pearson Education, Inc.

Chapmann, H.D. (1965). Cation Exchange Capacity. In: Methods of Soil Analysis, Part 2: Chemical and Microbiological Properties. Soil Science Society of America.

Hillel, D. (1998). Environmental Soil Physics. Academic Press.

Huang, P.M., Li, Y., & Sharratt, B. (2015). Soil Properties and Their Relationships to Crop Production. In: Advances in Soil Science. Edited by D. R. J. B. G. Satin. CRC Press.

Jansson, P.E., & Ahlström, A. (2005). Soil Properties and Processes. In: Soil and Environmental Quality. Edited by A. B. E. Ferrari. CRC Press.

McBratney, A.B., Santos, M.L., & Minasny, B. (2003). On Digital Soil Mapping. Geoderma, 117(1-2), 3-52.

Rawls, W.J., & Brakensiek, D.L. (1985). Prediction of Soil Water Properties for the Design of Drainage and Irrigation Systems. Soil Science Society of America Journal, 49(4), 1018-1025.

Soil Survey Staff. (2014). Keys to Soil Taxonomy. 12th Edition. USDA-Natural Resources Conservation Service.

Soil Survey Staff. (2017). Soil Survey Manual. USDA-Natural Resources Conservation Service.

Multiple Choice Questions

1. What is the phenomenon called when water behaves as if its surface is covered with a stretched elastic membrane?

   a) Cohesion b) Adhesion

   c) Surface tension d) Polarity

2. What is the C:N ratio of cultivated soils?

   a) 20:1 to 30:1 b) 8:1 to 15:1
   c) 100:1 to 400:1 d) 4:1 to 9:1

3. Which method is applicable to the field for quick textural determination of soil?

   a) Pipette method b) Feel method
   c) Elutriation method d) Decantation method

4. What is the maximum amount of hygroscopic water absorbed by 100g of dry soil under standard conditions called?

   a) Available water capacity b) Hygroscopic coefficient
   c) Maximum water holding capacity d) Soil moisture tension

5. What is the primary characteristic of sandy soils?

   a) At least 70% sand b) At least 80% silt
   c) At least 15% clay d) At least 50% organic matter

6. What is the relationship between the organic matter and nitrogen content in soil?

   a) There is no relationship
   b) Organic matter decreases nitrogen content
   c) There exists a relationship
   d) Organic matter increases nitrogen content

7. What is the particle size of fine clay?

   a) Less than 0.0002 mm b) 0.0002 - 0.0006 mm
   c) 0.0006 - 0.002 mm d) 0.002 - 0.006 mm

8. What is the effect of wetting and drying on soil structure?

   a) It has no effect
   b) It causes soil to swell and shrink
   c) It increases soil density
   d) It makes soil more acidic

9. What is the C:N ratio of straw?

   a) 20:1 b) 100:1
   c) 400:1 d) 30:1

10. What is the primary role of microbes in soil?
    a) They decompose organic matter
    b) They compact the soil
    c) They increase soil acidity
    d) They reduce soil fertility

Answer Key

1	c	2	b	3	b	4	b	5	a	6	c	7	a
8	b	9	b	10	a								

4

Soil-Water-Plant Relationship

Different kinds of soil water and soil water movement; Water in relation to plant growth and crop production; Soil texture, structure, infiltration and permeability; Bulk density and particle density of the soil; Concept and theories of water retention in soils; Consumptive use and $E_{T;}$ Soil moisture characteristics curves; Different losses of water (seepage and run off) and Soil moisture measurement, stress and plant growth

Soil-Water-Plant Relationships and Irrigation Scheduling Concepts

A good understanding of soil-water-plant relationships is the foundation for understanding irrigation concepts.

Water reaches the soil through irrigation, precipitation, or from groundwater drawn up into the root zone through "capillary action". To be available to plants, it must be held by the soil strongly enough so it is not lost to gravity before the plant gets it, but not so strongly that the plant cannot extract it from the soil. At the same time, the roots must have sufficient aeration to function properly.

Roots are responsible for obtaining almost all of the plant's water and nutrients. These substances move through the soil during and after irrigation/fertigation and rainfall. In addition, root systems are continually growing and branching to seek out water and nutrients. Root hairs are very fine structures that extend at right angles to the roots. Water and nutrients predominantly enter the root system via specialized cells near the tips of root hairs.

The xylem is the cellular pathway through the plant that transports water from the roots to the leaves for use in photosynthesis and to cool the plant by evaporation. The process by which water vapour is released to the atmosphere by the leaves and other parts of a living plant, resulting in evaporative cooling, is called transpiration. The xylem also transports water and nutrients to other parts of the plant. In a living plant, the xylem is completely filled with water, just as the veins and arteries of the human circulatory system are filled with blood. The plant needs a continual replacement of water lost through transpiration to keep the xylem full of water, and of course, this has to be accomplished without an organ analogous to a human heart. Tomato plants weighing 5 kg may only have a dry matter content of 1 kg but require 500 kg of water to produce.

In practice tomato crops that produce about 90 Mt/Ha require about 9 – 10 Megal / Ha in case furrow irrigation and 3.5 - 4.5 Megal for drip irrigation taking the rainfall figures into account.

Minor water deficits may result in a proportional growth reduction, but if the stress occurs at a critical growth period (eg during pollination), the loss of yield may be considerable and greatly out of proportion to the amount of stress.

Water Storage in the Soil

Soil moisture tension is a measure of how strongly the water is held by the soil particles. It is also called soil suction and is expressed in units of pressure (bars or atmospheres or kilopascals). In soils that are saturated, the soil moisture tension is zero and therefore the energy required to extract this water is zero. The roots essentially "grab" the free water, which is predominantly moving in a downward direction due to the force of gravity.

After significant precipitation or irrigation, water drains from the soil by gravity, initially from the largest pores where it is held least securely. Air fills the void in the pores as the water drains out. The remaining water is in a thinner layer, closer to the soil-particle surfaces. Once the layer is thin enough to be held securely by the soil particles so that rapid drainage by gravity is prevented, we say that the soil has reached "field capacity". This represents the approximate upper limit of soil water available for plant use. Soil moisture tension (or suction) at this point is about 10 centibars.

Water Movement in Soils

The rate of water movement in the soil depends on the existing soil and water conditions. The flow rate is predominantly determined by the size of the large pores and their proportion in the soil. Since coarse textured soils have large pores and a high proportion of large pores, saturated flow is more rapid than in a medium or fine textured soil.

Unsaturated flow takes place when soil pores are not filled with water. It takes place during and after irrigation. This flow occurs in the water films coating the soil particles. The rate of flow depends primarily on the thickness of the films in fine and medium pores and involves water flowing from one point to another because of a difference in relative wetness. In other words, water moves from a region where water films are thicker to a region where water films are thinner. Flow can be up, down or sideways. Unsaturated flow is often called capillary flow because it occurs due to the attractive forces between the water molecules and soil particles.

These concepts help explain the different wetted patterns that are created by subsurface drip irrigation systems in coarse (sandy), medium and fine-textured soils.

Different wetted patterns created by subsurface drip irrigation in three types of soil. In sand, there is more saturated flow (due to larger pores) and poor hydraulic conductivity in unsaturated flow conditions, leading to a narrower, deeper pattern. In clay soils, unsaturated flow and better hydraulic conductivity result in a wider, shallower wetted pattern in clay soils. To achieve a continuous wetted strip down the row, emitter (dripper) spacings must be narrower in sand compared to clay.

Water management and Irrigation Scheduling

Irrigation scheduling involves deciding when to irrigate and how much water to apply. The size of the root zone being explored by the crop and the water-holding capacity of the soil in the root zone must be taken into consideration when scheduling irrigations. We are interested in the ***available*** water-holding capacity (the amount of water between field capacity and the permanent wilting point), which varies dramatically depending on soil texture, soil structure, organic matter levels and the presence of hard pans. Table 1 shows the ranges of available water holding capacities for coarse, medium and fine-textured soils that are well structured.

Table 4.1: Ranges of available water for three soil textural groups

Soil textural group	**Available water (mm of water per meter of soil)**
Coarse (sand, loamy sand, sandy loam)	57 - 106
Medium (fine sandy loam, loam, silt loam)	117 - 183
Fine (clay loam, silty clay loam, clay)	173 - 183

Plant growth generally ceases well before permanent wilting, so only a portion of the available water is extracted between irrigations. With furrow or flood irrigation, water is applied when the amount of moisture in the soil reaches the ***allowable depletion***.

This specific level of soil moisture depends on crop characteristics, soil type and weather. With drip or micro irrigation, it is possible to always keep the soil moisture above the allowable depletion with frequent irrigations, as illustrated in Figure 5. However, it is still important to define this moisture level for your particular crop, root volume and soil type to ensure the moisture within the wetted soil volume created by drip irrigation is not depleted beyond the allowable depletion.

After planting depending on regions and climatic condition it is important to use water management as a tool to ensure root development to deeper layers in order to give these a better opportunity to exploit a larger portion of the soil. This will contribute to a more stable and stronger crop that will overcome stress intervals between irrigations. In the end this will lead to higher yields and might contribute to higher solids levels depending on the local condition.

When you multiply available water (mm/m) by the allowable depletion (%), you arrive at an estimate of **readily available water (RAW)**.

Frequent irrigation with drip irrigation keeps moisture in the ideal range, well above the allowable depletion. With surface irrigation, water is generally applied when soil moisture reaches the allowable depletion (a defined percentage of available water, based on crop, soil type and weather conditions).

The influence of weather can be accounted for in terms of **evapotranspiration (ET)**. Evapotranspiration is defined as the loss of water to the atmosphere due to a combination of evaporation from the soil surface (and in some cases the surface of the crop's foliage) and transpiration by the crop. Reference evapotranspiration can be calculated from atmospheric data collected via a weather station, using what is known as a modified Penman equation. This equation takes into account the relative influence of temperature, relative humidity, wind speed, solar radiation and atmospheric pressure. This calculation gives an estimate of ET by a "standard" or reference crop. Different crops transpire at different rates depending on their physiology, stage of growth and other factors. When the reference crop is an actively growing green grass cover, of uniform height (8 – 15 cm tall), completely covering the ground surface and adequately supplied with water, the calculated **reference evapotranspiration** figure is abbreviated **ETo**.

When ETo is multiplied by a **crop coefficient(Kc)**, the **daily crop water use (ETcrop)** can be estimated:

ETcrop = ETo x Kc

Crop coefficients are available for the various growth stages of many crops. It should be remembered that the crop coefficients used with ETo will be different from the **crop factors (Kp)** used with **evaporation from a "Class A" pan (Epan)**.

Using ET and the Water Budget Method for Irrigation Scheduling

A widely promoted procedure for irrigation scheduling is the Water Budget approach. It involves monitoring all of the additions of water to and losses of water from the soil in the crop's root zone and is based on maintaining a favorable soil water level. It is especially useful with high-frequency irrigation where known amounts of water are applied.

In its simplest form, the equation used with the Water Budget method is

Irrigation application + effective rainfall = crop water use + irrigation losses + leaching requirement

Crop water use, irrigation losses, the leaching requirement (if there is one) and effective rainfall are estimated to arrive at the amount of water to be applied.

Daily crop water use is calculated by multiplying daily reference crop water use (ETo or Epan) by the appropriate crop coefficient (Kc) or crop factor (Kp) for the stage of growth.

ETcrop = ETo x Kc

or

ETcrop = Epan x Kp

With drip irrigation, crop water use is generally replaced on a daily basis or every second or third day. This allows the soil moisture to be kept in an ideal range at all times. At some crop stages, it may be ideal to have the soil in the 0 – 25% depleted range (i.e. fairly close to field capacity). At other stages it may be recommended to keep the soil moisture in a more depleted state, for example to keep vegetative growth under control or to promote a quality attribute during ripening (eg. improved soluble solids in processing tomatoes).

With annual crops like processing tomatoes, the volume of soil being explored by the root system must be taken into account, and this volume will increase as the crop grows during the season. The volume of the root zone must be considered when making irrigation timing decisions (eg daily versus longer intervals). It makes no sense to replace 3 days' worth of crop water use if in doing so, you've let the soil moisture in the wetted pattern around the roots be drawn down beyond the allowable depletion, thereby causing stress.

If the wetted volume is allowed to expand beyond the boundary of the root system due to over-watering, water is being wasted and it is possible that valuable nutrients are also being pushed outside the effective root zone. Soil moisture monitoring is the best way to ensure the correct wetted volume is re-established after each irrigation.

Irrigation Losses and Leaching Requirement

The application efficiency of your irrigation system must be taken into account when deciding how much water to apply each time you irrigate. With drip and micro-irrigation, this is usually specified at the design stage. For example, if the flow variation is +/- 7%, you must add an additional 7% to ensure all parts of the field receive adequate water. In addition, there may also be a need to apply a small amount of water (eg 5% more on top of the extra 7% added to account for flow variation) to maintain a satisfactory salt balance in the soil.

Effective Rainfall

Effective rainfall, the portion that contributes to evapotranspiration, is one of the more difficult components to estimate in an irrigation water budget. The grower or consultant must make reasonable estimates of the following variables:

- Total rainfall;
- Root zone water content at the time of the rainfall event and the amount of water held in the root zone at field capacity;
- Surface runoff, which may be considerable if the surface soil seals off quickly, making infiltration rates too slow to take in rainfall from a downpour that lasts for a relatively short period;
- Losses to deep percolation.

It is best to make a direct check of the soil moisture level following a significant rainfall event. As a general rule, ignore any falls of rain less than 5 mm.

Field Checking to Confirm Water Budget Estimations

Finally, it is strongly recommended that you carry out routine field checking to ensure you have carried out the components of the water budget procedure accurately. A field check usually involves using a reliable method of soil moisture monitoring to verify the accuracy of the estimates that went into your water budget calculation. For example, if the crop factor or coefficient you are using is too low, the soil will be drier than what has been predicted by the water budget method and signs of plant stress may be noticed. If the crop factor or coefficient is too high, the soil will be wetter than expected.

Keep in mind that crop factors and crop coefficients are average figures. Local conditions such as exposure, windbreaks, soil characteristics and cultural practices can change the amount of irrigation required. By using both the water budget approach and soil moisture monitoring, growers will enhance their ability to make more reliable and efficient decisions when it comes to irrigation scheduling.

Incorrect usage of water can have a detrimental effect on product quality. To avoid excessive or insufficient water usage, methods of systematically predicting the crop requirement for water should be utilised. Where possible irrigation should be adjusted based on predicted rainfall, plant water use and evaporation. Daily rainfall records for outdoor production may be used to assist in planning irrigation requirements. Growers are recommended to obtain access to regular meteorological forecasts to support irrigation planning.

All growers should maintain records of irrigation water usage. These records should include

- Weather station records: rainfall, evaporation, temperatures
- Water application per day

Based upon risk assessments, irrigation water sources should be analysed at least once a year for microbial, chemical and mineral pollutants. The analysis should be compared against official standards and adverse results acted upon.

Soil-Water-Plant relationships

Soil physical properties have got tremendous influences on irrigation. Soil is a three-phase system comprising of the solid phase made of mineral and organic matter and various chemical compounds, the liquid phase called the *soil moisture* and the gaseous phase called the *soil air*. The main component of the solid phase is the soil particles, the size and shape of which give rise to pore spaces of different geometry. These pore spaces are filled with water and air in varying proportions, depending on the amount of moisture present. The presence of solid particles, liquid (soil solution) and gas (soil air) constitute a complex polyphasic system. The volume composition of the three main constituents in the soil system varies widely. A typical silt loam soil, for example, contains about 50 per cent solids, 30 per cent water and 20 per cent air. In addition to the three basic components, soil usually contains numerous living organisms such as bacteria, fungi, algae, protozoa, insects and small animals which directly or indirectly affect soil structure and plant growth.
The most important soil properties influencing irrigation are its infiltration characteristics and water holding capacity. Other soil properties such as soil texture, soil structure, capillary conductivity, soil profile conditions, and depth of water table are also given consideration in the management of irrigation water.

Water Relations of Soil

The mineral and organic compounds of soil from a solid (though not rigid) *matrix,* the interstices of which consists of irregularly shaped pores with a geometry defined by the boundaries of the matrix (Fig. 7.5). The pore space, in general, is filled partly with soil air and liquid vapour and partly with the liquid phase of soil water. Soil moisture is one of the most important ingredients of the soil. It is also one of its most dynamic properties. Water affects intensely many physical and chemical reactions of the soil as well as plant growth.

The properties of water can be explained by the structure of its molecule. Two atoms of hydrogen and one atom of oxygen combine to form a molelargely determined by that of the oxygen ion. The two hydrogen ions take up practically no space. Water molecules do not exist individually. The hydrogen in the water serves as a connecting link from one molecule to the other.

Soil serves as the storage reservoir for water. Only the water stored in the root zone of a crop can be utilized by it for its transpiration and buildup of plant tissues. When ample water is in the root zone, plants can obtain their daily water requirements for proper growth and development. As the plants continue to use water, the available supply diminishes, and unless more water is added, the plants stop growing and finally die. Before the stage is reached when crop growth is adversely affected, it is necessary to

Irrigate again. The amount of water to be applied to each irrigation, and the frequency of irrigation are dependent on the properties of the soil and the crop to be irrigated.

Kinds of Soil Water: When is added to a dry soil either by rain or irrigation, it isdistributed around the soil particles where it is held by adhesive and cohesive forces; it displaces air in the pore spaces and eventually fills the pores. When all the pores, large and small, are filled, the soil is said to be saturated and is at its maximum retentive capacity. The following re the three main classes of soil water:

1. **Hygroscopic Water:** Water held tightly to the surface of soil particles by adsorption forces;
2. **Capillary Water:** Water held by forces of surface tension and continuous films around soil particles and in the capillary spaces;
3. **Gravitational Water:** Water that moves freely in response to gravity and drains out of the soil.

Forces for Water Retention

Adhesion is the attraction of solid surfaces for water molecules. Adhesion is operative only at the solid-liquid interface and hence the film of water established by it is very thin. Cohesion is the attraction of water molecules for each other. This force makes possible a marked thickness of thefilms of water established by hydration until they attaining microscopic size. As the film gets thicker and thicker the forces of gravity act and water flows downward through the large pores. Such water is loosely held. Thus, when a soil is near saturation it is easy to remove an increment of water, but as moisture becomes less and less in the soil, the greater will be the force required to remove a unit amount of moisture.

When a dry soil samples is exposed to water vapour, it will take up moisture. The amount adsorbed depends on the nature and magnitude of the surface exposed the temperature and the degree of humidity. The moisture thus adsorbed is the water of hydration, water of adhesion, or commonly the hygroscopic water. When the air saturation is 100 per cent the maximum amount of such moisture will be acquired.

The capillary water is held between tensions of about 31 atmospheres and one-third atmosphere. Between 31 and 15 atmospheres, capillary adjustment is very sluggish. Comparatively easy movement does not occur until the water film thickens and pressures near one-third atmosphere are reached. As a result of its energy relations, the capillary water is the only fluid water bearing solutes that remains in the soil for any length of time, if drainage is satisfactory. Thus, it functions physically and chemically as the soil solution. The principal factors influencing the amount of capillary water in soils are the structure, texture and organic matter. The finger the texture of the mineral soil particles, the greater is likely its capillary capacity. Granular soil structure produces higher capillary capacity. Presence of organic matter increases the capillary capacity.

Water held in the soil at tensions of one-third atmosphere or less will respond to gravity and move downward, hence the name gravitational water. The water thus affected is that present in the non-capillary (large) pores. Of the three forms of water, only capillary and gravitational water are of interest to the irrigationists since hygroscopic water is not available to plants.

Infiltration

Movement of water into soils: The movement of water from the surface into the soil is called infiltration. The infiltration characteristics of the soil are one of the dominant variables influencing irrigation. Infiltration rate is the soil characteristic determining the maximum rate at which water can enter the soil under specific conditions, including the presence of excess water. It has the dimensions of velocity.

Infiltration is the term applied to the process of water entry into the soil generally (but not necessarily) through the soil surface and vertically downward. This process is of great practical importance since its rate determines the amount of run-off over the soil surface.

In other words, infiltration refers to the entry and downward movement of water in to the soil surface. Infiltration is a surface characteristic of a soil.

Infiltration Rate

It is the rate at which the water enters from the surface to the soil. Initially the infiltration rate is more but afterwards it decreases because the soil gets wet. According to the rate of entry of water from surface to the soil, infiltration rate is grouped in to four categories.

1. **Very Slow:** soils with less than 0.25cm per hour e.g. - very clay soils.
2. **Slow:** infiltration rate of 0.25cm to 1.25cm per hour e.g. Soils with high clay.
3. **Moderate:** infiltration rate of 1.25 to 2.5cm per hour. e.g. - sandy loam/silt loam soils.
4. **Rapid:** infiltration rate is more than 2.5cm per hour e.g. deep/sandy silt loam soils.

Factors Affecting the Rate of Infiltration

- **Compactness of soil surface:** A compact soil surface permits less infiltration whereas more infiltration occurs from loose soil surface.
- **Impact of rain drop:** the force (speed) with which the rain drop falls on the ground is said to be impact of rain drop. Ordinary size varies from 0.5 to 4mm in diameter. The speed of raindrop is 30ft per second and force is 14 times its own weight. When impact of raindrop is more then it causes sealing and closing of pores (capillaries) especially in easily dispensable soils resulting in infiltration rate
- **Soil cover:** Soil surface with vegetative cover has more infiltration rate than bare soil because sealing of capillary is not observed.
- **Soil Wetness:** If soil is wet, infiltration is less. In dry soil, infiltration is more.
- **Soil temperature:** Warm soil absorbs more water than cold soils.
- **Soil texture:** In coarse textured soils, infiltration rate is more as compared to heavy soils. In coarse textured soil, the numbers of macro-pores are more. In clayey soils, the cracking caused by drying also increases infiltration in the initial stages until the soil again swells and decreases infiltration.
- **Depth of soil:** Shallow soils permit less water to enter into soil than too deep soils. A coarse surface textured, high water stable aggregates, more organic matter in the surface soil and greater number of micro pores, all help to increase infiltration. As it is a dynamic and quite variable character of soil, it can be controlled by management practices. Cultivation practices that loosen the surface soil make it more receptive for infiltration e.g. course organic matter mulches increases infiltration.

The actual rate at which water is entering the soil at any given time is termed the *Infiltration velocity*. The infiltration rate decreases during irrigation. The rate of decrease is rapid initially and the infiltration rate tends to approach a constant value. The nearly constant rate that develops after some time has elapsed from the start of irrigation is called the *basic infiltration rate.*

pF of Soils

Scholfield (1935) suggested the use of the logarithm of soil moisturetension and gave the symbol pF of this logarithm which is an exponential expression of a free-energy difference (based on the height of a. water column above free-water level in cm). The pF function, analogous to the acidity-alkalinity scale pH, is defined as the logarithm to the base 10 of the numerical value of the negative pressure of the soil moisture expressed in cm of water.

$pF = \log 10^{h}$

in which

h = soil moisture tension in cm of water

If the osmotic tension is negligible, *i.e.,* at low salt concentration, the pF of the soil moisture may nearly equal the logarithm of the capillary tension expressed in cm of water.

Soil Water Retention and Movement

At relatively low suction (less than 1 bar), soil aggregation and at higher suction (1015 bars) the clay minerals regulate soil water retentivity in soil (Chibber, 1964; Tamhane and DuttaBiswas, 1965; Pharande et aI., 1969.) The water content at any particular suction followed the decreasing order: black soil (smectite), alluvial soil (illite) and laterite soil (kaolinite) (Ali and Biswas, 1971). Water retention and release for crop use in soil can be improved by application of FYM, straw, lime or SSP (Biswas *et al.*, 1967; Biswas *et al.*, 1971; Bhagat and Acharya, 1989). The values of organic carbon, silt + clay and CEC correlated significantly with water retention (Gupta *et al.*, 1984). Soil water movement processes such as infiltration and percolation of rain or irrigation water into soils and n upward flux from ground water containing plant nutrients and salts are govern by hydraulic conductivity K of soil, being its water content on volume basis. Darcy's law, which governs the movement of water in soil, takes the form

$Q = -k\,(-).\ DH / dZ$

Where 'q' is the flux of water, 'H' the hydraulic head (pressure head + gravity head) and 'Z' is the position coordinate. Saturated hydraulic conductivity K (s) as relevant to waterlogged rice soil, is invariant at a given temperature but depends on soil properties. Its values may be in the range of 1-10 cm h-1 in coarse textured soils and 10.1 cm ha-l or even less in heavy soils. But in *rabi*season, water moves under unsaturated conditions and such movement is guided by K(-) depending on water content. Using soil water pressure head values (H) measured by tensiometer and soil water content (-) by gravimetric or neutron probe method, K(-) can be directly determined in the field (Nielsen et al., 1973; Patra and Misra, 1985). K(-) for coarse and fine textured soils of Orissa has been formulated as :

K = 0.362 expo 48.17 (0-0.335) and

K = 0.205 expo 67.15 (0-0.468)respectively, where K is expressed in cm h^{-1}.

Dongale (1987) reported that under zero tillage, infiltration rate after 180 minutes increased from 4.4 cm h^{-1} in rice mono-cropped soil to 11.0 cm h^{-1} in rice -legume double-cropped soil. Tillage increased the intake of water further.

Classification of Soil Water

When water is added to a dry soil either by rainfall or irrigation, it is distributed around the soil particles where it is held by the forces of adhesion and cohesion. Adhesion is the attraction of water molecules towards solid particles and cohesion is the attraction of water molecules towards each other.

There are three main classes of soil water

- **Gravitational water:** A soil sample saturated with water and left to drain the excess out by gravity holds on to a certain amount of water. Water moves freely in response to gravitational force and drains out of the soil. This water is not available for plants use as it drains off rapidly from the root zone.
- **Capillary water:** The water content retained in the soil after the gravitational water has drained off from the soil is known as the capillary water. This water is held in the soil by surface tension. Plant roots gradually absorb the capillary water and thus constitute the principle source of water for plant growth. Capillary water is held between tension of 31 and 1/3 atmosphere.
- **Hygroscopic water:** The water that an oven dry sample of soil absorbs when exposed to moist air is termed as hygroscopic water. It is held as a very thin film over the surface of the soil particles by adsorption forces. This water is not available to plants. The tension with which water is held in soil surface is from 10,000 to 31 atmospheres.

Water Movement in the Soil

Water movement in the soil comprises three phases: Infiltration, redistribution and withdrawal. Immediately after application of irrigation or receipt of rain, water enters into the soil. It gradually redistributes into different layers. Water moves from higher potential to lower potential. The movement may be lateral, upwards or downwards depending on difference in water potential in different parts of the soil.

Movement of water within the soil influences water supply to plant roots and also contributes to underground water table. Water movement in the soil in three distinct ways: saturated flow, unsaturated flow and vapour movement.

Saturated flow: Saturated flow of water occurs when all pores, are filled with water either due to rain or irrigation or water logged situation. The flow of water is due to a gradient in metric potential from one region to another. The direction of flow is from higher potential to lower potential. The major force in driving water in saturated soil is gravity and naturally the major direction is downward movement. Lateral movement also occurs mainly due to metric potential.

Unsaturated flow: Immediately after irrigation or rain, saturated flow occurs and water enters into different soil layers. As drainage proceeds in a soil, larger pores are emptied of water. The unsaturated movement of water is restricted in micropores. The major driving force is metric potential and the direction of flow is from higher to lower potential.

Vapour movement: When the soil becomes dry, water from micropores is also emptied and water is mainly present in the form of water vapour. Water vapour moves from one zone to another due to vapour pressure gradient from higher vapour pressure to lower vapour pressure area. Water vapour moves from higher temperature region to cooler region. The amount of water supplied through vapour movement to plants is very small but is important for the survival of plants under severe drought conditions.

Soil Moisture Characteristics

Soil moisture tension is not necessarily an indication of the moisture content of neither the soil nor the amount of water available for plant use at any particular tension. These are dependent on the texture, structure and other characteristics of the soil and must be determined separately for each soil. Generally sandy soils drain almost completely at low tension, but fine textured clays still hold a considerable amount of moisture even at such high tensions that plant growing in the soil may wilt. Moisture extraction curves, also called moisture characteristic curves, which are plots of moisture content versus moisture tension, show the amount of moisture a given soil holds at various tensions. Knowledge of the amount of water held by the soil at various tensions is required, in order to understand the amount of water that is available to plants, the water that can be taken up by the soil before percolation starts, and the amount of water that must be used for irrigation.

Soil Moisture Stress

Soil-moisture tension as discussed in the preceding paragraphs is based on pure water. Salts in soil water increase the force that must be exerted to extract water and thus affect the amount of water available to plants. The increase in tension caused by salts is from *osmotic pressure*. If two solutions differing in concentration are separated by amembrane impermeable to the dissolved substance, such as a cell membrane in a plant root, water moves from the solution of lower concentration to the one of higher concentration. The force with which water moves across such a membrane is called osmotic pressure and is measured in atmospheres.

Plant growth is a function of the soil moisture stress which is the sum of the soil moisture tension and osmotic pressure of soil solution. In many irrigated soils, the soil solution contains an appreciable amount of salts. The osmotic pressure developed by the soil solution retards the uptake of water by plants. Plants growing in a soil in which the soil-moisture tension is, say, I atmosphere apparently can extract enough moisture for good growth. But if the osmotic pressure of the soil solution is, say, 10 atmospheres, the total stress is 11 atmospheres and the plants cannot extract enough water for good growth. Thus, for successful crop production in soils having appreciable salts, the osmotic pressure of the soil solution must be maintained as low as possible by controlled leaching and the soil moisture tension in the root zone is maintained in a range that will provide adequate moisture to the crop.

Soil Moisture Constants

Soil moisture is always being subjected to pressure gradients and vapour pressure differences that cause it to move. Thus, soil moisture cannot be said to be constant at any pressure. However, it has been found experimentally that certain moisture contents described below are of particular significance in agriculture and these are often called soil moisture 'constants'.

Saturation capacity: When all the pores of the soil are filled with water, the soilis said to be under saturation capacity of *maximum water holding capacity.* The tension of water at saturation capacity is almost zero and it is equal to free water surface.

Field capacity: The field capacity of soil is the moisture content after drainage ofgravitational water has become very slow and the moisture content has become relatively stable. This situation usually exists one to three days after the soil has been thoroughly wetted by rain or irrigation. The terms field capacity, field-carrying capacity, normal moisture capacity and capillary capacity are often used synonymously. At field capacity, the large soil pores are field with air, the micro pores are filled with water and any further drainage is slow. The field capacity is the upper limit of available moisture range in soil moisture and plant relations. The soil moisture tension at field capacity varies from soil to soil, but it generally ranges from 1/10 to 1/3 atmospheres.

Filed capacity is determined by ponding water on the soil surface in an area of about 2 to 5 sq m and permitting it to drain for one to three days, with surface evaporation prevented. Evaporation may be prevented by spreading a polythene sheet or a thick straw mulch on the ground surface. One to three days after the soil is thoroughly wetted, soil samples are collected with an auger from different soil depths at uniform intervals throughout the wetted zone. The moisture content is determined by the gravimetric method.

Moisture equivalent: Moisture equivalent is defined as the amount of waterretained by a sample of initially saturated soil material after being subjected to a centrifugal force of 1000 times that of gravity for a definite period of time, usually half an hour. To determine the moisture equivalent, a small sample of soil is whirled in a centrifuge with a centrifugal force of 1000 times that of gravity. The moisture remaining in the sample is determined. This moisture content when expressed as moisture percentage on over dry basis, gives the value of the moisture equivalent. In medium textured soils, the values of field capacity and moisture equivalent are nearly equal. In sandy soils, the field capacity exceeds the moisture equivalent. In very clayey soils, the field capacity is generally lower than the moisture equivalent.

Permanent wilting percentage: The permanent wilting percentage, also knownas permanent wilting point or wilting co-efficient, is the soil moisture content at which plants can no longer obtain enough moisture to meet transpiration requirements; and remain wilted unless water is added to the soil. At the permanent wilting point the films of water around the soil particles are held so tightly that roots in contact with the soil cannot remove the water at a sufficiently rapid rate to prevent wilting of the plant leaves. It is a soil characteristic, as all plants whose root systems thoroughly permeates the soil will wilt at nearly the same soil moisture content when grown in a particular soil in a humid atmosphere.

The moisture tension of a soil at the permanent wilting point ranges from 7 to 32 atmospheres, depending on soil texture, on the kind and condition of the plants, on the amount of soluble salts in the soil solution, and to some extent on the climatic environment. Since this point is reached when a change in tension produces little change in moisture content, there is little difference in moisture percentage regardless of the tension taken as the permanent wilting point. Therefore, 15 atmospheres is the pressure commonly used for this point.

The **wilting range** is the range in soil-moisture content through which plants undergo progressive degrees of permanent or irreversible wilting, from wilting of the oldest leaves to complete wilting of all leaves. At the permanent wilting point, which is the top of this range, plant growth ceases. Small amounts of water can be removed from the soil by plants after growth ceases, but apparently the water is absorbed only slowly and is enough only to maintain life until more water is available. The moisture content at which the wilting is complete and the plants die is called the *ultimate wilting*. Although the difference in the amount of water in the soil between the two points may be small, there may be a big difference in tension. At the ultimate wilting point soil-moisture tension may be as high as 60 atmospheres.

The most common method of determining the permanent wilting percentage is to grow indicator plants in containers, usually in small cans, holding about 600 grams of soil. Sunflower plant is commonly used as the indicator plant. The plants are allowed to wilt and are then placed in a chamber with an approximately saturated atmosphere to test them for permanent wilting. The residual soil moisture content in the container is then calculated which is the permanent wilting percentage. The determination of moisture content at 15 atmosphere tension which is the usually assumed value of permanent wilting point can be done by the pressure membrane apparatus (Richard, 1947).

Available water: Soil moisture between field capacity and permanent wiltingpoint is referred to as readily available moisture. It is the moisture available for plant use. In general, fine-textured soils have a wide range of water between field capacity and permanent wilting point than coarse textured soils. In contrast, sandy soils with their larger proportion of non-capillary pore space release most of their water within a narrow range of potential because of the predominance of large pores. It iIllustrates the three kinds of soil water and the difference in available water between typical sandy loam and siltloam soils. Tablebelow present the range of available water holding capacities of different soil textural groups. For irrigation system design, the total available water is calculated for a soil depth based on the root system of a mature plant of the crop to be grown.

Range of available water holding capacity of soils

Soil Type	% moisture based on dry weight of soil		Depth of available water /m depth of soil
	Field Capacity %	Permanent Wilting Point %	
Fine Sand	3-5	1-3	2-4
Sandy Loam	5-15	3-8	4-11
Silt Loam	12-18	6-10	6-13
Clay Loam	15-30	7-16	10-18
Clay	25-40	12-20	16-30

Movement of Water within Soils

The movement of water within the soil controls not only the rate of infiltration but also the rate of supply of moisture to plant roots and the rate of underground flow to springs and streams and recharge of ground water. Water in the liquid phase flows through the water filled pore space under the influence of gravity. In the films surrounding soil particles (under unsaturated conditions), it moves under the influence of surface tension forces. Water also diffuses as vapour through the air-filled pore spaces along gradients of decreasing vapour pressure. In all cases, the movement is along gradients of decreasing water potential.

Terminology

Water intake: The movement of irrigation water from the soil surface into andthrough the soil is called water intake. It is the expression of several factors, including infiltration and percolation.

Percolation: Percolation is the downward movement of water through saturatedor nearly saturated soil in response to the force of gravity. Percolation occurs when water is under pressure or when the tension is smaller than about ½ atmosphere. *Percolationrate* is synonymous with infiltration rate with the qualitative provision of saturated ornear saturated conditions.

Interflow: Interflow is the lateral seepage of water in a relatively pervious soilabove a less pervious layer. Such water usually reappears on the surface of the soil at a lower elevation.

Seepage: Seepage is the infiltration (vertically) downward and lateral movementsof water into soil or substrata from a source of supply such as a reservoir or irrigation canal. Such water may reappear at the surface as wet spots or seeps or may percolate to join the ground water or may join the subsurface flow to springs or streams. Seepage rate depends on the wetted perimeter of the reservoir or the canal and the capacity of the soil to conduct water both vertically and laterally.

Permeability

Qualitative: It is the characteristic of a pervious medium relatingto the readiness with which it transmits fluids.

Quantitative: The specific property governing the rate or readiness with whicha porous medium transmits fluids under standard conditions. According to this definition, equations used for expressing flow, which take into account the properties of the fluid, should give the same soil permeability value for all fluids which do not alter the medium.

The term *intrinsic permeability* is used as a permeability factor independent of the fluid. It must, however, be remembered that the factors which tend to change the permeability of the soil matrix to water will influence this value and prevent its use unless they can be measured or evaluated separately.

Hydraulic conductivity

Hydraulic conductivity is the proportionality factor k inDarcy's law ($v=ki$, in which v is the effective flow velocity and i is the hydraulic gradient). It is, therefore, the effective flow velocity at unit hydraulic gradient and has the dimensions of velocity (LT^{1}). The values of k depend on the properties of the fluid with the porous medium, such as swelling of a soil. A soil that has high porosity and coarse open texture has a high hydraulic conductivity value. For two soils of the same 'total' porosity, the soil with small pores has lower conductivity than the soil with large pores because of the resistance to flow in small pores. A soil with pores of many sizes conducts water faster if the large pores form a continuous path through the profile. In fine-textured soils, hydraulic conductivity depends almost entirely on structural pores. In some soils, particles are cemented together to form nearly impermeable layers commonly called *hardpans.* In other soils, very finely divided or colloidal material expands on absorbing water to form an impervious gelatinous mass that restricts the movement of water.

Hydraulic Head

Hydraulic head is the elevation with respect to a standard datumat which water stands in a riser pipe or manometer connected to the point in question in the soil. This will include elevation head, pressure head, and also the velocity head, if the terminal opening of the sensing element is pointed upstream. For non-turbulent flow of water in soil the velocity head is negligible. In unsaturated soil a porous cup must be used for establishing hydraulic contact between the soil water and water in a manometer. Hydraulic head has the dimensions of length (L).

Hydraulic Gradient

Hydraulic gradient is the rate of change of piezometric orhydraulic head with distance. Hydraulic gradient of ground water records the head consumed by friction in the flow in unit distance since in ground water flow the velocity heads are generally negligible.

Hydraulic equilibrium of water in soil

It is the condition for zero flow rate ofliquid or film water in the soil. This condition is satisfied when the pressure gradient force is just equal and opposite of the gravity force.

Classification of Soil Water or Kinds of Soil Water

When water is added to dry soil either by rain or irrigation, it is distributed around the soil particles, where it is held by adhesion and cohesive forces. It displaces air in the pore spaces and eventually fills the pores. When all the pores, large and small are filled, soil is said to be saturated and it is at its maximum retentive capacity.

Although the soil water cannot be sharply demarcated, yet for sake of understanding and as per utility of water to plant it is mainly classified into following categories.

- Hygroscopic water
- Capillary water a) Inner capillary b) outer capillary
- Gravitation water
- Water vapour

Hygroscopic Water

It is that part of soil water which is very tightly held on the surface of soil particles in very thin film by adsorption forces such as adhesion and cohesion. It is mostly in vapour form and forces with which it is held on surface of soil particles is estimated about 10,000 atmosphere towards the inner side and about 31 atmosphere at the outer side of hygroscopic water film. (One atmosphere at sea level is about 15 pounds per square inch, which means the force holding the water at one atmosphere is equal to about 15 pounds pre square inch or 1023 centimeters of water column height). This water is not any use to the plants.

Capillary Water

It is the water held by the forces of surface tension and continuous film around soil particles and in the capillary spaces. When soil particles absorb water even after the hygroscopic coefficient is reached, additional water is also held around the particles in the form thin film. This retension of water film continues until the film becomes quite thick and micro pores inside the soil mass get filled with water. A stage is then reached when the force of gravity becomes stronger and any further addition of water is pulled down by gravity and flows down as free water.

The capillary water is that water, which is held in the soil in excess of hygroscopic water but is up to the point where the gravity pull begins to move the water down wards, when free drainage conditions exist in the soil.

Capillary water is rather loosely held water (from 31 atmosphere to 1/3 atmosphere tension) and is capable of movement within the soil. The plant food nutrients are dissolved in it and therefore, it is most useful water for plants. The capillary water itself can be divided into two parts though there is no clear-cut line division.

Inner Capillary Water

It is that part of capillary water, which is nearest to the hygroscopic water and is in the form of a thinner film, held more tightly and moves rather very slowly than outer capillary water.

Outer Capillary Water

It is that part of capillary water which is not very tightly held in the soil and there after moves readily from place to place. It is the most useful water for plants as it is very quick available.

A soil which has a finer texture and granular structure indicating larger proportion of micro pores than macro pores holds more amount of capillary water than a single grained sandy soil having more percentage of macro pores. Soil rich organic matter content also holds much grater quantity of capillary water.

Gravitational Water

It is that part of soil water, which moves freely in response to gravity and drains out of the soil. When the maximum capillary capacity of a soil gets satisfied and further addition of water comes under the force of gravity. This water starts moving as free water through the macropores and it is called gravitational water.

It is superfluous and as such, it is of no use to the plants. Gravitational water is held at zero atmosphere tension. When down ward movement of gravitational water is more, some plant nutrients are leached out and when it is slow, it will adversely affect the aeration of soil.

Vapour Form

In this category, the water is present in gaseous form in the soil atmosphere but it is not directly used by plants and is therefore, not important unlike the first three kinds.

Kinds of soil water (classification of soil water)

- Hygroscopic water (Water of adhesion)
- Capillary water (water of cohesion)
- Gravitational water
- Water Vapour

Absorption and Movement of Water in Soil

The movement of water from the soil surface into and through the soil is called water intake. It is the expression of several factors including infiltration and percolation.

Permeability

It is defined as the characteristic that determines how fast air and water move through the soil describes what is known as permeability.

Once the water has entered into the top layer, its subsequent slow or rapid movement within the soil indicates its rapid or slow permeability. The permeability basically depends upon pore size distribution in the soil. Larger the number of macro pores (non-capillary pores), the greater is the permeability. The movement of water becomes slow in subsoil layers due to their compactness and low organic matter content but with deep-rooted plants, there is an increased permeability even in such sub soil layers. Permeability increases with the increasing fine texture.

Permeability depends up on

- **Number of micro pores:** More the number of macro pores higher is the permeability.
- **Soil aggregates:** Larger the size of capillary pores, greater is the permeability.
- **Depth of soil:** Permeability decreases with the depth, as the sub soil layers are more compact and have less organic matter.
- **Coarseness of soil texture:** In coarse textured soil, permeability is more, however fine textured soil is less.
- **Salt concentration:** Salt concentration affects permeability adversely. If the sodium is high in water; it would cause ready dispersion of soil and thus reduces permeability.
- **Soil moisture status:** Permeability decreases as the soil becomes drier and increases when soil becomes wet.
- **Organic matter content:** more organic matter in the soil results in more permeability.

The permeability is considered slow, if it is less than 2.5 cm per hour, moderate if it is about 5.0cm per hour. Like infiltration, permeability can be also controlled to a extent by suitable management practices. Continuous tillage reduces permeability, while the growth of deep-rooted crops like pulses or legumes, grasses and tress increases permeability. The permeability of soil varies with its moisture status and usually decreases as the soil becomes drier because air enters in to soil and reduces the permeability.

Percolation

The down ward movement of water through saturated or nearly saturated soil due to the forces of gravity is known as percolation. Percolation occurs when water is under pressure or when the tension is smaller than about 1/3 atmosphere.

Percolating water goes deep into the soil until it meets the free water table. Percolation studies are important for two reasons-

1. Percolating water is only source of recharge of ground water, which can be again be profitably used through springs and wells for irrigation.
2. Percolating water carries plant nutrients like Calcium, Magnesium deep into lower layers and depositing them beyond the reach of roots of common field crops. In sandy or open textured soils, there is a rapid loss of water through percolation.

Factors depends for Percolation

i) **Climate:** If the rainfall is more than evaporation, then there will be appreciable amount of percolation. In dry region, percolation is almost negligible.

ii) **Nature of soil:** Sandy soils permit more percolation as these occupy large number of macro-pores. The macro-pores serve as the main channels of the gravitational flow. However, clayey soil permits less water to percolate.

Capillary Movement

Once the flow due to gravitational forces has been ceased (stopped), the water moves in the form of thin or capillary film from a wet region to dry region. This type movement goes through the finer or micro-pores and it continues until the thickness of moisture film surrounding the soil particles is equal to both the regions (wet and dry regions). Capillary may be in all directions i.e. it may be downward, lateral or upwards from a low tension to high-tension area, since thicker film have lower tension; water from thicker film around the soil particles flows to thinner film. The greater the difference between the thicknesses of the film, the quicker is the capillary movement up to certain point and as difference narrows, the movement of water film also becomes slow and may cease (stop).

Forces Causing Water Movement and Retention of Water in Soil

Force Causing Water Movement:The forces which cause the water movement in soil are:

1. **Gravitational force or gravity tension:** The flow of water due to gravity is very marked when the soil is in saturated condition and generally, the direction of such flow is downward although a little lateral flow takes place. The large pores i.e. macro-pores serve as the main channels for gravitational flow.
2. **Capillary force or capillary tension:** In the soil, water is held by the forces of surface in the capillary spaces and around the soil particles. The movement of water under unsaturated soil conditions is due to force of surface tension. Once the flow due to gravitational force has ceased the water moves in the form of thin or capillary film from a wet region to a dry region through finer or micro-pores. The surface or capillary tension is responsible for the capillary movement of water to all directions from low tension to high tension.
3. **Vapour tension:** If the soil is not water logged, the movement of water vapour may take place to a very little extent from soil layers which gets more heated towards the cooler soil layers particularly when difference between their temperatures are very wide.
4. **Osmotic pressure:** The movement of water takes place due to difference in osmotic pressure of the soil solution. the situation is only observed in only saline soil which has excessive salts.

In all these four forces, the gravitational and capillary forces are important because their significance in the movement of water in the soil is more. However, vapour transfer and osmotic pressure are less important because of their negligible significance in case of normal soils.

Retention of water in soil

Water that enters in the soil is retained by means of the following three forces:

i) **Force of adhesion:** It is the attraction of solid surface of water molecules (It is the attraction of unlike materials to each other). Due to the force of adhesion, the water molecules are attached to the surface of soil particles and thus a thin film of water is tightly held around the soil particles. Finer the soil particles, greater the surface area and consequently, the water film is held or retained more tightly.

ii) **Force of cohesion:** It is attraction between similar molecules of like characteristics. Cohesion is attraction of water molecules for each other. When more water is added to the moist soil, the cohesive force comes into action and the freshly added molecules get attracted towards already existing water molecules. This results in thickening of water film around the soil particles.

iii) Soil colloids: (Clay or humus particles): The water is also retained in the soil due to soil colloids like clay or humus particles. The water thus retained in the soil is called imbibitional moisture.

Such retention of moisture is different in different soils. Fine textured soils having greater aggregation and more organic matter or humus retain much more quantity of water than those coarse textured single grained soils which are poor in organic matter.

Soil Moisture Constant

Water contents under certain standard conditions are referred as soil moisture constants.

Under field conditions, water content of soil is always changing constantly with time and depth of soil and is not static or constant. However, the concept of soil moisture constants greatly facilitates in taking decision in irrigation.

Important Soil Moisture Constant: While studying soil water and discussing its availability or other wise to plant, some specific terms called as soil moisture constants are used. A brief explanation of some important and commonly used terms is given below and the methods of expressing them are indicated in the table below:

Table 4.2: Soil Moisture Constants

Appearance of soil	Type of Soil	Soil Moisture Constant	Moisture Tension in Atmosphere
Wet soil	Gravitational water	Maximum water	0.001
Moist soil	Available water	Field capacity	0.33 (1/3)
	Water held in micro pores	Wilting point	15
Dry soil	Unavailable water tightly held	Hygroscopic coefficient	31
		Air dry	1000
		Oven dry	10,000

Important Soil Moisture Constants

1. **Oven dry weight:** This is the basis for all soil moisture calculations. The soil is heated in an oven at 105 degree Celsius until it loses no more water and final weight is recorded as oven dry weight. Equivalent moisture tension at this stage is 10,000 atmospheres.
2. **Air-dry weight:** Unlike oven dry weight, this is a variable constant. Soil exposed in humid atmosphere will have a higher weight than the same soil, if put in dry atmosphere. Under average conditions, moisture at air dryness is held with a force of about 1000 atmosphere.

3. **Hygroscopic coefficient:** It is the maximum quantity of water absorbed by any soil in a saturated atmosphere (i.e. at 99 percent relative humidity) at 25 degree Celsius temperature. The hygroscopic coefficient varies with the type of soil, its texture and organic matter content. This constant is equal to a force of about 31 atmospheres and determined by placing the soil in a saturated atmosphere at 25oC temperature. Water held by the soil at this constant is not available to plants because it is mostly in vapour form but it is useful to certain bacteria.
4. **Permanent Witling Point (PWP):** The wilting point is also known as a wilting coefficient or permanent wilting point or permanent wilting percentage. After using the water from outer capillary portion, the plant roots begin to utilize although with difficultly the inner capillary water. Thus, as the moisture film becomes thinner, it is held more and more tightly and it is difficult for plant roots to remove each successive portion of the water film. But later on, a stage is reached at which plants cannot obtain enough water to meet transpiration requirement and remain wilted even under saturated atmosphere, unless water is added to soil. The soil moisture constant at this stage (wilting is called as wilting co-efficient or permanent wilting percentage. Water at this constant is with force of a 15 atmosphere. The wilting co-efficient differs in different soils. It is as low as 4 to 6 percentage in sandy soils and as high as about 16 to 20 percent in clayey soils which are rich in organic matter. The wilting point is a lower limit of available soil moisture.
5. **Field Capacity (F.C.):** Field capacity is the moisture content in percentage of a soil on oven dry basis, when it has been completely saturated and down ward movement of has practically ceased.

   With 2 to 3 days after a heavy rains or irrigation, the gravitational or free water is drained. The moisture content at this stage in the soil is said to be at field capacity. The field capacity is the upper limit of available soil moisture range in the soil moisture and plant relations. The moisture tension at this stage is about 1/3 atmosphere. The fine textured granular soil with high organic matter content more soil moisture than sandy soil at field capacity.
6. **Moisture equivalent:** According to the modified technique, moisture equivalents is the amount of moisture in percentage on oven dry weight basis held by 30 grams of dry soil when subjected to 1000 times the gravitational force in a centrifuge for 30 minutes.

   For practical purpose, field capacity may be considered as equal to the moisture equivalent. The value (moisture content may be considered as equal to the moisture equivalent are nearly equal in loamy soil but for sandy soils, the moisture equivalent is slightly higher than filed capacity.

7. **Maximum capillary capacity:** When water is added to the soil whose field capacity is already reached, that water goes on thickening the moisture film. A stage is then reached after which any further additional of water will get percolated down by the force of gravity. This is the point of maximum capillary capacity.
8. **Maximum water holding capacity:** Any further addition of water to the soil after its maximum capillary capacity is reached will start moving down by force of gravity, if it is a well drained soil but when drainage is restricted, maximum amount of water can be held until all micro and macro pores are filled with water. This stage is called the maximum water holding capacity. It is only in case of poorly drained soils or soils having hard pan near the surface that maximum water is retained in the soil for a long period.

The values of different soil moisture constant (moisture percent) differ according to soil type. The values for these moisture constant for some the soils are given below.

Table 4.3: Moisture constants for few typical Indian soils (in percent of oven dry soil)

Soil type	Air dry moisture	Hygroscopic co-efficient	Wilting Coefficient	Moisture equivalent	Maximum water holding capacity
Heavy black	3.8	20.7	29.9	53.2	79.7
Medium black	2.1	13.3	20.6	45.6	66.6
Alluvial	1.6	7.6	13.5	40.4	48.7
Sandy	0.5	1	5.3	21.8	25.2
Laterite	0.8	2.8	5.5	32.9	39.6

Available and Unavailable Soil Water

It is evident that all the water present in the soil profile is not available for the use of plants. Even the capillary water which is considered to be loosely held by the soil particles is not utilized by plants.

Three tentative divisions of the soil water may be made on the basis of availability

i) Unavailable water
ii) Desirably available water
iii) Superfluous or excess of water not needed by plants.

Table 4.4: Available and Unavailable Water

Type of water	Atmospheric Pressure	Status
Oven Dry	10000	Unavailable Water
Air Dry	100	
Hygroscopic Co-efficient	31	Difficultly Water
Wilting Point	15	
Field Capacity	0.33	Available Water
Ground Water	0.001	Unavailable Water

Unavailable soil water

Types of water are not available to the plants are

a) Hygroscopic water

b) Fraction of inner capillary

c) Water vapour

Water below the hygroscopic co-efficient is held so tenaciously above 31 atmosphere that is unavailable to plants. The water held between the hygroscopic co- efficient (31 atmosphere) and the wilting point (15 atmospheres) is inner capillary water. Its movement is extremely sluggish and is only difficultly available to plants. Only certain type of plants under arid conditions make its use. So also some bacteria and fungi use the inner capillary water. It includes whole of the hygroscopic water plus a part of inner capillary water being below the wilting point.

Available or Desirably Available Water

The range of water between the limits of field capacity and wilting point (co-efficient) is considered as the desirably or available water. The soil moisture between field capacity (1/3 atmosphere) and wilting point (15 atmosphere) is readily available moisture.

Superfluous Water

It includes gravitational water (excess of field capacity). This water is also unavailable to the use of plants because it is lost due to deep percolation. The preference of superfluous water in soil for longer period is harmful to plant growth.

Moisture Absorption by Crops

Absorption of water is not dependent of process but it is related to transpiration. Absorption is controlled by rate of water loss in transpiration at least when water is readily available to the roots. Absorption and transpiration are linked by the continuous water column in xylem system of plants. Due to the loss of water in transpiration, it produces the energy gradient which causes the movement of water from soil in to the plants and from plants to atmosphere. In the maintenance of water column in xylem, the cohesive and adhesive properties of water play important role. Moisture enters in to plant roots by process of osmosis (movement of liquid through semi permeable membrane caused by unequal concentration on the two sides). The concentration of soluble material in cell sap of roots is increased because of loss of water through transpiration. When concentration of soluble material in cell sap within roots is greater than the soil moisture, the water passes in the roots to equalize the concentration. A more correct view to consider the concentration of water molecule in cell sap reduced because of quantity of soluble substances present and hence the number of water molecules in the soil solution is greater. As a result more water molecules strike against cell wall and water passes into the roots from the zone of higher concentration of water to a zone of lower concentration of water.

When the concentration of soluble substances in the soil moisture exceeds that cell sap, situation will be reserved and water will pass out of the roots to the soil. Plants growing in saline soils with high concentration of soluble salts absorb water with difficulty due to high osmotic pressure of the soil solution. The absorption of water by plants is closely related with transpiration. The sun provides energy for vaporization of water from leaves. Loss of water from leaf cells cause an increase in interior osmotic pressure which causes water to move in to them from xylem vessels. The xylem vessels of leaf are continuous with that of stem and roots and cause a tension created by loss of water from leaf to be transmitted to roots. Increased osmotic pressure in root cells occurs and uptake of water is encouraged. The absorption of water takes place in terminal portion of roots but the maximum absorption takes place in the zone of root hairs, 1 to 10cm behind root tip. In other words, water is absorbed mainly through roots hairs. Root absorbs water both passively and actively.

Passive absorption takes place when water is drawn into the roots by negative pressure in the conducting tissues created by transpiration.

Under the conditions during which there is little transpiration, the roots of many plants absorb water by spending energy that is called active absorption. Under normal conditions of transpiration, the contribution of active absorption to the water supply of plant is negligible and it is usually less than 10 percent of total absorption.

Certain plants are able to absorb moisture from the atmosphere when soil is at permanent wilting point. This is known as aerial absorption or negative transpiration. Direct absorption of water by leaves that are wetted by rain, dew or overhead irrigation can help to resaturate dehydrated leaf tissue.

The leaves are borne throughout the stem in all plants which are mainly responsible for the loss of water. The leaf surface shows small pores surrounded by two cells. The pores are called stoma and cells surrounding them are called guard cells. The stoma (stomata) regulates the loss of water as vapour and exchange of CO2 in leaf and other organs. It is thus the efficiency of these structures which possibly determine water loss from plant. The efficiency of the stomata depend on their size and number per unit area.

References

Gardner, W. R. (1960). "Dynamics of Soil Water." Soil Science Society of America Journal, 24(3), 517-521. A seminal paper discussing the dynamics of soil water and its relationship with plant water availability.

Hillel, D. (2004). Soil and Water: Physical Principles and Processes. Academic Press. This book provides an in-depth understanding of the physical properties of soil and water interactions and their effects on plant growth.

Kirkham, M. B. (2005). Principles of Soil and Plant Water Relations. Academic Press. This text covers the fundamental principles governing the interaction between soil water and plant physiology.

Lal, R. (2008). Soil and Water Conservation for Food Security. Springer. It discusses the importance of soil and water conservation practices for sustainable agriculture and food security, emphasizing the soil-water-plant relationship.

López, M. V., & Gonçalves, S. M. (2010). "Interactions Between Soil Water and Plant-Acquired Nutrients in Sandy Soils." Plant and Soil, 339(1-2), 151-162. This study investigates the interplay between soil moisture levels and nutrient uptake by plants in sandy soil environments.

Nielsen, D. R., & Wendroth, O. (2003). Spatial and Temporal Variability in Soil and Crop Water Relations. Water Resources Publications. This book examines the variability of soil and crop water relations, incorporating concepts of spatial and temporal interactions.

Passioura, J. B. (2002). "Soil Structural Matters." Soil and Tillage Research, 66(2), 121-129. Discusses how soil structure affects water availability and plant growth, linking soil properties with plant performance.

Schroeder, J. (1985). "Water Relations of Plants." Annual Review of Plant Physiology and Plant Molecular Biology, 36(1), 15-37. An overview of how plants interact with water, including the role of soil in these relations.

Talsma, T. (1975). "Soil-Water Relations: Methods and Strategies." Soil Science, 120(1), 1-10. This article provides insight into methodologies for studying soil-water relationships and their implications for plant health.

Zhang, L., & Chen, X. (2009). "Soil-Water-Plant Relationships in a Semi-Arid Region." Soil Science Society of America Journal, 73(3), 873-877. A research paper highlighting specific case studies focusing on the interactions among soil, water, and plant systems in semi-arid areas.

Multiple Choice Questions

1. What is the primary function of plant roots in relation to irrigation?
   a) To provide structural support
   b) To absorb water and nutrients
   c) To release oxygen into the soil
   d) To prevent soil erosion
2. What term is used to describe the moisture content at which soil holds water against the force of gravity?
   a) Hygroscopic Water
   b) Gravitational Water
   c) Capillary Water
   d) Field Capacity
3. Which irrigation method generally uses less water than furrow irrigation?
   a) Surface Irrigation
   b) Drip Irrigation
   c) Flood Irrigation
   d) Sprinkler Irrigation
4. The water absorbed by the soil during irrigation that moves freely due to gravity is called:
   a) Hygroscopic Water
   b) Capillary Water
   c) Gravitational Water
   d) Available Water
5. What is the term for the percentage of water that can be held in soil between field capacity and the permanent wilting point?
   a) Osmotic Water
   b) Readily Available Water (RAW)
   c) Gravimetric Water
   d) Saturated Water
6. Which of the following is not a factor influencing the rate of water infiltration into soil?
   a) Soil texture
   b) Temperature
   c) Soil color
   d) Soil compactness
7. What is 'evapotranspiration'?
   a) The process of photosynthesis
   b) The volume of water entering soil from rain
   c) The loss of water from soil and plants to the atmosphere
   d) The transportation of water within a plant
8. In which type of soil does capillary movement typically occur more effectively due to its finer texture?
   a) Sandy Soil
   b) Clay Soil
   c) Loamy Soil
   d) Silt Soil

9. Which method is **not** commonly used for estimating soil moisture?
   a) Gravimetric (oven-dry)
   b) Time Domain Reflctometry
   c) Neutron moisture meter
   d) Class A pan evaporation
10. What happens if a crop experiences minor water deficits at a critical growth period?
   a) Growth improves with stress
   b) Yield loss may be significant
   c) No effect on crop yield
   d) The crop grows faster

Answer Key

1	b	2	d	3	b	4	c	5	b	6	c	7	c
8	b	9	d	10	b								

5

Water Budgeting and Moisture Estimation

Water budgeting and its importance, methods of soil moisture estimation, evapo-transpiration and crop water requirement and effective rainfall Water budgeting and its importance, irrigation scheduling approaches

Water budgeting

Allocation of the water receipt including anticipated within the crop period and itsdetailed account of expenditure for efficient and profitable farm management is called as water budgeting.

Water budgeting may be for an irrigation system planned by irrigation engineers; may be for a canal or for an area (block) or may be for a farm according to the need andplan by responsible persons who plan the irrigation efficiency.

Importance of water budgeting

Efficient utilization of available recourse (water) for bringing more area under irrigation:

1. To increase the productivity of a region / farm.
2. To increase cropping intensity of a region / farm
3. To tide over some dry-spells
4. To reduce excess irrigation and losses caused thereby
5. To avoid run off losses

Irrigation scheduling

Irrigation scheduling is defined as frequency with which water is to be applied based on needs of the crop and nature of the soil.

Irrigation scheduling is nothing but number of irrigations & their frequency required to meet the crop water requirement.

Irrigation scheduling may be defined as scientific management techniques of allocating irrigation water based on the individual crop water requirement (ETc) underdifferent soil and climatic condition, with an aim to achieve maximum crop productionper unit of water applied over a unit area in unit time.

Based on the above definition, the concept made is

"If we provide irrigation facility the agricultural production and productivity will go up automatically" Irrigation scheduling is a decision making process repeated many times in each year involving when to irrigate and how much of water to apply? Both criteria influence the quantity and quality of the crop. It indicates how much of irrigation water to be used and how often it has to be given.

Effect of application of right amount and excess amount of water

Excess irrigation is harmful because

1. It wastes water below root zone
2. It results in loss of fertilizer nutrients
3. It causes water stagnation and salinity
4. It causes poor aeration
5. Ultimately it damages the crops

However, Irrigation scheduling has its own meaning and importance according tothe nature of the work.

Irrigation Engineers

Irrigation scheduling is important to cover more area with available quantity of water or to satisfy the whole command from head to tail reach in the canal or riversystem.

Soil Scientists

It is important that the field should not be over irrigated or under irrigated as bothwill spoil the chemical and physical equilibrium of the soil.

Agronomists

It is very much important to get higher yield per unit quantity of water in normal situations and to protect the crop to get as much as possible yield under drought situation by means on supplying water in optimum ratio and minimizing all field losses.

Importance of irrigation scheduling

How much and how often water has to be given depends on the irrigation requirement of the crop.

Irrigation requirement (IR) = Crop water requirement (CWR) – Effective rainfall (ERF)It can be expressed either in mm/day/ or mm/month

If the crop water requirement of a particular crop is 6 mm per day, it means every day we have to give 6 mm of water to the crop. Practically it is not possible since it is time consuming and laborious. Hence, it is necessary to schedule the water supply by means of some time intervals and quantity. For example the water requirement of 6 mm/ day can be scheduled as 24 mm/for every 4 days or 30 mm/for every 5 days or 36 mm/forevery 6 days depending upon the soil type and climatic conditions prevailing in that particular place. While doing so we must be very cautions that the interval should not allow the crop to suffer for want of water.

Using ET and the Water Budget Method for Irrigation Scheduling

A widely promoted procedure for irrigation scheduling is the Water Budget approach. It involves monitoring all of the additions of water to and losses of water from the soil in the crop's root zone and is based on maintaining a favorable soil water level. It is especially useful with high-frequency irrigation where known amounts of water are applied.

In its simplest form, the equation used with the Water Budget method is:

Irrigation application + effective rainfall = crop water use + irrigation losses + leaching requirement

Crop water use, irrigation losses, the leaching requirement (if there is one) and effective rainfall are estimated to arrive at the amount of water to be applied.

Daily crop water use is calculated by multiplying daily reference crop water use (ETo or Epan) by the appropriate crop coefficient (Kc) or crop factor (Kp) for the stage of growth.

ETcrop = ETo x Kc

or

ETcrop = Epan x Kp

With drip irrigation, crop water use is generally replaced on a daily basis or every second or third day. This allows the soil moisture to be kept in an ideal range at all times. At some crop stages, it may be ideal to have the soil in the 0 – 25% depleted range (i.e. fairly close to field capacity). At other stages it may be recommended to keep the soil moisture in a more depleted state, for example to keep vegetative growth under control or to promote a quality attribute during ripening (eg. improved soluble solids in processing tomatoes).

With annual crops like processing tomatoes, the volume of soil being explored by the root system must be taken into account, and this volume will increase as the crop grows during the season. The volume of the root zone must be considered when making irrigation timing decisions (eg daily versus longer intervals). It makes no sense to replace 3 days' worth of crop water use if in doing so, you've let the soil moisture in the wetted pattern around the roots be drawn down beyond the allowable depletion, thereby causing stress.

If the wetted volume is allowed to expand beyond the boundary of the root system due to over-watering, water is being wasted and it is possible that valuable nutrients are also being pushed outside the effective root zone. Soil moisture monitoring is the best way to ensure the correct wetted volume is re-established after each irrigation.

Irrigation Losses and Leaching Requirement

The application efficiency of your irrigation system must be taken into account when deciding how much water to apply each time you irrigate. With drip and micro-irrigation, this is usually specified at the design stage. For example, if the flow variation is +/- 7%, you must add an additional 7% to ensure all parts of the field receive adequate water. In addition, there may also be a need to apply a small amount of water (eg 5% more on top of the extra 7% added to account for flow variation) to maintain a satisfactory salt balance in the soil.

Field Checking to Confirm Water Budget Estimations

Finally, it is strongly recommended that you carry out routine field checking to ensure you have carried out the components of the water budget procedure accurately. A field check usually involves using a reliable method of soil moisture monitoring to verify the accuracy of the estimates that went into your water budget calculation. For example, if the crop factor or coefficient you are using is too low, the soil will be drier than what has been predicted by the water budget method and signs of plant stress may be noticed. If the crop factor or coefficient is too high, the soil will be wetter than expected.

Keep in mind that crop factors and crop coefficients are average figures. Local conditions such as exposure, windbreaks, soil characteristics and cultural practices can change the amount of irrigation required. By using both the water budget approach and soil moisture monitoring, growers will enhance their ability to make more reliable and efficient decisions when it comes to irrigation scheduling.

Incorrect usage of water can have a detrimental effect on product quality. To avoid excessive or insufficient water usage, methods of systematically predicting the crop requirement for water should be utilised. Where possible irrigation should be adjusted based on predicted rainfall, plant water use and evaporation. Daily rainfall records for outdoor production may be used to assist in planning irrigation requirements. Growers are recommended to obtain access to regular meteorological forecasts to support irrigation planning.

All growers should maintain records of irrigation water usage

- These records should include:
- Weather station records: rainfall, evaporation, temperatures

Water application per day

Based upon risk assessments, irrigation water sources should be analysed at least once a year for microbial, chemical and mineral pollutants. The analysis should be compared against official standards and adverse results acted upon.

India's water budget

Total geographical area = 328M.ha.Average annual rainfall = 1190mm

In million hectare metre = 1190 x 328 = 392 M ha m Contribution from snowfall = 8 M ha m

Total = 400 m ha m.

The rainfall below 2.5 mm is not considered for water budgeting, since it will immediately evaporate from surface soil without any contribution to surface water or ground water.

When rainfall occurs, a portion of it immediately evaporates from the ground or transpires from vegetation, a portion infiltrates into the soil and the rest flows oversurface as run off.

There are on an average 130 rainy days in a year in the country out of which the rain during 75 days considered as effective rain. The remaining 55 days are very light and shallow which evaporates immediately without any contribution to surface or ground water recharge. Considering all these factors it is estimated that out of 400 million hectare meter of annual rainfall 70 million hectare meter is lost to atmosphere through evaporation and transpiration, about 115 million hectare meter flows as surface run-off and remaining 215 million hectare meter soaks or infiltrates into the soil profile

Land utilization pattern of India

Total geographical area	= 328.00 M ha
Net area reported	= 307.47 M ha
Area under forest	= 65.90 M ha
Area under nona gricultural use, barren and uncultivablewaste	= 100.45 M ha
Net Area sown	= 141.12 M ha
Net area irrigated	= 31.20 M ha
Gross area sown	= 164.00 M ha
Gross area irrigated	= 80.50 M ha

Land use pattern of West Bengal (Area in hectares)

1. Area not available for cultivation 17,53,328
2. Other uncultivated land excluding current fallows 1,40,701
3. Area under forest 11,74,977
4. Current fallows 3,19,246
5. Net cropped / sown area 52,94,702
6. Gross cropped area 95,32,607
7. Reporting area (excluding Kolkata) 86,82,954
8. Total geographical area 88,75,200

World Irrigation Statistics

Countries	Area irrigated in million hectares
Australia	1.150
Botswana	0.002
Brazil	0.141
Burma	0.753
Canada	0.627
Ethiopia	0.030
France	2.600
India	37.640
Indonesia	3.797
Iran	4.000
Iraq	3.107
Israel	0.153
Japan	3.390
Pakistan	11.970
USSR	9.900
USA	16.932
China	74.000

Field Checking to Confirm Water Budget Estimations

Finally, it is strongly recommended that you carry out routine field checking to ensure you have carried out the components of the water budget procedure accurately. A field check usually involves using a reliable method of soil moisture monitoring to verify the accuracy of the estimates that went into your water budget calculation. For example, if the crop factor or coefficient you are using is too low, the soil will be drier than what has been predicted by the water budget method and signs of plant stress may be noticed. If the crop factor or coefficient is too high, the soil will be wetter than expected.

Keep in mind that crop factors and crop coefficients are average figures. Local conditions such as exposure, windbreaks, soil characteristics and cultural practices can change the amount of irrigation required. By using both the water budget approach and soil moisture monitoring, growers will enhance their ability to make more reliable and efficient decisions when it comes to irrigation scheduling.

Incorrect usage of water can have a detrimental effect on product quality. To avoid excessive or insufficient water usage, methods of systematically predicting the crop requirement for water should be utilised. Where possible irrigation should be adjusted based on predicted rainfall, plant water use and evaporation. Daily rainfall records for outdoor production may be used to assist in planning irrigation requirements. Growers are recommended to obtain access to regular meteorological forecasts to support irrigation planning.

Methods of Soil Moisture Estimation Laboratory & Field Methods

By measuring soil moisture at regular interval and at several depths within the root zones, information can be obtained as to the rate at which moisture is being used by the crops at different depths. This provides the base for determining when to irrigate and how much water to be applied.

For practical purpose, irrigation should be given when about 50 percent of available moisture in the root zone is depleted. The amount of water to be applied is directly related to the water already present in the soil. The methods of measuring soil moisture are divided in to:

A. **Direct method:** Measurement of moisture content in the soil (wetness)

B. **Indirect methods:** Measurement of water potential or stress or tension under which water is held by the soil.

A. Direct methods

I. **Gravimetric methods:** In the gravimetric method, basic measurement of soil moisture is made on soil samples of known weight or volume. Soil sample from the desired depths are collected with a soil auger. Soil sample are taken from desired depth at several locations of each soil type. They are collected in air tight aluminum containers. The soil samples are weighed and they are dried in an oven at 105 oC for about 24 hours until all the moisture is driven off. After removing from oven, they are cooled slowly to room temperature and weighed again. the difference in weight is amount of moisture in the soil. The moisture content in the soil is calculated by the following formula -

$$\text{Moisture content on weight basis} = \frac{\text{Wet weight - Dry weight}}{\text{Dry weight}} \times 100$$

PROBLEMS: Wet weight of a soil sample with can is 210 gms and weight with can is 180 gms weight of empty container is 40 gms calculated moisture content of soils sample?

Solution

Weight of wet soil sample = wet weight – weight of empty can

= 210-40

= 170

Dry weight of soil sample = Dry weight – weight of can

=180-40

=140

$$\text{Moisture content (\%)} = \frac{\text{Wet weight of soil - Dryweight of soil}}{\text{Dry weight of soil}} \times 100$$

$$= \frac{170 - 140}{140} \times 100$$

$$= \frac{30}{140} \times 100$$

= 21.4%

II. **Volumetric method:** Soil sample is taken with a core sampler or with a tube auger whose volume is known. The amount of water present in soil sample is estimated by drying it in the oven and calculating by following formula.

Moisture content = Moisture content (%) by weight x Bulk Density (%) by volume.

PROBLEM: Undisturbed soil sample was collected from a field, two days after irrigation when the soil moisture was near field capacity. The inside dimension of core sampler was 7.5 cm diameter and 15 cm deep. Weight of core sampling cylinder weight of the core-sampling cylinder was 1.56 kg. Determine the available moisture holding capacity of soil and the water depth in centimeter per meter depth of soil.

Solution

Weight of moist soil = 2.76-1.56 = 1.20kg

Weight of oven dry soil = 2.61-2.56 = 1.05kg

$$\text{Moisture content} = \frac{1.20 - 1.05}{1.05} \times 100$$

= 14.28%

Volumeof core sampler = ----------------------------X d2 x h

= -----------X7.5X7.5X15

4

= 662 cu. Cm

$$\text{Apparent specific gravity} = \frac{\text{Wt. of dty soil in grams}}{\text{Volume of soil in cu. Cm}}$$

$$= \frac{1.05}{662} = 1.58$$

Available moisture= Ap. Sp. Gr. X moisture content

= 1.58 X 14.28

= 22.56 cm / m depth of soil

The method is though accurate and simple it is used mainly for experimental purpose. Sampling, transporting & repeated weighing give errors. It is also laborious and time consuming. The errors of the gravimetric method can be reduced by increasing the size and number of samples. however the sampling disturbs the experimental plots and hence many workers prefer indirect methods.

III. **Using Methyl Alcohol:** Soil sample is mixed with a known volume of methyl alcohol and then measure the change in specific gravity of school with a hydrometer. This is a shot cut procedure but it is no in common use.

IV. **Using calcium chloride:** Soil sample is mixed with a known amount of calcium chloride. calcium chloride reacts with water and removes it in the form of acetylene gas. The moisture is determined has come in common use.

B. Indirect methods

In those methods, no water content in the soil is directly measured but the water potential or stress or tension under which the water is held by the soil is measured. The most common instrument used for estimating soil moisture by indirect method is:

1. Tensiometer
2. Gypsum block

3. Neutron probe
4. Pressure plate and pressure membrane apparatus

In all these methods, the reading from above instruments and corresponding soil moisture content is determined by oven drying method are plotted on a graph. Subsequently, these calibration curves are used to know soil moisture content from the reading of these instruments.

1. **Tensiometer:** Tensiometer is also called irrometers since they are used in irrigation scheduling. Tensionmeters provide a direct measure of tenacity (tension) with which water is held by soil. It consist of 7.5 cm porous ceramic or clay cup, a protective metallic tube, a vacuum gauge and a hollow metallic tube holding all parts together. At the time of installation, the system is filled with water from the opening at the top and rubber corked when set up in the soil. moisture from cup moves out with drying of soil, creating a vacuum in the tube which is measured with the gauge. Care should be taken to install tensiometer in the active root zone of the crop. When desired tension is reached, the soil is irrigated. The vacuum gauge is graduated to indicate tension values up to one atmosphere and is divided in to fifty divisions each of 0.2 atmosphere value. The tensiometer works satisfactory up to 0.85 bars of atmosphere.

Merits of Tensiometer

1. It is very simple and easy to read soil moisture in situ.
2. It is very useful instrument for scheduling irrigation to crops which require frequent irrigations at low tension.

Limitations

Sensitivity of a tensiometer is only up to 0.85 atmospheres while available soil moisture range is up to atmosphere and hence is useful more on sandy soils wherein about 80% of available water is held within 0.85 ranges.

2. **Gypsum Blocks:** Gypsum blocks or plaster of Paris resistance units are used for measurement of soil moisture is situ. These were first invented by Bouycos and Mick in 940. the blocks are made of various materials like gypsum, nylon fiber, glass, plaster of Paris or combination of these materials. The blocks are generally rectangular shaped. A pair or electronics is usually made of 20 mesh stainless steel wire screen soldered to copper lead wire. The common dimensions of screen electrodes are 33.75 cm long & 0.25 cm wide. The usual spacing between the electrodes is 2 cm. A similar block is 5.5 cm long, 3.75 cm wide and 2 cm thick.

Working principle: It works on principal of conductance of electricity. When two electrodes A and B are placed parallel to each other in a medium and then electric current is passed, the resistance to the flow of electricity is proportional to the moisture content in the medium. Thus, when the block is wet, conductivity is high and resistance is low. Generally these read about 400 to 600 ohms resistance at field capacity and 50,000 at wilting point. the readings are taken with portable Wheatstone Bridge Bouycos water Bridge operated by dry cells.

While placing the gypsum block in soil, care should be taken that the blocks must have close contact with undisturbed soil. After placing, the blocks get wetted with soil moisture due to capillary movement. Pure gypsum block sets in about 30 minutes. The gypsum block is sensitive to soil to moisture from 1.0 atm tension to 20.0 atm. How ever, the gypsum blocks are not reliable in wet soils.

3. **Pressure membrane and pressure plate apparatus:** Pressure membrane and pressure plate apparatus (developed primarily by Richards) is generally used to estimate field capacity, permanent wilting point and moisture content at different pressures. The apparatus consists of air tight metallic chamber in which porous ceramic pressure plate is placed. The pressure plate and soil samples are saturated and are placed in the metallic chamber. The required pressure, say 0.33 bar or 15 bars is applied through a compressor. The water from the soil sample which is held at less than the pressure, Applied trickles out of the outlet till equilibrium against applied pressure is achieved after that, thc soil samples are taken out and oven dried for determining the moisture content.
4. **Neutron meter (neutron scattering method):** Soil moisture can be estimated quickly and continuously another with neutron moisture meter without disturbing the soil. Another advantage is that soil moisture can be estimated from large volume of soil. This meter scans the soil to about 15 cm. diameter around the neutron probe in wet soil and 50 cm in dry soil. it consists of a probe and a scalar or rate meter. The probe contains fast neutron source, which may be a mixture of radium and beryllium or Americium and beryllium. Access tubes are aluminum tubes of 50 to 100 cm length and are placed in the field where moisture to be estimated.

Neutron probe is lowered into access tube to the desired depth. fast neutrons are released from the probes, which scatter into the soil. When neutrons encounter nuclei of hydrogen atom of water, their speed is reduced. the scalar or the rate meter counts the number of slow neutrons, which are directly proportional to water molecules. Moisture content of soil can be known from the calibration curve with counts of slow neutrons.

Limitations: The two drawbacks of the instruments are that it is expensive and moisture content from shallow top layers cannot be estimated. The fast neutrons are also slowed down by other source of hydrogen (present in the organic matter). Other atoms such as chlorine, boron and iron also slow down the fast neutrons, thus overestimating the soil moisture content.

5. **Gama Ray absorption method:** It is the technique of measurement of changes in soil water content by change in amount of gamma radiation absorbed. The amount of radiation passing through soil depends on soil destiny which varies chiefly with change in water content. This is suitable where change in bulk destiny is very small.
6. **Feel and appearance method:** A practical estimate of moisture content is obtained by the feel and appearance of soil samples taken from the desired depths. the soil sample is squeezed in the hand and its feel and appearance are taken into consideration. In this method, actual moisture content is not determined.
7. **Soil moisture characteristic curve:** The energy status of water and amount of water in the soil are related with the soil moisture characteristic curve. As the energy status of water decreases (moisture towards more negative values) soil water content also decreases. In other words, as soil moisture content deceases, more energy has to be applied to extract moisture from the soil. the relation between suction (externally applied force) and water content of the soil are represented graphically by a curve which is known as a soil are moisture characteristic curve.

Hysteresis

The relation between energy status and moisture content can be obtained in two ways, (i) in absorption by taking an initially saturated soil sample and applying increasing suction to dry the soil gradually and (ii) in absorption by gradually wetting an initially dry soil. The measurement of energy status and moisture content during this process are taken and plotted on graph. The curves obtained through desorption and sorption is different for the same soil sample. the moisture content at a given suction is greater in desorption than in absorption and this phenomenon is known as hysteresis.

III. **Using Methyl Alcohol:** Soil sample is mixed with a known volume of methyl alcohol and then measure the change in specific gravity of school with a hydrometer. This is a shot cut procedure but it is no in common use.

IV. **Using calcium chloride**: Soil sample is mixed with a known amount of calcium chloride. calcium chloride reacts with water and removes it in the form of acetylene gas. The moisture is determined has come in common use.

B. **Indirect methods:** In those methods, no water content in the soil is directly measured but the water potential or stress or tension under which the water is held by the soil is measured. The most common instrument used for estimating soil moisture by indirect method is:

1. Tensiometer
2. Gypsum block
3. Neutron probe
4. Pressure plate and pressure membrane apparatus

In all these methods, the reading from above instruments and corresponding soil moisture content is determined by oven drying method are plotted on a graph. Subsequently, these calibration curves are used to know soil moisture content from the reading of these instruments.

1. **Tensiometer:** Tensiometer is also called irrometers since they are used in irrigation scheduling. Tensionmeters provide a direct measure of tenacity (tension) with which water is held by soil. It consist of 7.5 cm porous ceramic or clay cup, a protective metallic tube, a vacuum gauge and a hollow metallic tube holding all parts together. At the time of installation, the system is filled with water from the opening at the top and rubber corked when set up in the soil. moisture from cup moves out with drying of soil, creating a vacuum in the tube which is measured with the gauge. Care should be taken to install tensiometer in the active root zone of the crop. When desired tension is reached, the soil is irrigated. The vacuum gauge is graduated to indicate tension values up to one atmosphere and is divided in to fifty divisions each of 0.2 atmosphere value. The tensiometer works satisfactory up to 0.85 bars of atmosphere.

Merits of Tensiometer

1. It is very simple and easy to read soil moisture in situ.
2. It is very useful instrument for scheduling irrigation to crops which require frequent irrigations at low tension.

Limitations

Sensitivity of a tensiometer is only up to 0.85 atmospheres while available soil moisture range is up to atmosphere and hence is useful more on sandy soils wherein about 80% of available water is held within 0.85 ranges.

2. **Gypsum Blocks:** Gypsum blocks or plaster of Paris resistance units are used for measurement of soil moisture is situ. These were first invented by Bouycos and Mick in 940. the blocks are made of various materials like gypsum, nylon fiber, glass, plaster of Paris or combination of these materials. The blocks are generally rectangular shaped. A pair or electronics is usually made of 20 mesh stainless steel wire screen soldered to copper lead wire. The common dimensions of screen electrodes are 33.75 cm long and 0.25 cm wide. The usual spacing between the electrodes is 2 cm. A similar block is 5.5 cm long, 3.75 cm wide and 2 cm thick.

Working principle: It works on principal of conductance of electricity. When two electrodes A and B are placed parallel to each other in a medium and then electric current is passed, the resistance to the flow of electricity is proportional to the moisture content in the medium. Thus, when the block is wet, conductivity is high and resistance is low. Generally these read about 400 to 600 ohms resistance at field capacity and 50,000 at wilting point. the readings are taken with portable Wheatstone Bridge Bouycos water Bridge operated by dry cells.

While placing the gypsum block in soil, care should be taken that the blocks must have close contact with undisturbed soil. After placing, the blocks get wetted with soil moisture due to capillary movement. Pure gypsum block sets in about 30 minutes. The gypsum block is sensitive to soil to moisture from 1.0 atm tension to 20.0 atm. How ever, the gypsum blocks are not reliable in wet soils.

3. **Pressure membrane and pressure plate apparatus:** Pressure membrane and pressure plate apparatus (developed primarily by Richards) is generally used to estimate field capacity, permanent wilting point and moisture content at different pressures. The apparatus consists of air tight metallic chamber in which porous ceramic pressure plate is placed. The pressure plate and soil samples are saturated and are placed in the metallic chamber. The required pressure, say 0.33 bar or 15 bars is applied through a compressor. The water from the soil sample which is held at less than the pressure, Applied trickles out of the outlet till equilibrium against applied pressure is achieved after that, the soil samples are taken out and oven dried for determining the moisture content.

4. **Neutron meter (neutron scattering method):** Soil moisture can be estimated quickly and continuously another with neutron moisture meter without disturbing the soil. Another advantage is that soil moisture can

be estimated from large volume of soil. This meter scans the soil to about 15 cm. diameter around the neutron probe in wet soil and 50 cm in dry soil. it consists of a probe and a scalar or rate meter. The probe contains fast neutron source, which may be a mixture of radium and beryllium or Americium and beryllium. Access tubes are aluminum tubes of 50 to 100 cm length and are placed in the field where moisture to be estimated.

Neutron probe is lowered into access tube to the desired depth. fast neutrons are released from the probes, which scatter into the soil. When neutrons encounter nuclei of hydrogen atom of water, their speed is reduced. the scalar or the rate meter counts the number of slow neutrons, which are directly proportional to water molecules. Moisture content of soil can be known from the calibration curve with counts of slow neutrons.

Limitations: The two drawbacks of the instruments are that it is expensive and moisture content from shallow top layers cannot be estimated. The fast neutrons are also slowed down by other source of hydrogen (present in the organic matter). Other atoms such as chlorine, boron and iron also slow down the fast neutrons, thus overestimating the soil moisture content.

5. **Gama Ray absorption method:** it is the technique of measurement of changes in soil water content by change in amount of gamma radiation absorbed. The amount of radiation passing through soil depends on soil destiny which varies chiefly with change in water content. This is suitable where change in bulk destiny is very small.
6. **Feel and appearance method:** A practical estimate of moisture content is obtained by the feel and appearance of soil samples taken from the desired depths. the soil sample is squeezed in the hand and its feel and appearance are taken into consideration. In this method, actual moisture content is not determined.
7. **Soil moisture characteristic curve:** The energy status of water and amount of water in the soil are related with the soil moisture characteristic curve. As the energy status of water decreases (moisture towards more negative values) soil water content also decreases. In other words, as soil moisture content deceases, more energy has to be applied to extract moisture from the soil. the relation between suction (externally applied force) and water content of the soil are represented graphically by a curve which is known as a soil are moisture characteristic curve.

Hysteresis

The relation between energy status and moisture content can be obtained in two ways, (i) in absorption by taking an initially saturated soil sample and applying increasing suction to dry the soil gradually and (ii) in absorption by gradually wetting an initially dry soil. The measurement of energy status and moisture content during this process are taken and plotted on graph. The curves obtained through desorption and sorption is different for the same soil sample. the moisture content at a given suction is greater in desorption than in absorption and this phenomenon is known as hysteresis.

Crop Water Requirement

The demand for water to crop usually comes from evaporation and transpiration, which is commonly known as consumptive use, associated with unavoidable conveyance losses during application and some special needs during land preparation and transplanting. The supply of water to meet the crop needs usually come from effective rainfall, irrigation water application and contribution from storage moisture at the soil profile.

Equationally, WR=IR+ER+S where WR= Water Requirement, IR= Irrigation Application, ER= Effective Rainfall and S= Storage moisture in soil profile

So, IR=WR-(ER+S)

To calculate the water requirement of the crop, it is essential to know the existing moisture status of the soil at sowing and harvesting on which the crop is growing. It needs the knowledge of root system and rooting habit of the particular crop and procedure to determine the moisture content in the different soil profile up to root zone depth of the crop. The crop usually extracts moisture for essential activities through root system. Water comes up to the tip of the leaves to complete the food production processes and goes down to the root again.

So before going to apply irrigation, it is very important to know

- The crop which the farmer wants to grow with its rooting habits;
- The physiological critical growth stages; and
- The existing moisture status of the soil profiles up to root zone depth.

Water Balance

It is very important to harvest excess rain water as runoff during medium to heavy rainfall period for its utilization in crop production during rainless period and development of rainfed agriculture through scientific management of rain water, particularly in the East African countries. The gains of water would come from total rainfall over period of application of water as irrigation. The losses constitute a total runoff (R), downward draining (D), Evaporation (E) and Transpiration (T). Thus the balance sheet shows as Surplus and Deficit over a particular period and it uses to give a changed situation of stored moisture in the soil profiles as well as in the plants.

Equationally,

Gains = P (rainfall) + I (Irrigation); Losses = Runoff (R)+ Down ward Drain (D) + Evaporation (E) + Transpiration (T); Gains-Losses = S + V where,

S and V are changed moisture status in soil profiles and plants, respectively.

Water Use for Crop Production

To get proper understanding on water use, it needs experimentation and observation on the basis of scientific information on these aspects mentioned above. Only the research findings on water requirement of crops and irrigation water use for crop production would lead to formulate a guide lines on effective utilization of irrigation potential whatsoever it has it has been created. The economic and judicious water application also ensured by these guidelines;

So the entire principles of irrigation, its theories and practices, depends on

1. How much water to apply? i.e. Measurement of irrigation water;
2. When to apply the supplemental water as irrigation? i.e. proper timings and scheduling of irrigation application; and
3. How best to apply irrigation? i.e. methods of water application to the crop for sustainable production.

References

Allen, R. G., & Pruitt, W. O. (1986). "Estimating gross irrigation requirements." Journal of Irrigation and Drainage Engineering, 112(3), 174-183. doi:10.1061/(ASCE)0733-9437(1986)112:3(174)

Allen, R. G., Pereira, L. S., Raes, D., & Smith, M. (1998). Crop Evapotranspiration: Guidelines for Computing Crop Water Requirements. FAO Irrigation and Drainage Paper No. 56. Food and Agriculture Organization of the United Nations.

Chang, H. H., & Hsu, H. (1987). "A water budget model for a watershed in southeastern Wisconsin." Journal of Hydrology, 94(1-2), 213-229. doi:10.1016/0022-1694(87)90204-2

Famiglietti, J. S., & Rodell, M. (2013). "Water in the Balance." Science, 340(6138), 1300-1301. doi:10.1126/science.1236410

Jiong, W., & Xue, Y. (2016). "Estimating groundwater storage changes and the water budget in the North China Plain." Hydrology and Earth System Sciences, 20(6), 2267-2279. doi:10.5194/hess-20-2267-2016

Koster, R. D., & Milly, P. C. D. (1997). "The effect of seasonal soil moisture on the atmospheric circulation." Geophysical Research Letters, 24(24), 3257-3260. doi:10.1029/97GL03015

Mpetsheni, N. P., & Ngwenya, B. T. (2020). "Water budgeting for irrigated agriculture: case study of the Madikwe River catchment." Water, 12(7), 1985. doi:10.3390/w12071985

Ritchie, J. T. (1972). "Modeling plant and soil systems." Soil Science Society of America Journal, 36(5), 752-758. doi:10.2136/sssaj1972.03615995003600050006x

Van Loon, A. F., & Van Beek, L. P. H. (2019). "A global water budget tool for cross-sectoral water resource management." Environmental Research Letters, 14(3), 035004. doi:10.1088/1748-9326/aaf276

WMO. (2017). Guide to Hydrological Practices. World Meteorological Organization. Retrieved from https://library.wmo.int/doc_num.php?explnum_id=3674

Multiple Choice Questions

1. What is water budgeting primarily concerned with?
   a) Increasing crop yield directly
   b) Allocating water resources efficiently
   c) Financial management in agriculture
   d) Organic certification procedures
2. How does irrigation scheduling contribute to farm management?
   a) By increasing the labor costs involved
   b) By determining how and when water is applied based on crop needs
   c) By eliminating the need for fertilizer
   d) By focusing solely on soil types without regard for crops
3. What is a key benefit of efficient water budgeting?
   a) Reducing excess irrigation and associated losses
   b) Increasing crop prices in the market
   c) Ensuring all farmers have access to water
   d) Eliminating the need for crop rotation
4. What does excess irrigation NOT lead to?
   a) Wasting water below the root zone
   b) Loss of fertilizer nutrients
   c) Improved soil aeration
   d) Water stagnation
5. Which of the following is a direct method for measuring soil moisture?
   a) Gypsum blocks
   b) Tensiometer
   c) Neutron probe
   d) Gravimetric methods
6. What is the purpose of maintaining a favorable soil water level in the Water Budget Method?
   a) To avoid the use of fertilizers
   b) To ensure proper crop growth with minimal water loss
   c) To reduce the frequency of irrigation in all conditions
   d) To eliminate the need for soil testing

7. Effective rainfall is difficult to estimate due to which of the following factors?
   a) Total rainfall and soil characteristics
   b) Evaporation rates alone
   c) Only surface runoff
   d) Seasonal planting schedules
8. What is one of the recommended actions after a significant rainfall event?
   a) Increase irrigation immediately
   b) Ignore the rainfall if it is below 5mm
   c) Measure soil moisture to assess how much water was absorbed
   d) Apply additional fertilizers regardless of water levels
9. How is the irrigation requirement (IR) calculated?
   a) By subtracting crop yield from effective rainfall
   b) By adding effective rainfall to crop water requirement
   c) By subtracting effective rainfall and storage moisture from water requirement
   d) By measuring atmospheric conditions only
10. Which of the following instruments can provide a easure of tension with which water is held by soil?
   a) Gypsum blocks
   b) Pressure plate apparatus
   c) Tensiometer
   d) Neutron probe

Answer Key

1	b	2	b	3	a	4	c	5	d	6	b	7	a
8	c	9	c	10	c								

6

Dry-land, Rainfed Agriculture and Watershed Management

Introduction

India is the seventh largest having 329.73 million ha of geographical area and second most populous country having more than a billion population in the world. There is no other substitute other than to use every square inch of land and every drop of water effectively and efficiently. Though there was markedly increase in annual food grain production from 51 (1950-51) million tones t0 206 million tones (2000-01), still the death on hunger is visualized.

The demand for food is increasing with ever-increasing population growth rate. So there is no option other than to increase food production. There is very little or no scope of horizontal expansion due to land limitation, the question of vertical expansion is the only way to fit with the situation. Genetically improved crop variety, hybrids, mechanized agriculture, higher rate of fertilizers and chemicals application may give one or two more green revolution, but the question lies how far it can sustain greenness?

Adoption of improved crop cultivars, cropping system, increasing cropping intensities, precision farming can hold good to get out of this situation where whole system is to be considered.

The cropping intensities in the states like Assam, Bihar, MP, Orissa, UP and West Bengal are 137.7, 137.8, 125.6, 154.6, 148.6 and 159.6 percent, respectively (FAI,1994-95). This indicated that a vast majority of lands remain fallow during dry season due to lack of proper technologies, which are under rainfed conditions, neglected. Except West Bengal, more than half of the arable land remains fallow during rainless period. Increasing cropping intensities in rainfed uplands is not easily achievable but there is great potential of intensification of cropping in rainfed lowland by utilizing natural resources and conserved residual moisture. Pulses like black gram, green gram, lentil, lathyrus, chick pea and oilseeds like linseed, groundnut, rapeseed could not only increase the cropping intensities but also come forward to increase national productivity.

Table Increasing demand of food at the end of 9th and 10th 5-year plans along with production trends

Crops	Production (in million tones)		
	1998-99	**Ending 2001-02**	**Ending 2006-07**
Rice	84.74	94.00	103.50
Wheat	71.01	75.07	84.30
Coarse Cereals	30.90	32.60	34.40
Total Cereals	186.65	202.30	222.20
Pulses	15.89	18.20	21.00
Food grains	203.54	220.50	243.20
Oilseeds	25.60	28.00	33.80
Sugarcane	295.70	287.00	352.80
Vegetables	87.53	93.60	110.70
Fruits	44.04	53.70	70.50

Source: Kumar, 1996 and The Hindu Survey of Indian Agriculture 2000

Rainfed lowland ecology

Topographically, the cultivated lands are divided in to three broad categories viz: up, medium and lowlands. These classifications are based on the relative elevations o the fields and determine the depth and duration of submergence of these lands during a year. Rainfed lowlands offer very limited scope for growing crop where rice is compulsory options to the farmers for this situation and environment.

Table Total lowlands rice area in different states of the country

States	Total rice area	Rainfed lowland rice area
	(in million ha)	
Assam	2.32	0.99
Bihar	5.17	2.90
Madhya Pradesh	4.83	3.12
Orissa	4.37	2.30
Uttar Pradesh	5.54	2.34
West Bengal	5.20	2.96
East (Total)	24.43	14.61
Haryana	0.56	
Himachal Pradesh	0.09	
Jammu & Kashmir	0.28	
Punjab	1.65	
North (Total)	2.58	

Andra Pradesh	3.53	0.01
Karnatak	1.16	0.23
Kerala	0.74	0.44
Tamil Nadu	2.52	0.05
South (Total)	7.95	0.73
Gujrat	0.57	0.13
Maharasthra	1.52	0.62
Rajasthan	0.16	
West (Total)	2.25	0.75
Other small areas	0.95	0.36
All India (Total)	41.16	16.45

About 16.45 million ha of rainfed lowland, rice is grown under several constraints. In Eastern India, comprising of Assam, West Bengal, Bihar, Orissa, eastern MP and eastern UP, rainfed lowland rice occupies more than 50% of the total area of 27.43 million ha with an average yield of 1.2 to 1.5 tones/ha as against national average of 1.9 tones/ha. The flood plains of Brahmaputra and its tributaries in Assam, the Ganges and its tributaries in eastern UP, Bihar and West Bengal and the Mahanadi river system in Orissa constitute about 50% of the total rainfed lowlands in this zone.

Generally, three physiographic-hydrologic typesf rice lands are recognized by pluvial, phraetic and fluxial (Mooremann and Breeman, 1978). Since pluvial lands have free drainage, the lowlands are either phraetic or fluxial. In phraetic lands, the water requirement of rice crop is met both by rainwater and groundwater. The water table remains shallow at least during most part of the growing/cropping season. Water availability in the root zone during the growing cycle is influenced by the nature and distribution of natural rainfall, inherent soil characteristics and its management. In phreatic lands, rice could be grown with less drought risk. Fluxial rice lands receive water fully or partly from surface flow runoff or floodwater from rivers, streams. These lands remain inundated during the cropping season. The duration of flooding preludes the growing of other upland crops in the rainy season. In fluxial rice lands, floodwater availability depends on the rainfall pattern and the runoff from the catchment. Floodwater is not dependable source of water for growing rice in drought years. Both shallow and deep-water rice also suffers. In wet years, excessive rains in the catchment area cause widespread flooding, damaging the standing crop, Rice may be drowned or uprooted and land damaged by scouring.

Soil Characteristics

Out of 15 agro-climatic zones of the country delineated by the Planning Commission, five zones (II) eastern Himalayas comprising Assam, West Bengal, Manipur, Meghalaya, Nagaland, Arunachal Pradesh, Tripura and Sikkim, (III) lower Gangetic plains region comprising West Bengal, (IV) middle Gangetic plains with eastern UP and Bihar, (VIII) eastern plateaus and hills comprising MP, Orissa and part of West Bengal and (IX) the northern part of east coast plains and hills comprising coastal Orissa constitute eastern India. The climates in these zones varies from humid to sub-humid (rainfall: 1840-3528 mm), dry sub-humid to moist sub-humid (1302-1607 mm), dry sub-humid to moist sub-humid (1271-1436 mm) and semi-dry to dry sub-humid (780-1287 mm), respectively (Annoymous, 1998; Vaswani,1990). The average minimum and maximum temperature in these zones are in the ranges of 6-10 and 33-41 degree C. Alluvial soils (Entisols and inceptisols), coastal and deltaic alluvial soils (Inceptisols and Alfisols), red soils (Alfisols) and laterite sois (Ultisols) occupy a considerable area of this part of the country (Murthy, 1978). In addition to these, salt affected soils, mixed red & yellow and mixed red &blacj soil also occurs in this tract.

Soils of lowlands areas of eastern India are usually finer in texture, massive in structure, slightly acidic to neutral in reaction and relatively more fertile than that of other land situations. Rice is harvested during mid-November to end of December depending upon the land physiography and rainfall pattern. Immediately after rice harvest, there remains adequate residual soil moisture and a shallow groundwater table in these lowlands, which can sustain a short duration pulse or oilseed crop (Misra and Panda,1994).

In a typical lowland at CRRI, Cuttack clay content in soils ranged from 35 to 41% at 0-4 cm depth, which slightly declined with increase in depth. The total available water holding capacity of soil in 1 m profile was 17-22 cm. In lowlands at Balasore, clay content was in the range of 22-23% in 0-20 cm soil layer, which suddenly drop down to 6% at 20-30 cm and then increased to 49% at 1 m depth. The total available water in 1 m profile was 17.3 cm.

Dryland Farming

Dryland farming is agriculture dependent upon the vagaries of weather, especially precipitation. In its broadest aspects, dryland farming is concerned with all phases of land use under semiarid conditions. Not only how to farm but how much to farm and whether to farm must be taken into consideration. Above all else, dryland farming must emphasize the capture and efficient use of precipitation. Dry land farming is cultivation of crops in regions with annual rainfall of more than 750 mm. Inspite of prolonged dry spells, crop failure is relatively less frequent. These are semiarid tracts with a growing period between 75 and 120 days. Moisture conservation practices are necessary for crop production. However, adequate drainage is required especially for vertisols. Main emphasis is on soil and water conservation, sustainable crop yields and limited fertilizer use according to soil moisture availability.

Areas under dryland

India has 143 million ha of its geographical area under cropping, out of which 113 m ha has the potential to be irrigated, thus, 30 m ha are rainfed agriculture, even if the full potential of irrigation has been utilised. Thus, dryland agriculture occupies larger land than the wet agriculture.

Importance of dryland in Indian agriculture

- It supports 40% of the population and occupies 66% of land. The ratio needs to be rectified, by increasing the productivity of these region
- For the balanced development of country, dryland agriculture must be emphasized.
- lack of irrigation facilities, complete dependency on monsoon rainfall and reflecting a typical cropping pattern of coarse grains millets and oilseeds, pulses, cotton etc.
- All dryland agriculture qualify to be called as rainfed but it is not so vice – versa, as the fundamental difference is the agro – ecological properties and the cropping pattern between the two.
- Dryland agriculture refers to the region or the type of farming which is operated in sub humid to arid conditions with inefficient hydrology.
- India has 1/3rd of its geographical area is under humid conditions, while 2/3rd is subhumid or arid conditions.
- Dryland agriculture is based on the crops e.g. cotton, groundnut, oilseeds, pulses, tobacco.
- Dryland agriculture mostly produces hardy and nutritious crops like Jowar, bajra, ragi, pulses, oilseeds, cottonseeds, sunflower, safflower. Thus, the nutrition / malnutrition problems in poor areas can be fought with the help of dryland agriculture.
- Dryland agriculture involves cash cropping, E.g. Jatropha cultivation can reduce the magnitude of petroleum crisis since liquid produced by it can be added to petroleum products without reducing the efficiency.
- Dryland agriculture has the potential to **produce fodder and cattle feed.** Thus, it can help in white revolution. Also, here the pasture lands are more extensive and cattle breeds have greater per-capital yield.

- Sheep-raising has greater prospect in semi-arid areas.
- Solution to Improve Efficiency and Productivity of Dryland Agriculture
- Dryland areas have a significant contribution in pulses, oilseeds, coarse grain crops, and cotton. Optimal use of dryland technologies for increased agricultural production can help to fill the demand-supply gap in pulses and oilseeds production and improve the economic status of farmers. The various methods to improve the efficiency and productivity of dryland agriculture are summed below:
- Water-harvesting: It means capturing rain where it falls or capturing the runoff in your own village or town. And taking measures to keep that water clean by not allowing polluting activities to take place in the catchment. Water harvesting can be undertaken through a variety of ways.
- Capturing runoff from rooftops
- Capturing runoff from local catchments
- Capturing seasonal floodwaters from local streams
- Conserving water through watershed management
- Organic farming is a system that avoids or excludes use of synthetic inputs like pesticides, fertilizers, hormones, etc. and relying on techniques like crop rotation, organic wastes, farm manure, rock additives and crop residues for plant protection and nutrient utilization.

Strategies for development of dryland farming

1. Selecting an appropriate cropping system tailored to the area is a crucial strategy, which becomes more profitable with effective management of sowing time.
2. Employing proper tillage, fertilizer management, weed control, and implementing plant protection measures also play essential roles in boosting productivity.
3. The use of drought-tolerantcrop varieties is another important technique, as these varieties can endure prolonged periods of drought more effectively than other types.
4. Soil condition can be Improved planting cover crops (planted to cover the soil, not for harvesting). They can reduce erosion, improve soil health, enhance water availability, and help control pests and diseases. Cover crops also increase resilience to drought conditions and erratic rainfall.

5. Mulching is a common dryland technique to conserve moisture in the soil by preventing evaporation.

Initiatives for development of dryland farming

1. In the 1950s, the Government realised the significance of dryland agriculture and decided to improve stability and productivity of drylands through research and development (R&D) efforts. Various R&D centres at different dryland locations started work for developing appropriate soil and water conservation practices.
2. In 1970, the Indian Council of Agricultural Research (ICAR) initiated the All India Coordinated Research Project for Dryland Agriculture (AICRPDA) in Hyderabad, establishing 23 collaborating centers across the nation. It marked the start of location specific adaptive research in dryland agriculture.
3. In 1985, ICAR established Central Research Institute for Dryland Agriculture (CRIDA) at Hyderabad to further strengthen basic and adaptive research. CRIDA has successfully developed a large number of technologies in rainwater management, watershed development, soil health management, cropping systems, etc.
4. The ICAR launched a flagship network project called the National Innovations in Climate Resilient Agriculture (NICRA), primarily to develop and promote climate-resilient technologies in agriculture with a special focus on rainfed regions.
5. The NICRA project has developed several climate-resilient technologies which include climate resilient varieties of different crops, resilient intercropping systems, crop diversification with alternate crops, agro-forestry systems, in-situ moisture conservation, farm pond, integrated framing systems, etc.
6. The Government is also implementing a scheme on rainfed area development under the National Mission for Sustainable Agriculture. The scheme focuses on integrated farming systems for enhancing productivity and minimising risks associated with climatic variability.
7. Schemes such as Per Drop More Crop, Soil Health Card & Paramparagat Krishi VikasYojana are also contributing to the improvement of dryl & agriculture.

Dryland farming

Dry land farming is cultivation of crops in regions with annual rainfall of more than 750mm. Inspite of prolonged dry spells, crop failure is relatively less frequent. These are semiarid tracts with a growing period between 75 and 120 days. Moisture conservation practices are necessary for crop production. However, adequate drainage is required especially for vertisols. Main emphasis is on soil and water conservation, sustainable crop yields and limited fertilizer use according to soil moisture availability.

Rainfed farming

Rainfed farming is crop production in regions with annual rainfall of more than 1150 mm. Crops are not subjected to soil moisture stress during the crop period. These are humid regions with growing period of more than 120 days. Emphasis is on disposal of excess water, maximum crop yield, high levels of inputs and control of water erosion.

Introduction to Rainfed agriculture

- In agriculture, we can find different types of crops depending on the availability of water.
- **Rainfed agriculture refers to the area under various types of crops where the cultivation is dependent upon the monsoon rainfall.**
- Rain-dependent areas can be broadly split into two: '*dry lands*', which receive less than 750 mm of rain a year; and rainfed areas, which receive more than 750 mm.
- Rainfed agriculture is practiced under a wide variety of soil type, agro-climatic and rainfall conditions ranging from 400 mm to 1600 mm per annum.

Dryland Agriculture in India

- India has 143 million ha of its geographical area under cropping, out of which 113 m ha has the potential to be irrigated, thus, 30 m ha are rainfed agriculture, even if the full potential of irrigation has been utilised.
- Presently, India has 93 m ha (recent data = 100 m ha) under irrigation. Thus, 43 m ha are under rainfed conditions.
- **Dryland agriculture refers to the region or the type of farming which is operated in sub humid to arid conditions with inefficient hydrology**, lack of irrigation facilities, complete dependency on monsoon rainfall and reflecting a typical cropping pattern of coarse grains millets and oilseeds, pulses, cotton etc.
- All dryland agriculture qualify to be called as rainfed but it is not so vice – versa, as the fundamental difference is the agro – ecological properties and the cropping pattern between the two.

Need for Dryland Agriculture in India

- India has **1/3rd of its geographical area is under humid conditions, while 2/3rd is subhumid or arid conditions**. Thus, dryland agriculture occupies larger land than the wet agriculture.
- It supports 40% of the population and occupies 66% of land. The ratio needs to be rectified, by increasing the productivity of these region
- For the balanced development of country, dryland agriculture must be emphasized.
- **Dryland agriculture is based on the industrial crops** e.g. cotton, groundnut, oilseeds, pulses, tobacco. For development of agro processing industries and to make Indianagriculture more export- oriented greater emphasis on dryland agriculture is inevitable.
- Agriculture has forwarded and backward linkages with industry and thus, if industrial crops are produced in greater quantum, meso-scale, household, cottage industries could be developed in rural areas and rural dev processes can be propelled. It will also result in industrialisation and bring about economic growth in the country.
- Dryland agriculture mostly **produces hardy and nutritious crops** eg. Jowar, bajra, ragi, pulses, oilseeds, cottonseeds, sunflower, safflower. Thus, the nutrition / malnutrition problems in poor areas can be fought with the help of dryland agriculture.
- Dryland agriculture involves cash cropping, E.g. **Jatropha cultivation can reduce the magnitude of petroleum crisis** since liquid produced by it can be added to petroleum products without reducing the efficiency.
- Dryland agriculture has the potential to **produce fodder and cattle feed**. Thus, it can help in white revolution. Also, here the pasture lands are more extensive and cattle breeds have greater per-capital yield.
- Sheep-raising has greater prospect in semi-arid areas.

Solution to Improve Efficiency and Productivity of Dryland Agriculture

- Dryland areas have a significant contribution in pulses, oilseeds, coarse grain crops, and cotton. Optimal use of dryland technologies for increased agricultural production can help to fill the demand-supply gap in pulses and oilseeds production and improve the economic status of farmers. The various methods to improve the efficiency and productivity of dryland agriculture are summed below:
- **Water-harvesting:** It means capturing rain where it falls or capturing the runoff in your own village or town. And taking measures to keep

that water clean by not allowing polluting activities to take place in the catchment. Water harvesting can be undertaken through a variety of ways.

- Capturing runoff from rooftops
- Capturing runoff from local catchments
- Capturing seasonal floodwaters from local streams
- Conserving water through watershed management

- **Agronomical practices on scientific basis** e.g. Crop rotation, intercropping etc
- **Soil preparation:** No soil is ideal hence it necessitates the preparation of soil before and after the cultivation. Soil fertility may be lost due to continuous farming; for the replenishment of soil contents, it is prepared prior to sowing of seeds. In agriculture, ploughing, levelling, and manuring are the three steps of soil preparation.
- **Organic farming:** Organic farming is a system that avoids or excludes use of synthetic inputs like pesticides, fertilizers, hormones, etc. and relying on techniques like crop rotation, organic wastes, farm manure, rock additives and crop residues for plant protection and nutrient utilization.
- **Watershed management:** Watershed management implies effective conservation of soil and water resources for sustainable production with minimum non-point resources (NFS) pollutant losses. It involves the management of land surface and vegetation so as to conserve the soil and water for immediate and long-term benefits to the farmers, community, and society as a whole.

The various measures adopted under soil and water harvesting is

1. Vegetative barriers
2. Building of contour bunds along contours for erosion
3. Furrow/Ridges and Furrow ridge method of cultivation across the slope.
4. Irrigation water management through drip and sprinkler methods.
5. Planting of horticultural contour species on bunds.

Ecological conservation techniques in farming: Ecological farming includes all methods, including organic, which regenerate ecosystem services like: prevention of soil erosion, water infiltration and retention, carbon sequestration in the form of humus, and increased biodiversity. Many techniques are used including no-till, multispecies cover crops, strip cropping, terrace cultivation, shelterbelts, pasture cropping, etc.

Watershed Management

In India, National Watershed Development Project for Rainfed Areas (NWDPRA) was launced to generate all available & possible natural resources and its effective utilization for sustainble agriculture and the effective management of soil-water resources to uplift the socio-economic status of the rural people wherein emphasis was applied on rain water conservation for development the ares under rainfed conditions. With a view to develop suitable agro-technology for soil water conservation, cropping systems, agro-forestry and livestock production and for its effective transfer under massive scale, the Operational Research Projects for Resources Development and Management on Watershed Basis was formulated to meet the location-specific need along witth need based research programme. At present, out out of 141 million hectare net area sown, about 100 million hectare of land was remained rainfed (Tripathi *et. al.*,1990) which contributed about 42% of the total food grain production in the country. Most of the land occupied under rainfed conditions was predominantly based on monocropping where after harvest of rainy season crop(s), nothing could be grown and remained fallow during winter months due to lack of irrigation water or adequate rainfall in these areas (Zaman,1993). Growing second crop(s) with necessary soil and water conservational practices was possible with varying degree of success in such areas (Zaman and Mallick, 1987). Watershed was appeared as integrated management of all possible natural resources like soil-water-plant-animal and human to maximize agricultural production on sustainable basis (Zaman,1993). The performances of various watershed located at different agro-climatic zones in India revealed that the production of food, fodder, fuel, fibre and fruits along with the animal components increased considerably (Bhardwaj and Singh,1990). The achievement of all these watersheds operated in the country were quantified on the basis of results obtained under the Operational Research Projects for Resource Development and Management. With a view to increase the agricultural production on sustainable basis and to bring more area under crop cultivation specially introducing second crops during winter months under appropriate soil and water convervational measures along with increasing or creating irrigation potential collecting run-off, harvesting and recycling of excess rain water, the programme was undertaken at location-specific watershed under red and laterite tract.

Thus the programme on watershed management was found successful and accepted as pivotal for improvement of crop yields in rainfed areas that left the tremendous expansion possibilities including development of agro-forestry, agri-horticulture, silvipasture and animal husbandry along with fisheries.

Selection of suitable crop(s) and its appropriate varieties, introduction of new crops with proper sequences at different land situations the areas through proper utilization of all possible natural resources increased the crop as well as land productivity. The watershed management programme was proved both labour and capital intensive and benefit cost ratio was found as high as 7:1. So the programme offered high avenues for employment generation in the rural areas which augmented the average income of all the benefiaciaries of the adopted villages. Barron *et. al,* (2003), Zaman (2005) stated that watershed refers to a natural geo-hydrological unit that drains at a common point. In other word it may be described as an area of land from which the runoff water flows through a definite of drain channel, stream or river. This definite point is known as outlet.

The size and shape of Watershed

The size and shape of any watershed may vary depending upon the morphology of the land surface. Based on the total area of contributing run-off water to be drained at any outlet, watersheds are subdivided into different categories:

a	Catchment	Drainage area of a primary river which flows to sea having area more than 1 lakh ha
b	Sub-catchment	Drainage area of a secondary river or tribute having area more than 40,000 ha
c	Watershed	Drainage area of a tertiary stream having area 4,000-40,000 ha
d	Sub-watershed	Drainage area of a quaternary stream having area 2,000-4,000 ha
e	Mini-watershed	Drainage area of a streamlet having area 1,000-2,000 ha
f	Micro-watershed	Drainage area of a streamlet having area 500-1000 ha

Spectacular improvement in agricultural production through green revolution in several parts of India has been achieved due to high yielding varieties and application of irrigation and fertilizers and the improvement is not sufficient for feeding the increasing population unless a breakthrough in crop production technology takes place particularly in rainfed and dry areas which comprise about 65% of the cultivable land of our country. The increase crop production in those areas is linked with proper soil resource management along with water resource development.

The vast rainfed and dry areas are usually associated with soil degradation, aridity, water scarcity, poor soil fertility, uncontrolled grazing, decline of forest, poor socio-economic condition of the human resources. Development of such areas in recent years are undertaken through proper physical and biological management and referred as watershed management. Basically the purpose of integrated watershed management is to follow improved land management for preventing soil loss and enhancing crop productivity, reducing run-off, in-situ and ex-situ water harvesting and conservation, maintain ecological balance between land-water-plant-animal kingdoms (Zaman,2002).

The main principles of watershed management based on resource conservation, resource generation and resource utilization are as follows:

a) Utilization of land as per capability
b) Production of productive top soil
c) In-situ and Ex-situ water conservation
d) Increasing cropping intensity
e) Safe utilization of marginal lands
f) Enhancing crop productivity
g) Stabilization of farm income
h) Improvement of infrastructural facility
i) Setting up of small scale agro industries
j) Improvement of animal kingdom
k) Upliftment of socio-economic status

Behera and Panda (2006), Bosch (2008) and Zaman (2003) stated that watershed management is a multidimensional developmental approach for steady uplift of masses living in that area. The main objectives of watershed development programme can be described by a word "POWER" which means

P = Production of food-fodder-fuel-fibre-fish-milk

= Pollution control

= Prevention of floods

O = Over exploitation of natural resources to be minimized

W = water conservation and storage

E = Erosion control

= Economic stability

= Ecosystem safety

= Employment

R = Ground water recharge

= Reduction of draught through management

The above said objectives can be achieved by proper planning & implementation of an integrated watershed development programme. The developmental planning should ensure proper blending of people's requirement, suggestions, technical information about the area to be treated, technically variable solutions and methodology for bringing about desired improvement. The proper planning are usually carried out by the following ways:

A. Collection of soil-climate-geological data base

i) Detailed soil survey of the watershed area has to be carried out to find out soil type, thickness, reaction, horizon differentiation, slope;

ii) Rainfall and its frequency, distribution and duration along with other meteorological parameters;

iii) Existing water resources like pond, dug well, shallow tube well, % of irrigated areas;

iv) Existing drainage pattern, channel system, rate and quality of run-off, geomorphology;

v) Physiography like slope, relief, size and orientation of the area.

Preparation of thematic maps

With the help of 1: 50,000 remote sensing imagery obtainable from NRSA, Hyderabad and based map of the area through GIS techniques following thematic maps to be prepared:

1. Drainage map
2. Relief map
3. Soil map
4. Slope map
5. Geomorphological map
6. Land use map

B. Analysis of secondary data

a) **Natural resources:** secondary data about the existing natural resources like crop productivity, cropping pattern, and varieties used, fertilizer and pesticide use, livestock population, various land uses are to be collected from various sources and Government publications. Certain random field verifications are required to ensure the quality of data. All these data collected to be analyzed to yield much information.

b) **Human resources:** various data on socio-economic aspect like male female ratio, status of education, number of schools, prevailing caste and its percentage to total population are to be analyzed.

c) **Prioritization of watershed for development:** prioritization of watershed is a criterion to select a watershed where developmental work is utmost necessary for economical uplift of that area. Many models are in use to prioritize a watershed. Among these models CSWSI model is commonly followed. In this model various physical and biological resources are used as indicators. A simple formula is used as,

The cumulative SI of different indicators along with its weightage and rank give the prioritization value. More higher is the SI value, the watershed needs development. Some indicators and its weightage and ranks are given below:

Table SI= Weightage of indicators X Rank

Indicator		Range of parameters			Rank
		Low	Medium	High	
1.	Varieties	Local (0.6)	Improved (0.3)	HYV (0.1)	9
2.	Cropping intensity	100-110 (0.6)	110-120 (0.3)	>120 (0.1)	1
3.	Crop yield (optimum yield)	<50% (0.6)	50-90% (0.3)	>90% (0.1)	10
4.	Irrigation	<30% (0.6)	30-50% (0.3)	>50% (0.1)	4
5.	Marginal farmers	<25% (0.1)	20-50% (0.3)	>50% (0.6)	7

On completion of watershed planning, the developmental work should be initiated in the following ways:

a) Work should be carried out from ridge to valley;
b) Major structure to be constructed;
c) Both mechanical and vegetative measures for effective moisture conservation should go side by side;
d) Long term soil conservation work should be initiated;
e) Awareness programme should be initiated;
f) Co-operative societies to be constructed; and
g) Knowledge based support compared to finance based support should be initiated.

Various state government departments and NGO's are engaged in planning and development of many identified micro-watersheds particularly in drought prone area in a big way. However, sustainability of this development mostly depends on the adoptability of the farmers' friendly technologies co-operative attitude by the beneficiaries and dedication of the implementing agencies.

Water resource conservation

The average annual rainfall in India is about 1170 mm. It is distributed unevenly and erratically over the country's geographical area of 329 million hectares. This account for the total annual precipitation including the snowfall is around 4000 billion cubic-meter. There is large variation in the distribution of rainfall from region to region, season to season and year to year. Rainfall mainly occurs in three to four months during the monsoon season covering only 100 hours in a year. The normal rainfall is as low as 100 mm in the western parts of Rajasthan to over 11000 mm at Mausynram in Meghalaya. The spatial and temporal variation of rainfall is so complex and wide that led to flooding in some areas and drought situation in other places. The increasing disparity between the water availability and growing demands for various sectors needs water conservation on an urgent basis.

It is to be mentioned here that Agriculture is the major user of water and it consumes about 80 per cent of the current level of water utilization. However, with the increasing demand from other competing sectors, the allocation of water is likely to reduce by 10 to 15 per cent in the next two decades. In addition, shrinking water resources due to falling groundwater tables, silting of reservoirs and pollution of water resources from untreated domestic waste, industrial effluents and agro-chemicals are threatening the acute water shortages in the coming years as well as the sustenance of existing levels of agricultural production. Besides, potential agricultural productivity in many areas is not realized due to poor irrigation management. The water use efficiency is as low as 35 to 40 per cent at the field level. Concomitantly, the problems on the quality of groundwater and rapid decline in water table depth due to unscientific exploitation are also noted in many parts of the country which has already been designated as the dark and grey blocks. Thus a proper assessment of the availability of surface and sub-surface water is the cornerstone for proper planning, development and management of this scarce and vulnerable resource. The highest scope of conserving of the precious water resource lies with the agricultural sector followed by other sectors of importance. In addition, it needs to be conserved in all sectors of its uses through various measures from the point it falls as rainfall till it drains into sea for maintaining self-sufficiency in food production and in order to keep India as an economic country with the improved living style.

Aims of water conservation

Water conservation basically aims at matching demand and supply of water. The strategies for water conservation may be demand oriented, supply oriented and management oriented. The demand oriented strategies mostly depend on the field of water use i.e. domestic, irrigation, industrial, navigation and environmental requirements. The supply oriented strategies such as creation of storages, long distance transfer and control of water loss through evaporation are generally common to all the fields of water use. The management oriented strategies include mainly the operational and maintenance aspects for judicious and economic water utilization.

Dimensions of water conservation

Water conservation can be achieved in following three ways:

i) *Water resources conservation* – efficient management of rainwater through storage, allocation and transfer for optimal use and preservation of the quality including its supporting ecosystem.

ii) *Water use conservation* – water supply and distribution with minimum losses and consumption through prevention of wastage.

iii) *Technology and cropping package* – efficient use of water through adoption of water saving technologies and cropping patterns.

Water Conservation Strategies and Technology Options

Some of the conservation measures & technologies available in the agricultural sectors are discussed in the following headlines:

Creation of surface storages

Creation of surface storages is a water conservation measures as it prevents the water flowing waste into the oceans without being put to beneficial use. India with the existing/on-going/proposed major, medium and minor reservoirs in combination has storage capacity of 400 BCM comparing to the ultimate potential of 690 BCM which indicates the expansion of surface storage capacity is critically important to enhance the water availability to cater the needs of various sectors. Greater attention has also been paid to provide better soil condition for retention of monsoon rains in the soil as groundwater storage through adequate soil and water conservation measures on watershed basis.

Conclusion

Despite numerous constraints, rainfed, watershed and dryland farming can be successful by considering local climate and soil conditions, selecting appropriate crops, and utilizing suitable technologies. By adopting an integrated farming model of prper utilization of this vast majority of areas have the potential to cultivate multiple crops within a single season, incorporating horticultural or livestock elements for added benefits.

References

Amani, M. & Rahmani, A. (2020). "Integrated Watershed Management: A Tool for Sustainable Development in Arid Regions." Land Use Policy, 92, 104405.

Bandyopadhyay, S., & Bhatnagar, A. (2017). "Soil and Water Conservation Practices in Rainfed Agriculture: A Review." International Soil and Water Conservation Research, 5(1), 47-60.

Bhatia, A., & Dey, M. (2017). "Role of Agroforestry in Watershed Management in Dryland Areas." Forest Ecology and Management, 396, 38-50.

FAO (Food and Agriculture Organization). (2018). The Future of Food and Agriculture: Trends and Challenges. FAO.

Kaur, S., & Sidhu, D. (2019). Sustainable Rainfed Agriculture: Innovations and Technologies. Springer Nature.

Ogbonna, A., & Nwachukwu, I. N. (2021). "Low-Cost Technologies for Rainfed Land Management in Semi-arid Regions." Agricultural Water Management, 242, 106408.

Quinn, K., & Le Roux, C. (2019). Climate-Smart Agriculture: Enhancing Resilience to Climate Change in Rainfed Production Systems. Wiley-Blackwell.

Ranjan, R., & Singh, R. (2019). "Water Management in Rainfed Agriculture: Techniques and Practices." Agronomy for Sustainable Development, 39(1), 18.

Sharma, D. R., & Soni, P. (2018). Watershed Management for Sustainable Agriculture and Rural Development. New India Publishing Agency.

Thorat, D. B., & Chinchmalatpure, A. S. (2020). "Water Harvesting and Watershed Management in Dryland Areas." Journal of Soil and Water Conservation, 75(4), 238-246.

Multiple Choice Questions

1. What is the geographical area of India?

   a) 329.73 million ha b) 250 million ha
   c) 500 million ha d) 400 million ha

2. By which year did India's annual food grain production increase to 206 million tonnes?

   a) 2000 b) 2001
   c) 2002 d) 2003

3. Which of the following states has the highest cropping intensity according to the text?

   a) Bihar b) Telangana
   c) West Bengal d) Uttar Pradesh

4. What type of agricultural practices are emphasized for improving yield in dryland areas?

   a) Monocropping b) Organic farming
   c) Crop rotation d) Both b and c

5. How much of India's geographical area is under rainfed agriculture?

   a) 143 million ha b) 100 million ha
   c) 30 million ha d) 50 million ha

6. In which year was the All India Coordinated Research Project for Dryland Agriculture initiated?

   a) 1960 b) 1970
   c) 1985 d) 1995

7. Which crops are primarily associated with dryland agriculture in India?
   a) Rice and Wheat
   b) Pulses and Oilseeds
   c) Fruits and Vegetables
   d) All of the above
8. What does the acronym NICRA stand for?
   a) National Initiative for Climate Research in Agriculture
   b) National Innovations in Climate Resilient Agriculture
   c) National Implementation for Climate Resilient Agriculture
   d) None of the above
9. What percentage of India's agricultural water consumption is accounted for by the agriculture sector?
   a) 50%
   b) 60%
   c) 80%
   d) 90%
10. What is the main objective of watershed management according to the text?
   a) Increase monocropping
   b) Enhance agricultural production sustainably
   c) Reduce employment opportunities
   d) Minimize crop yield

Answer Key

1	a	2	a	3	d	4	d	5	a	6	b	7	b
8	b	9	c	10	b								

7

Effective Utilization of Rainfed Lowland

Rice farming under different subsystems of lowland ecology

1. **Rainfed lowland with limited water at seeding and excess water at later growth stages:** In rainfed lowland areas, the rice is direct seeded in dry soil in April-May before onset of monsoon (June) and seeds germinate with the showers received after sowing. Drought at the seedling stage kills seedlings and results in poor stand establishment. Furthermore, heavy rains in catchment areas of river basins result in water logging and flooding at late growth stages. The water hinders tillering and results in poor panicle number less than 200/m^2 as against 350 /m^2 of high yielding rice cultivars.
2. **Rainfed lowlands with limited/excess water at transplanting and excess water at later stages:** When on-set of the monsoon is delayed, transplanting delayed because the farmer has to wait for adequate rains for land preparation. Monsoon rain submerges the transplanted crop, resulting in poor stand. In some areas, the land gets inundated before transplanting and farmers generally transplant aged and tall seedlings to keep the plant parts above the water. Plant survival in deeper water depends on rise in water level and on the elongation ability and submergence tolerance of the plant. Varieties that elongate fast develop weak culms and may lodge when the water level recedes. In this situation tillering is poor and panicle per unit area are less.
3. **Assured irrigated areas with excess water during crop growth due to poorly drained condition:** Rice grown in the river deltaic areas of Krishna, Godavari and Indo-Gangetic rivers suffer from water stagnation even though some areas are having irrigation facilities. Heavy rains in the catchment cause flooding and stagnation in rice fields because of congested drainage or lack of drainage caused by topographically saucer-shaped land. These conditions develop during peak monsoon (July-September). The late planted crops, being at an early stage of growth, suffersfro water stagnation. In these areas, transplanting early in the season is necessary for plants to grow enough to escape submergence.

4. **Rainfed areas with excess water through out the crop growth:** Some lowland areas as get inundated with the first few monsoon breaks before seedlings and transplanting operation take place. Such areas fall in to two categories: (i) those in which water deptdose not exceeds 1 m and (ii) those in which depth is exceeds 1 m. In former case, seedlings of 75 cm are normally used for transplanting. In later case, deep-water rice varieties are grown and the crop is seeded well in advance of the rains.
5. **Rainfed lowlands with flash flood:** The lowland areas where water depth varies from 0-15 cm during most of the crop growth period, transplanted rice may be submerged for a few days to 2 weeks due to flash floods caused by over-flood from river or due to excess rainfall in shorter period of time. Some cultivars, which could tolerate floods to an extent and recover, might be suitable if floodwater recedes within 2 weeks.
6. **Rainfed coastal saline areas with excess water at later growth stage:** Large part of Orissa, West Bengal, Karnatak, Tamil Nadu, Goa and Maharasthra suffer from tidal inundation. The coastal saline areas having varied characteristics, the most common being low-lying topography, inadequate drainage, fairly high water table of poor quality (5-8 m mhos/cm) and moderate to heavy soil texture. About 2 million ha of rice lands are saline and waterlogged (30-70 cm) and in these areas rice yields vary from 0.5 to 1.0 tones/ha. Cyclonic storms bring seawater to coastal areas through rivers and estuaries during May and June. The lands, even not inundated, it does not support crops because of high salt content. The salt rises to the surface through capillaries and accumulates.
7. **Rainfed lowlands with acid sulfate soils and tides:** In Kerala and West Bengal, most of the acid sulfate soils are found. Crops failure due to salinity and acidification during droughts, intrusion of salt water and deep flooding are common. Observed problems include low pH, toxicity from Al, Fe and H_2S partial pressure of CO_2, salt injury, toxicity of organic acids, deficiency of P and micro-nutrients and low nutrient status (Bloomfield and Coulter,1973); Ponnamperuma *et. al.*, 1973). The potential acid sulfate soils in tidal swamps will not acidify as long as the tidal effects are strong enough to prevent prolonged aeration. The soil remains productive only if wet conditions are maintained (Sinh *et. al.*, 1982).

Available water resources

The annual global flow of water ranges from 4000 to 4700 million ham, out of which 1400 m ham is utilizable for irrigation and other purpose. The present level of utilization of this available potential is just around 25%., out of which irrigation constitute about three-forth (Harswarup Singh, 1990). As per estimate of National Commission of Agriculture (1976), te average annual rainfall over the geographical area of the country is 370 m ham (112 cm). An amount of 20 m ham could be added, which is brought in form of catchment areas lying outside the country. It is estimated that about 85 m ham of water soaks into the soils (contributing to profile storage/groundwater). And 130 m ham evaporates into atmosphere. The remaining 180 m ham constitutes the average annual flow in different rivers, of which only 70 m ham is utilizable owing to the topographical, hydrological, climatic, soil and other limitations (Harswarup Singh, 1990). The available data on groundwater potential in AP shows the groundwater balance for development to be round 2.7 million ha. Distribution of groundwater reveals that the canal command districts of coastal areas as well as Rayalseema and Telengana, 33 in Rayalseema and 15 on the coastal districts largely in the non-ayacut areas (Siddique,1995), if policy decision is taken on groundwater development in command areas, an additional areas of 0.5 to 1.0 million ha in Andra Pradesh, Tamil Nadu, West Bengal, Orissa, Bihar and Assam could be brought under crop cultivation during *rabi* season, and summer rice could be grown. Even other crops like groundnut, maize, potato and linseed could successfully be grown in this rainfed lowland ecology after rainy season rice harvested on availability of profile moisture.

Cropping system under rainfed lowland ecology

The cropping system in these areas is highly influenced by rainfall pattern and amount of precipitation, soil type, topography within lowlands, time of rice harvesting and resourcefulness of the farmers. Despite their adequate skill for managing crops, farmers could be provided tools at the levels of crop germ plasm, crop management and cropping system to minimize losses due to vagaries of nature. With the objective to produce more quality food grain with less water in the forthcoming decade, rainfed agriculture has its pivotal role in meeting the daunting challenge.

Profile moisture in various soils

The lowland rice growing imposes high demand on time and energy for conversion of lowland soil to freely drained upland soil. This turn around time required for conversion to upland environment discourages the use of rice fallows. The vast majority of rainfed lowland rice of eastern India is kept fallows due to this reason. The raising of non-rice crops on residual moisture in lowland rice fallows depends on;

1. The rapidity with which the soil dries up and become amenable for tillage in a crumb soil structure of a garden soil;

2. Availability of irrigation after rice crop harvested;
3. Severity of aridity and mildness of temperature in post-rice season;
4. Sensitivity of the crops to a saturated soil condition; and
5. Length of time the soil remains moist.

Table 7.1: Profile moisture storage of different soils:

Soil type	Storage (cm)
Clay	76.20
Silt loam	60.96
Sandy loam	45.72
Fine sand	30.48
Coarse sand	15.24

Harvesting time of *kharif* rice

Utilization of residual moisture under rainfed lowland through various crops depends on the time of ***kharif*** rice harvest. Chandra and Manna,1988 reported that rice crop occupied in lowland field even up to December and thus crop planning could be done to sow other crop in December-January.

Table 7.2: Time of *kharif* rice harvest in different states

State	Time of harvest
Orissa	Mid-December
Bihar	Last week of November
Eastern UP	First week of December
Eastern MP	Second week of December
West Bengal	Second week of November
Assam	Last week of October

Tillage operations

Under rainfed lowland situation, after *kharif* rice harvest, it is most important to establish next crop at earliest opportunity. There would be problem for land preparation in light and medium textured soils than that with heavy clay soils, which generally occurs in lowlands. Several pulses like green gram, black gram and lentil and oilseeds like linseed and sunflower could be raised under heavy soil condition through zero tillage and dibbling of seeds in rows. The time consuming and energy demanding tillage operations is totally dispensed with and the seed broadcast either in standing crop ready for harvest pr in the stubble soon after harvest is followed. The crop so sown used to thrive on receding moisture regime. The non-rice crop is a bonus in such farming system. Surprisingly, this opportunity could not be availed of in many situations, particularly in eastern India.

Crop options for the no-till farming on residual moisture are: sun hemp for hay where soils are heavy and the non-rice growing season is short, black and green gram when whether is mild, and rape seed when ambient temperature is low for grain legume. Research agenda for no-till residual moisture farming in lowland would entail, selection of benchmark sites, development of seeding techniques, screening of crops and suitable varieties and evaluation of soil moisture conserving techniques. Besides, there is need for screening the crops towards sensitivity to tilth regarding their germinibilty and growth. It was observed that sesame required fine tilth, sunflower cloddy seddbed and black gram and *khesari* required muddy fields. The crops may also be evaluated for root penetrability. Experiments conducted at CRRI shown that the reduced tillage or shallow tillage at 0-5 cm surface conserved adequate soil moisture to grow successful crop of groundnut. Study on soil moisture profile revealed that under non-tillage treatment, the profile progressively dried, where as under reduced tillage such drying was effectively checked due to soil mulch effect. Sharma *et. al.,* (1985) observed the shallow tillage of surface soil (3-5 cm depth) during dry season and use of 1.5 – 2 cm thick layer of wheat straw mulch on wet black soil reduced evaporation by 30% and signicantly increased water use efficiency and grain yield of non-irrigated wheat , wheat, safflower, chick pea, lentil, linseed, *taramira*, mustard and toria could be established well on the residual moisture in black soil with applied minimum tillage or animal drawn improved blade harrow, bullock-drawn sweeps with tool frame.

Sowing methods and prospect of post rainy season crops

In rainfed lowland areas, deep sowing (5-7 cm) provides good opportunity for better germination and stand establishment of pulse crops. Saraf *et. al.*, (1990) reported that for sustaining the productivity of chick pea sowing should be done taking opportunity of first available shower when soil would have sufficient moisture at seeding depth to ensure good germination. Pulses like chickpea, lentil, lathyrus, could be advocated in rice fallows under such situations. Seed treatment, such as soaking of seeds in potassium dihydrogen orthophosphate (KH_2PO_4) before sowing or even seed soaking in water facilitates the germination and early establishment. Chatterjee and Maiti (1983) reported that during period of November-February, the temperature is low and provided good scope for growing a number of winter crops under rainfed lowland conditions like lentil, gram, safflower, sunflower, linseed, niger and peas depending upon topographical and agro-climatic conditions. They also reported that where winter in mild and short with possibilities of receiving some rainfall during February-March, the sequence of rice-groundnut and rice-sesame appeared to be very profitable.

Soil water retention and movement

At relatively low suction (less than 1 bar), soil aggregation and at higher suction (1015 bars) the clay minerals regulate soil water retentivity in soil (Chibber, 1964; Tamhane and DuttaBiswas, 1965; Pharande et al., 1969.) The water content at any particular suction followed the decreasing order: black soil (smectite), alluvial soil (illite) and laterite soil (kaolinite) (Ali and Biswas, 1971). Water retention and release for crop use in soil can be improved by application of FYM, straw, lime or SSP (Biswas *et al.*, 1967; Biswas *et al.*, 1971; Bhagat and Acharya, 1989). The values of organic carbon, silt + clay and CEC correlated significantly with water retention (Gupta *et al.*, 1984).

Soil water movement processes such as infiltration and percolation of rain or irrigation water into soils and n upward flux from ground water containing plant nutrients and salts are govern by hydraulic conductivity K of soil, being its water content on volume basis. Darcy's law, which governs the movement of water in soil, takes the form

$$Q = -k (-). DH / dZ$$

Where 'q' is the flux of water, 'H' the hydraulic head (pressure head + gravity head) and 'Z' is the position coordinate. Saturated hydraulic conductivity K (s) as relevant to waterlogged rice soil, is invariant at a given temperature but depends on soil properties. Its values may be in the range of 1-10 cm h-1 in coarse textured soils and 10.1 cm ha-l or even less in heavy soils. But in *rabi*season, water moves under unsaturated conditions and such movement is guided by K(-) depending on water content. Using soil water pressure head values (H) measured by tensiometer and soil water content (-) by gravimetric or neutron probe method, K(-) can be directly determined in the field (Nielsen et al., 1973; Patra and Misra, 1985). K(-) for coarse and fine textured soils of Orissa has been formulated as :

$$K = 0.362 \text{ expo } 48.17 (0\text{-}0.335) \text{ and}$$

$$K = 0.205 \text{ expo } 67.15 (0\text{-}0.468)$$

respectively, where K is expressed in cm h^{-1}.

Dongale (1987) reported that under zero tillage, infiltration rate after 180 minutes increased from 4.4 cm h^{-1} in rice mono-cropped soil to 11.0 cm h^{-1} in rice -legume double-cropped soil. Tillage increased the intake of water further.

Mulching

Mulching is an important technique for conserving residual soil moisture especially in the no-tillage plots under rainfed lowland conditions. Besides conserving moisture, it also suppresses weed growth. Chatterjee and Sen

(1977) reported that mulching the crop fields with 5 cm thick straw could increase the yields of winter crops. They observed yield improvement with mulching in case of linseed, safflower and lentil because of increased water use efficiency.

Mandal *et at.* (1990) observed that a good crop of safflower can be harvested under late sown condition after the harvest of lowland paddy even without giving any irrigation with the help of residual soil moisture and with a lower dose of each of N, P_20_5 and K_20 @ 30 kg h^{-1} provided that the soil moisture is conserved with the help of straw mulch @ 7 t h^{-1} at sowing.

Paira Cropping

Paira cropping is another useful practice for raising winter crops in rainfed lowlands. Roysharma et al. (1982) reported that lentil give grain yield *of* 15.9, 16.3 and lOA q hI at Dholi (1976-80), Kanpur (1974-76) and Barielly (1974-75), respectively with inoculated seeds and top-dressing *of* 20 kg N and 30 kg P20S h-I (twice at 40 and 80 DAS) with paira cropping. Gangasaran (1990) reported that paira cropping *of* linseed is certain parts *of* U.P. and West Bengal. Linseed is broadcast in the standing crop *of* rice 10-20 days before harvest.

In an experiment on *paira* cropping, rice seedlings were transplanted on 15th July and chickpea, oilseed flax and lentil were broadcast into the standing rice crop on 25th November at normal and one and half times the normal seed rate without and with 20 kg N + 18 kg P h^{-1} for chickpea and lentil and 30 kg N + 9 kg P h^{-1} for flax. Oilseed flax intercrop with rice gave the highest rice yield equivalent of5.1 t h^{-1} when fertilizers were applied. Sowing rates had no significant effects on yield (Thakur,1990).

Physical improvement in soil structure

Physical soil health is an important consideration in seed germination and conservation. Utilization ofresidual soil moisture in *rabi* pulses and oilseeds after rice in rainfed lowlands is depended on physical health of the soil. Biswas (1982) has conclusively suggested three practices for improvement of soil physical condition leading to increased soil productivity *viz.* (1) Judicious use of organic manure, (2) use *of* phosphates and (3) legume cultivation in crop rotation. Biswas and Khosla (1971) have earlier reported that long term manuring with FYM or sun-hemp green manuring in red and laterite soils at Bhubaneswar and Ranchi improved water stable aggregates percentage, water storage capacity and available water content. The results obtained from All India Coordinated Project on long-term Fertilizer Experiments revealed that available water holding capacity ofsoil increased from 18.8% with N alone to 19.5% with combined use ofNPK + FYM for 11 years in 8.8% was reported in laterite soils of Bhubaneswar (Nambiar and Ghosh, 1984). Long-term fertilizer experiments conducted at CRRI for17 years showed that continuous combined application of inorganic N (44.8 kg/ha) and compost (9.2 t/ha/year) significantly improved soil aggregation (Bandyopadhya *et al.,* 1969). For sustainability in production, it is neither chemical fertilizers nor the organic manures alone but their integrated use is highly beneficial (Patnaik *et al.,* 1989).

Soil fertility improvement

All India Coordinated Project on Long-term fertilizer experiments clearly showed declining trend of productivity even with application of N, P and K fertilizer due to emergence of deficiency of Sand Zn, Available S (Morgan's extractable S) decreased to around 3-5 mg/kg in plots where DAP had been applied but increased in plots where SSP had been continuously applied. This effect was quite spectacular at Barrackpore, Jabalpur and Bhubaneswar (Nambiar and Abro, 1989). In the context of growing pulses and leguminous oilseeds after rice in rainfed lowlands, supply of P, S and Ca is very important to improve crop productivity.

Effective utilization of water

In lowland areas of eastern India, because of high rainfall there is a rise reserve of ground water at shallow depth. But, except in Uttar Pradesh and some parts of West Bengal, in rest of the eastern region, the ground water utilization is very poor (Verma and Agarwal, 1990; Rao, 1990; Patnaik, 1994). Although in semi-arid and arid regions of the country, indiscriminate use of ground water lowers the water table and is expected to eventually create the problems of drought, there is less chance of and deep tube wells should be developed depending on the position of the aquifers and quality of water. This may help in providing life saving irrigation to the crops grown after rice. Lowland areas of eastern India also provide great scope for storage of rainwater in ponds, ditches or trenches at one site or at comer of the field for adoption of the water-harvested structures (Sinhababu and Venkateswarlu, 1995).

Coastal saline soil management

Coastal soil salinity generally develops due to seawater inundation, capillary rise of saline ground water and its evaporation during the dry season and deposits of windblown salts on land surface. This can be minimized by providing sea dykes with one way sluice gate to allow drainage from land to sea but to check the reverse flow of sea water into land (Yadav *et at.,* 1979). These dykes need to be stabilized by growing *Casuarina*, cashew nut and salt tolerant grasses on them. Shelterbelt and windbreak plantations along seacoast are also recommended to check deposit of wind blown salts on arable land.

Salt tolerant rice variety could be grown during wet season. Lunishree recently developed at CRRI, Cuttack and CSR 6 is more suited to coastal salt affected lowlands. Salt tolerant crops such as sugar beet, linseed, barley and even wheat have been found to be promising during dry season. Most of the land in coastal saline areas remains fallow during dry season because of high salinity or non-availability of good quality irrigation water. Bandyopadhya and Sen (1975) have found that at few locations of Sunderban area, quality of ground water is good only at a depth of 180 m or below. A study on assessment of quality of surface water available in Orissa coast canal and at few definite sites along the Suvarnarekha and Salandi rivers closer to seas coast indicate that there is scope of adopting multiple cropping in some pockets of Balasore by using water from these sources during the first half of the dry season (Panda *et. at.*, 1980)

Cropping pattern followed in rainfed lowlands

An exhaustive compilation on agro-climatic zone specific research under National agricultural Research Project indicates the feasibility and prospects of the following rice based cropping systems in rainfed lowlands of different agro-climatic zones of eastern India (Ghosh, 1991)

Table 7.3: Important cropping pattern followed in rainfed lowlands of eastern India

Assam	Jute-rice-rapeseed Jute-rice-lentil Rice-black gram
Bihar	Rice-linseed Rice- Khesari (*Piara*crop) Rice-chickpea
Eastern MP	Rice-khesari Rice-linseed Rice-Black gram Rice-chickpea Rice-rapeseed –mustard/kulthi
Orissa	Rice-green gram Rice-black gram Rice-kulthi Rice-sesamum Rice-groundnut Rice-khesari Jute-rice-green gram/black gram Rice-field pea (*Piara* crop)
Eastern UP	Rice-Chickpea Rice-peas Rice-linseed Rice-lentil

West Bengal	Rice-lentil Rice-khesari (*Paira* crop) Rice-rape seed/mustard Rice-linseed Rice-safflower/sunflower

Crop response under residual soil moisture ecology

The various crops have been identified for cultivation under residual soil moisture condition under different agro-climatic regions.

Table 7.4: Rice based crop yield under residual soil moisture in irrigated ecology

Crop	Variety	Yield t/ha		Place of work	Source
		Residual moisture	Irrigated Ecology		
Mustard	Pus a Bold	0.37	0.37	Cuttack	Dinesh Chandra
	PusaBahar	0.37	1.03	Cuttack	1993
Raimash	PDR 14	0.53	1.07	Cuttack	-do-
Peanut	ICGS 44	0.71	1.36	Cuttack	Chandra *et al.*1988
Mungbean	PDM 11	0.11	0.10	Cuttack	-do-
Urdbean	PDU 54	0.15	0.18	Cuttack	-do-
Wheat	Swati	3.72	4.16	Bilaspur	Parihar *et al.*, 1955
Chickpea	JG74	2.02	1.88	Bilaspur	-do-
Indian mustard	Baruna	1.05	1.22	Bilaspur	-do-
Sunflower	Modem	1.91	2.11	Bilaspur	-do-

Irrigating these crops sometimes reduced the yields e.g. mung bean safflower and horse gram, Mung bean variety PusaBaisakhi, safflower varieties JSF-l and APRR-3 and horse gram (local varieties) were well suited under the residual moisture in rainfed lowland ecology at Cuttack.

Future Strategies of research and development

The extent of occurrence of rainfed lowland, particularly in eastern India is large. But cropping intensity in this ecosystem is low, most probably due to the lack of technology of efficient utilization of residual soil moisture during dry season. In fact, intensification of cropping with the utilization of this rice-natural reserve in rainfed lowlands of eastern India is possible and can provide food security to the growing population of this region. To achieve this objective, the following research and development strategies may be pursued.

1. Improvement in soil moisture storage and regulation of upward flux of water through soil structural manipulations;
2. Multi-locational evaluation of cropping systems, introduction of new crop species and varieties of different pulses and oilseeds in dry season

under rainfed lowland situations along with their screening for sensitivity to soil tilth and soil/air temperature and for root penetrability;

3. Genetic improvement in pulses, oilseeds and other crops with particular reference to residual soil moisture utilization;
4. Identification of specific limiting factors of crop production (bio-physical and socio-economic factors) and their mitigation options.
5. Integrated approaches of nutrient management, water management and plant protection measures.

References

M. Ray, Roy, D C. and Zaman, A. 2016. Evaluation of rice (Oryza sativa L.) – based cropping systems for increasing productivity, resource use efficiency and energy productivity in Coastal West Bengal. Indian Journal of Agronomy. 61(2): 131-137;

Maitra, S and Zaman, A. 2017. Brown manuring - an effective technique for yield sustainability of cereals crops. International Journal of Bio-resources Science 4(1): 1-5;

Maitra, S and Zaman, A., 2017. Organic farming in India: status, scope and potential, http://aditicert.net/newsletter/Organic-farming-in-India-Status-Scope-and-Potential.pdf:2017: 1-10;

Patra, S. and Zaman, A. 2017. Agricultural drainage in saline and non-saline waterlogged lands. In Crop and Water Productivity in Agriculture, BCKV Publication 2017. p 55-62;

Patra, S. and Zaman, A. 2017. Tapping the unrealized potential of residual soil moisture through mustard and rapeseed paira cropping in lowland rice ecosystem, Success stories of Water Management Project, Gayeshpur Centre (Series-1, Published by the ICAR in its Website: http://www.icar.org.in pp 114 – 117;

Sarkar, S., Sarkar, A. and Zaman, A. 2016. Yield, water use and economics of chickpea (Cicer arientum) as influenced by different levels of irrigation and mulches. Indian Journal of Agronomy, 61(4): 479-483;

Sarkar, S., Sarkar, A. and Zaman, A. 2017. Effect of irrigation and phosphorous levels on braod beans (Vicia faba L.) for improving growth, yield and water extraction pattern, Legume Research: An International Journal. 40(2): 257-263;

Sarkar, S., Sarkar, A. and Zaman, A. 2017. Effect of irrigation and phosphorous fertilization on growth, yield and nodulation of broad beans (Vicia faba L.) Indian Journal of Agricultural Research 51(1): 69-73;

Zaman, A. 2017. Water Management under rainfed ecology: Rainwater harvesting technologies for crop and water productivity. http://agroinformatics2017/3.blogspot. in. water management-zaman.html. 22: 1-6;

Zaman, A. and Maitra, S., 2017. Organic Agriculture: Prospects, Problems and Constraints, http://aditicert.net/newsletter/Organic-Agriculture%20-Prospects-Problems-and-Constraints.pdf: 2017: 1-13;

Zaman, A. and Patra, S. 2017. Water resource management for sustainable agriculture. In Crop and Water Productivity in Agriculture, BCKV Publication 2017. p 1-10;

Zaman, A., Manik Lal, P. Zaman and Md. Hedayetullah. 2016. Adoption of Improved water Management Technologies for sustaining agricultural productivity. International Journal of Agronomy and Crop Science, 1(1) 34-39;

Zaman, A., Zaman, P. and Maitra, S. 2017. Water resource development and management for agricultural sustainability. Journal of Applied and Advance Research, 2(2):73-77;

Zaman, A., Zaman, P., Hedayetullah, M. and Talukder M L. 2016. Effective utilization of lowland ecosystem for water and crop productivity. Journal of Ecology, 5 (1); 1-4;

Zaman, A., Zaman, P., Hedayetullah, M. and Talukder M L. 2016. Management of acid soils for sustainable crop production in eastern India. Journal of Crop Science & Technology, 4(1): 22-32;

Zaman, A., Zaman, P., Hedayetullah, M. and Talukder M L. 2016. Water resource development and management for agricultural sustainability in eastern India. Global Journal of Biological Research, 1(1): 33-36;

Multiple Choice Questions

1. In rainfed lowland areas with limited water at seeding, rice is typically:
   a) Transplanted immediately after monsoon
   b) Direct seeded in dry soil before monsoon
   c) Grown exclusively in flooded fields
   d) Harvested earlier in the dry season
2. During delayed monsoon onset, what challenge do farmers face during transplanting?
   a) Excessive soil erosion
   b) Poor seed germination due to drought
   c) Water logging and submergence
   d) High competition from weeds
3. In river deltaic areas, water stagnation leads to poor productivity primarily due to:
   a) Lack of rainfall b) Congested drainage systems
   c) Poor quality of seeds d) High temperatures
4. Which of the following statements about rainfed areas with flash floods is true?
   a) No cultivars can tolerate flooding
   b) Some plants can survive submersion for up to 2 weeks
   c) Flash floods occur only at the end of the growing season
   d) Floods are beneficial for all rice varieties
5. The coastal saline areas in India generally experience what issue affecting rice yields?
   a) High altitudes leading to poor drainage
   b) Tidal inundation and high salt content
   c) Constant droughts during the growing season
   d) Excessively dry soils from lack of irrigation

6. Which soil type in rainfed lowland ecosystems has the highest water storage potential?

   a) Sandy loam b) Coarse sand
   c) Clay d) Fine sand

7. What is a significant benefit of mulching under rainfed lowland conditions?

   a) It increases soil temperature
   b) It suppresses weed growth and conserves moisture
   c It enhances the salinity of the soil
   d) It decreases the soil's nutrient content

8. Paira cropping involves:

   a) Intercropping rice with deep-rooted crops
   b) Sequential planting of crops in different seasons
   c) Broadcasting winter crops in standing rice
   d) Only planting legumes after rice harvest

9. Which practice is NOT recommended for improving soil physical health in rainfed lowlands?

   a) Use of organic manure
   b) Continuous use of only chemical fertilizers
   c) Crop rotation including legumes
   d) Application of phosphates

10. The annual global flow of water available for irrigation purposes is estimated to be around:

   a) 1000 million ham b) 1400 million ham
   c) 2500 million ham d) 4000 million ham

Answer Key

1	b	2	c	3	b	4	b	5	b	6	c	7	b
8	c	9	b	10	b								

8

Scheduling of Irrigation

Irrigation Scheduling: Approaches and Advantages

Efficient irrigation scheduling aims to optimize crop yields while maximizing water use efficiency and minimizing soil damage. This involves determining the correct timing and amount of water to apply for each irrigation event. Several criteria can be used for scheduling irrigations, broadly classified into climatological, soil water regime, and plant-based approaches. The most suitable criterion depends on specific soil, plant, climate, and management factors.

Climatological Approach

Under conditions where soil water is readily available, the potential rate of water loss from a crop primarily depends on the atmosphere's evaporative demand. This approach uses potential evapotranspiration (PET) or cumulative pan evaporation (CPE) over short periods as an index for irrigation scheduling.

Penman (1948) defined PET as the water transpired by a short, uniformly green crop completely covering the ground and never lacking water. He also noted that PET is generally a fraction of pan evaporation under the same conditions. PET can be estimated using various techniques, including lysimeters, energy balance calculations, aerodynamic approaches, and empirical formulas. Based on PET or crop water use rates over short periods, irrigation can be conveniently scheduled. Lysimeters isolate the crop root zone, allowing accurate determination of water balance components by controlling and measuring difficult-to-assess processes.

Soil Water Regime Approach

This approach uses the available soil water held between field capacity and the permanent wilting point in the effective crop root zone as a guide for irrigation scheduling. Soil moisture tension, which reflects how tightly water is held by soil particles, can also serve as an indicator.

Feel and Appearance: This simple, traditional method relies on visual assessment and tactile examination of soil samples. Experience improves accuracy, and guidelines help estimate soil moisture content and determine the depletion of available soil moisture (DASM). Despite being subjective, this method can be satisfactory when combined with local knowledge.

Depletion of Available Soil Moisture (DASM): This method uses the permissible depletion level of available soil moisture in the effective root zone as an index for irrigation timing. For many crops, irrigating at 20-25% DASM is optimal during moisture-sensitive stages, while 50% DASM may be suitable during other periods.

Soil Moisture Tension: Soil moisture tension, measured using tensiometers at a specific depth in the root zone, can also guide irrigation scheduling. Tensiometers implicitly account for soil water storage properties and water use rates influenced by climate and plant vigor. Pairs of tensiometers, placed at different depths, help determine when to start and stop irrigation based on tension readings. This method is frequently used for orchards and vegetables in coarse-textured soils.

Plant indices method

1. Visual symptom,
2. Mini plot technique,
3. Plant growth,
4. Plant population,
5. RWC,
6. Plant water potential,
7. Canopy temperature,
8. Indicator plant and
9. Critical growth stages

Advantages of Irrigation Scheduling

Effective irrigation scheduling offers numerous benefits

- **Optimized Water Rotation:** Enables farmers to effectively manage water distribution among fields, minimizing crop water stress and improving yields.
- **Reduced Water and Labor Costs:** Minimizes irrigations, maximizing soil moisture use and lowering associated expenses.
- **Lower Fertilizer Costs:** Reduces nutrient loss through surface runoff and deep percolation (leaching).
- **Increased Net Returns:** Improves crop yields and quality, boosting profitability.

- **Minimized Water-logging:** Reduces drainage requirements, preventing water-logged conditions.
- **Controlled Root Zone Salinity:** Allows controlled leaching to manage salinity.
- **Additional Returns:** Frees up water for irrigating non-cash crops during dry periods.

A typical lysimeter installation in field condition

Lysimeters, used to directly estimate water balance components in a cropped field, offer valuable data for calibrating other methods. Their accuracy in detecting changes in soil water varies. Weighing lysimeters, which measure water loss directly through mass changes, can determine evapotranspiration with high precision (within a few hundredths of a millimeter) and over short time periods. Non-weighing lysimeters calculate evapotranspiration by subtracting drainage water collected from the bottom of the lysimeter from the total water input. A critical requirement for lysimeter use is ensuring that the vegetation within and immediately surrounding the device is perfectly matched in terms of height and leaf area index. However, lysimeters are expensive to build and require meticulous operation and maintenance, limiting their use to specific research applications.

A simpler alternative for irrigation scheduling involves using open pan evaporimeters. The transpiration rate of a crop is closely related to the evaporation from these devices, which integrate the effects of climatic parameters into a single measurement. For example, studies have shown that wheat requires 75-100 mm of cumulative pan evaporation (CPE) at Ludhiana, sugarcane needs 75 mm CPE in Maharashtra, green gram benefits from 180 mm CPE at Ludhiana, and sunflower requires 60 mm CPE at Bangalore. Prihar et al. (1974) advocated for scheduling irrigation based on the ratio between irrigation water depth (IW) and cumulative evaporation from a Class A pan evaporimeter minus precipitation (CPE) since the last irrigation. An IW/CPE ratio of 1.0 suggests irrigating with an amount of water equal to the evaporation loss.

Another approach to irrigation scheduling relies on observing plant indices. Plants exhibit signs of water stress, making them valuable indicators. Visual cues like wilting, drooping, leaf curling in maize, foliage color changes, and leaf angle adjustments in beans can signal the need for irrigation. The appearance of carotenoid or anthocyanin pigments, shortened internodes in sugarcane and cotton, stem elongation retardation in grapes, leaf abscission, lack of new growth, and redness at terminal growth points in almonds can also indicate water stress.

The "soil-cum-sand mini-plot technique," also known as profile modification, is another method. This technique involves creating a small area with reduced water-holding capacity by mixing sand with the soil. Plants in this mini-plot will exhibit wilting symptoms sooner than the rest of the field, serving as a warning to irrigate the entire crop. A 1x1 meter area is excavated to a depth of 1 meter, and the soil is mixed with approximately 5% sand by volume. The pit is then refilled with the mixture, compacting each 15 cm layer to maintain the original bulk density. When the mini-plot plants show wilting, it signals the need for irrigation.

Increasing plant population density in representative spots (1x1 meter areas) can also be used as an indicator. A plant population 1.5 to 2.0 times higher than optimum will deplete available water more rapidly, causing earlier wilting and signaling the need for irrigation;

Finally, monitoring the rate of growth can be a helpful approach. Plant growth is dependent on turgor pressure, which is influenced by soil water availability. Reduced stem elongation or fruit circumference growth rates can indicate water stress. For example, in cotton, inter nodal length elongation of 0.5-1.0 cm/day indicates approaching critical soil moisture levels. Similarly, in oranges, a fruit circumference growth rate below 0.2-0.3 mm/day suggests that irrigation is necessary.

Approaches to Irrigation Scheduling

Scientific irrigation scheduling aims to optimize crop yields and water use efficiency by determining the precise timing and amount of water to apply. This minimizes water losses and protects soil properties. Multiple criteria are used for irrigation scheduling, based on the soil, plant, climate, and management practices.

Advantages of Irrigation Scheduling

Proper irrigation scheduling offers several benefits:

- Optimizes water distribution among fields to minimize crop water stress and maximize yields.
- Reduces water and labor costs by decreasing the number of irrigations.
- Lowers fertilizer costs by minimizing runoff and deep percolation.
- Increases net returns through improved crop yields and quality.
- Reduces waterlogging by controlling drainage needs.
- Helps manage root zone salinity.
- Allows for irrigating additional crops during water-scarce periods.

Criteria for Irrigation Scheduling

Various criteria can be broadly classified into three categories: soil water regime, plant indicators and critical growth stages

1. **Soil Water Regime:** This method uses the available soil water between field capacity and permanent wilting point within the crop root zone as an indicator for scheduling irrigations. Soil moisture tension can also guide irrigation timing. One method of assessing soil water regime is **feel and appearance of soil**.
   - **Feel and Appearance of Soil:** This simple method involves visually and tactilely assessing soil moisture by hand. With experience, users can estimate soil moisture content and schedule irrigations based on the depletion of available soil moisture (DASM). While subjective and reliant on experience, this method can be useful when combined with local knowledge.
2. **Plant Indicators: Relative Water Content (RWC):** Introduced by Weatherly (1950), RWC measures a plant's actual water content relative to its water content at full saturation. It is calculated as:

   RWC reflects the balance between atmospheric evaporative demand and root water absorption. It serves as a reliable indicator of plant water stress, aiding in irrigation scheduling. For example, critical RWC levels for cotton and sesame are 72% and 75%, respectively; irrigation should be scheduled when plants reach these values. However, RWC measurement can be tedious and time-consuming.
3. **Plant Water Potential:** Plant water potential measures the energy status of water in a plant, similar to the tension of water in soil. It is a strong indicator of physiological processes. Leaf water potential can be measured using a pressure bomb/chamber in situ or a dye method in the laboratory. Critical plant water potential values for yield reductions are 1.2 to 1.25 MPa for cotton throughout its life cycle. For sunflower, these values are 1.0, 1.2, and 1.4 MPa at vegetative, pollination, and seed formation stages, respectively. For maize, irrigation is needed if canopy temperature exceeds ambient temperature by more than 0.7°C between 13:30 and 14:00 hours. Tomato shows no temperature variation sensitivity to water stress.
4. **Indicator Plants:** Some researchers suggest using indicator plants to guide irrigation scheduling. For example, irrigating wheat based on wilting symptoms in maize and sunflower can result in high grain yields.

5. **Critical Growth Stages:** Crop plants have growth phases that are particularly sensitive to water stress. The most critical stage is the one where water deficiency leads to the greatest yield loss. Irrigation at these critical stages can maximize water use efficiency, particularly when water is scarce. Scheduling irrigation based on these stages is simple for farmers but doesn't account for available soil water. Excessive irrigation without a significant water deficit can harm plants and reduce yields in some cases. This criterion is less suitable for long-duration crops (e.g., sugarcane, cotton), crops needing frequent irrigation (e.g., potato, rice), or situations with significant rainfall.

Irrigation Scheduling Approaches

There are three primary approaches to scheduling irrigation: soil moisture depletion, plant-based indices, and climatological methods.

1. **Soil Moisture Depletion:** This approach relies on monitoring the available soil moisture within the crop's root zone. Irrigation is triggered when the soil moisture drops to a predetermined threshold, typically around 50% depletion of the available water (the moisture between field capacity and wilting point). The goal is to replenish the soil moisture before the crop experiences stress and yield reduction. Soil moisture deficit, the difference between field capacity and current moisture levels, is measured using various methods that assess percentage availability, tension, or resistance.
2. **Plant-Based Indices:** This approach uses the plant's response to water stress as an indicator for irrigation. Visible signs like leaf drooping, curling, or changes in foliage color can signal water deficiency. However, these are qualitative indicators. More quantitative measures include monitoring cell elongation rates, plant water content, leaf water potential, and leaf diffusion resistance. Certain "indicator plants," like sunflowers in Hawaii (used for sugarcane irrigation), can also reflect soil moisture conditions.
3. **Climatological Approach:** This method uses climate data to estimate evapotranspiration (ET), the water lost from the soil and plant surfaces. Irrigation is scheduled when the cumulative ET reaches a specific level. The amount of water applied can equal the calculated ET or a fraction thereof. Common methods include:

- **IW/CPE Ratio Method:** A fixed amount of irrigation water (IW) is applied when the cumulative pan evaporation (CPE) reaches a predetermined level. A typical practice involves applying 5 cm of water at an IW/CPE ratio of 0.75 to 1.0.

- **Pan Evaporimeter Method:** This method uses the relationship between crop water use (CU) and evaporation from a pan to determine irrigation needs. The Class A pan and sunken screen pan are two common types.
    - **U.S. Class A Evaporation Pan:** This widely used pan measures evaporation from a water surface. The daily difference in water level, adjusted for rainfall, indicates evaporation.
    - **Sunken Screen Evaporimeter:** This pan, developed by Sharma and Dastane, is designed to more closely approximate evapo-transpiration rates. Evaporation values from this pan reportedly correlate well with actual crop water use. It consists of a screen-covered pan sunk into the ground with the water level maintained level with the soil surface. The amount of water needed to refill the pan is measured to determine evaporation.

1. **Evaporimeter Setup:** Evaporimeters are set up in pairs to detect leakages, with a minimum distance of 3 meters between them. The pan is cleaned regularly, painted white annually, and thoroughly inspected for leaks. They are situated in open fields, away from wind obstructions, and aligned perpendicularly to the main wind direction to prevent interference.
2. **Critical Growth Approach:** Certain crop growth stages are particularly sensitive to moisture stress, leading to irreversible yield loss if not managed properly. These stages, known as critical or moisture-sensitive periods, require adequate irrigation water. For crops like cereals, the critical stages are panicle initiation, flowering, and pod development. Sufficient irrigation water is supplied during these stages, while non-sensitive stages can be skipped, especially under limited water supply conditions.
3. **Plant Water Status Approach:** This modern approach involves using the plant's water content as an indicator for irrigation scheduling. Although not yet widely adopted, this method can provide valuable insights into soil moisture and climate factors

Simple Irrigation Scheduling Techniques

1. **Soil cum Sand Mini Plot Technique:** A 1-cubic-meter pit is dug in the middle of the field and mixed with 5% sand. After filling the pit, crops are grown as usual. Plants in the pit wilt earlier, signaling the need for irrigation.
2. **High Seed Rate Technique:** In elevated areas, crops are grown with four times the natural seed rate in a 1-square-meter plot. High plant density causes earlier wilting, indicating the need for irrigation.

3. **Feel and Appearance Method:** Soil moisture can be estimated by feeling the soil and forming it into a small ball. This method requires experience to accurately estimate moisture levels.
4. **Tensiometer:** Tensiometers measure the tension with which water is held in the soil. When the desired tension is reached, the soil is irrigated. Tensiometers are limited to sandy soils and low tension irrigation scheduling.
5. **Plant Indices:** Plants can serve as guides for irrigation scheduling. Indicators such as leaf dropping, curling, or rolling, and foliage color changes can signal the need for irrigation. However, these symptoms do not allow for quantitative moisture deficit estimation.
6. **Infrared Thermometer:** Canopy temperature is measured using an infrared thermometer, which displays the canopy temperature (Tc) and air temperature (Ta) difference (Tc-Ta). Negative Tc-Ta values indicate sufficient water, while zero or positive values suggest the need for irrigation.
7. **Remote Sensing:** For large, single-crop areas, remote sensing data can be used for irrigation scheduling. The reflectance of solar radiation by plants with adequate water differs from that of stressed plants, allowing for the detection of water stress and scheduling of irrigation.

Problem: Calculate cumulative evaporation required irrigation at 0.5, 0.6, 0.75, 0.8, 1.0 with 5 cm of irrigation water.

Solution

Example: Cumulative pan evaporation at IW/CPE ratio of 0.5=IW/CPE=0.5

$$= \frac{5}{\text{CPE}} = 0.5, \text{CPE} \times 0.5 = 5 \quad \text{CPE} = \frac{5}{0.5} = \frac{50}{5} 10\text{cm}$$

Irrigation of 5 cm is given when CPE is 10 cm

CPE at 0.6 ratio = 5/0.6 = 8.33 cm

CPE at 0.75 ratio = 5/0.75 = 6.66 cm

CPE at 0.8 ratio = 5/0.8 = 6.25 cm

In IW/CPE ratio approach, irrigation can also be scheduled at fixed level of CPE by varying amount of irrigation water.

Problem: Calculate the amount of water for each irrigation for scheduling irrigation at 0.5 and 0.8 IW/CPE with 10cm of CPE.

Solution

Amount of water to be given at IW/CPE ratio of 0.5=IW/10=0.5 IW=0.5 X 10= 5cm

Amount of water to be given at IW/CPE ratio of 0.8 =IW/10=0.8, IW=10 X 0.8=8cm

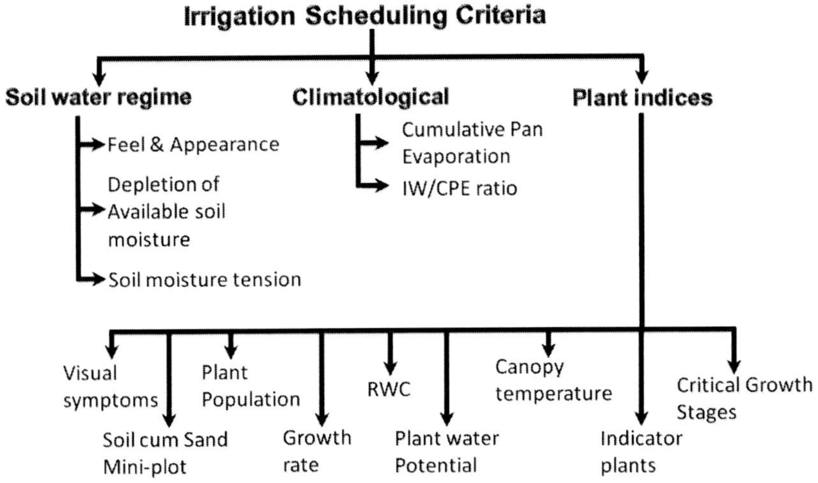

Crop	Critical growth period for water supply
Rice	Primordial development, Heading & Flowering
Ragi	Primordial initiation & Flowering
Maize	Tasseling, Silking& Pollination
Bajra	Heading & Flowering
Wheat	Crown root initiation, Shooting & Earing
Groundnut	Flowering, Peg penetration & Pod development
Sesame	Flowering to Maturity
Sunflower	Star formation, Flowering & Seed development
Safflower	Rosette, flowering and Seed development
Soybean	Flowering & Seed formation
Cotton	Flowering & Boll development
Sugarcane	Formative & Stem elongation
Tobacco	Rapid growth & Topping stage
Chillies	Flowering & Fruit development
Potato	Tuber initiation to Tuber maturity
Onion	Bulb enlargement to Ripening
Tomato	Flowering & Fruiting
Mango	Flowering & fruit development

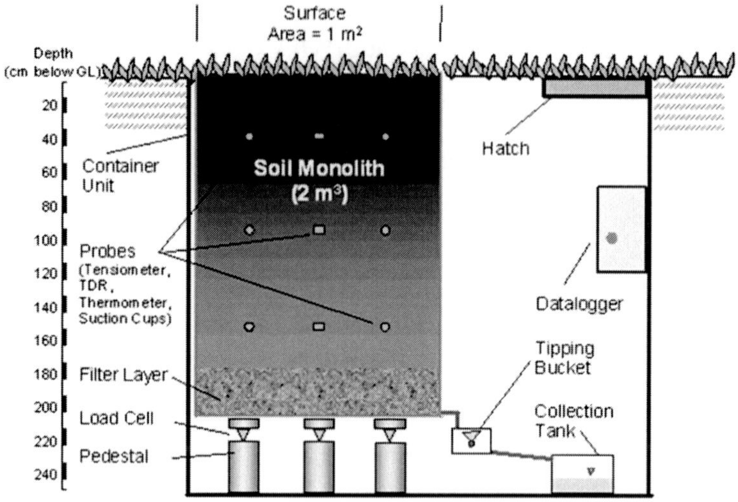

Schematic view of Lysimeter

Table Moisture sensitive stages of important crops.

Sr. No.	Crop	Important Moisture Sensitive Stages
1	Rice	Panicle Initiation, Flowering
2	Wheat	Crown Root Initiation, Jointing, Milking
3	Sorghum	Seedling, Flowering
4	Maize	Silking. Tasseling
5	Bajara	Flowering, Panicle Initiation
6	Nachani	Panicle Initiation, Flowering
7	Ground Nut	Rapid Flowering, Pegging, Early Pod Formation
8	Red Gram	Flowering & Pod Formation
9	Green Gram	Flowering & Pod Formation
10	Black Gram	Flowering & Pod Formation
11	Sugarcane	Formative Stage
12	Sesamum	Blooming stage to Maturity
13	Sunflower	Two weeks before & after flowering
14	Safflower	From rosette to flowering
15	Soybean	Blooming & seed formation
16	Cotton	Flowering & Ball Formation
17	Tobacco	Transplanting to Full Bloom
18	Chilies	Flowering
19	Potato	Tuber Initiation to Tuber Maturity
20	Onion	Bulb Formation to Maturity
21	Tomato	From the Commencement of Fruit Set

Threshold soil moisture tension values for

S.No.	Crop	Critical soil moisture Tension (Centibars)	Depth of maximum Rooting density (cm)
1.	Sugarcane	30–65	30
2	Maize	25–50	30
3	Cotton	40–50	30
4	Sorghum	50–65	60
5	Tobacco	30–40	30
6	Soybean	50–60	30
7	Wheat	40–50	30
8	Potato	25–50	30

Table Optimum IW/CPE ratios for scheduling irrigation in important crops

Crop	Optimum IW/CPE ratio
Groundnut	• 0.75 to 1.O IW/CPE ratio depending on crop developmental stages in Andhra Pradesh, Maharashtra & West Bengal
Sunflower	• 0.5 to 1.O IW/CPE ratio depending on crop developmental stages at Hyderabad & Kanpur
Wheat	• 1.O IW/CPE ratio at Ludhiana, Kanpur and Bikramganj
Bengal gram	• 0.4 IW/CPE ratio at Ludhiana
Mustard	• 0.4 IW/CPE ratio at Hissar
Maize	• 0.75 to 1.0 IW/CPE ratio depending on crop developmental stages at Delhi & Hyderabad
Sugarcane	• 0.5 to 1.0 IW/CPE ratio depending on crop developmental stages at Lucknow

However, the criterion does not take into consideration the critical growth stages for irrigation. It may also not hold well under the following situations

1. For crops requiring 1-2 irrigations only
2. Determining timing of first and last irrigation
3. Under low temperature/ frost/ salt stresses

Further, the optimum value of CPE for fixed/ variable depth of irrigation has to be worked out for different crops and locations

References

Allen, R. G., Pereira, L. S., Raes, D., & Smith, M. (1998). Crop Evapotranspiration – Guidelines for Computing Crop Water Requirements. FAO Irrigation and Drainage Paper No. 56. Food and Agriculture Organization of the United Nations.

Doorenbos, J., & Pruitt, W. O. (1977). Guidelines for Predicting Crop Water Requirements. FAO Irrigation and Drainage Paper No. 24. Food and Agriculture Organization of the United Nations.

Fereres, E., & Fontanelli, G. (2010). Irrigation Scheduling: A Key Tool for Water Management. In: Irrigation and Fertilization. Agronomy, 1(4), 22–33.
Hargreaves, G. L., & Samani, Z. A. (1982). Estimation of Potential Evapotranspiration. Journal of the Irrigation and Drainage Division, ASCE, 108(IR3), 225-230.
Jensen, M. E., Burman, R. D., & Allen, R. G. (1990). Evapotranspiration and Irrigation Water Requirements. ASCE Manuals and Reports on Engineering Practice No. 70. American Society of Civil Engineers.
Kisekka, I., & Ekhwan (2016). Improved Irrigation Scheduling Using Soil Moisture Sensors. Final report of the Water and Energy Program for the Midwest, University of Illinois.
Phene, C. J., & Lamm, F. R. (2001). Scheduling Irrigation with the Help of Soil Moisture Sensors. Proceedings of the 2001 ASAE Annual International Meeting, Paper No. 011092.
Steduto, P., Hsiao, T. C., Kang, S., & Fereres, E. (2012). Crop Yield Response to Water. FAO Irrigation and Drainage Paper No. 66. Food and Agriculture Organization of the United Nations.
Stegman, E. C., & Waller, J. (1998). Water Management for Crop Production. In: Irrigation in the Near Future. Kansas State University Agricultural Experiment Station and Cooperative Extension Service.
Zhu, J., & H. Y. (2015). A Review of Irrigation Scheduling Strategies. Water Resources Management, 29(8), 2857–2870. https://doi.org/10.1007/s11269-015-0940-1

Multiple Choice Questions

1. What is the primary goal of scientific irrigation scheduling?
   a) To reduce labor costs
   b) To optimize crop yield and water use efficiency
   c) To minimize soil erosion
   d) To increase crop variety
2. Which method involves assessing the feel and appearance of the soil to determine moisture content?
   a) Climatological approach   b) Soil moisture regime approach
   c) Indicator plant method   d) Plant indices method
3. What does the IW/CPE ratio method relate to in irrigation scheduling?
   a) The depth of rooting zones
   b) The cumulative pan evaporation
   c) The rate of plant growth
   d) The type of soil used
4. Which plant character is commonly used as an index for scheduling irrigation in maize?
   a) Leaf color   b) Root growth
   c) Leaf curling   d) Flowering time

5. What is the function of tensiometers in irrigation scheduling?
   a) To measure crop yield
   b) To display soil temperature
   c) To measure soil moisture tension
   d) To calculate evaporation rates
6. In which irrigation scheduling approach are visual symptoms like wilting and leaf curling used?
   a) Climatological approach
   b) Soil moisture regime approach
   c) Plant indices approach
   d) Empirical formulae approach
7. What is the critical soil moisture tension range for cotton mentioned in the text?
   a) 25 – 50 centibars b) 40 – 50 centibars
   c) 30 – 40 centibars d) 50 – 60 centibars
8. Which method is described as a "crude but effective" way to judge soil moisture through direct hand experience?
   a) Tensiometer method b) Feel and appearance method
   c) Pan evaporation method d) Lysimeter method
9. The potential evapotranspiration (PET) is primarily influenced by what factor?
   a) Soil type b) Crop variety
   c) Climatic conditions d) Water availability
10. What is the main disadvantage of using the visual plant symptoms method for scheduling irrigation?
   a) It requires advanced technology
   b) It cannot provide quantitative estimations of moisture deficit
   c) It increases labor costs
   d) It is only applicable to certain crops

Answer Key

1	b	2	b	3	b	4	c	5	c	6	c	7	b
8	b	9	c	10	b								

9

System and Method of Irrigation Water Application

Surface: border strip, furrow, basin and check basin

Sub-surface

Micro-irrigation: drip, sprinkler; Pichter and bottle

Definition of Irrigation Method

Application of irrigation water to cropped field by different types of layouts is called as irrigation methods. The methods of irrigation initially might have been started to check the over flow of water from one field to another. But today, it has become necessary to save the water by proper methods to arrest run-off loss, percolation loss, evaporation loss etc., and to optimize the crop water need. Hence, irrigation method can be defined as the way in which the water is applied to the cropped field without much application and other losses, with an objective of applying water effectively to facilitate better environment for crop growth.

Factors Influencing Irrigation Methods

Soil type: The soil physical properties such as texture, structure, porosity, infiltration rate, etc. influence the selection of irrigation methods. Heavy texture soil restricts water movement than light texture soil wherein water move freely to deeper sections due to high porosity. Single grain structure soil allows water freely to move downward compared to other structures

Soil depth: If soil is shallow which holds less water, leveling and forming bunds etc. to hold? Maximum water is to increase the irrigation interval. Similarly if the soil is deep, it holds more water and needs longer irrigation interval. Accordingly, the irrigation methods can be selected.

Topography of land: In undulating topography, it is very difficult to adopt normal methods of irrigation. The slope of the land also decides the methods to be adopted. If the land is more sloppy, basin method cannot be used. In this condition strip method can be used. For undulating topography instead of strip or basin method, sprinkler or drip methods can be used.

Climate: Rainfall, temperature, humidity, wind velocity, radiation, etc., influence the irrigation methods. For example, heavy wind affects sprinkler irrigation and temperature affects surface method of irrigation by high evaporation loss.

Water sources: The flow velocity, quantity and quality of available water are the other main factors, which decide the methods of irrigation to be adopted.

Crops to be grown: The value of the plant and geometry of the crop to be cultivated are the main criteria to decide the method of irrigation. For example, if the crop is a high value or cash crop or wide spaced crop, sprinkler or drip method of irrigation can be adopted. Irrigation water can be applied to the land in the following general ways.

By flooding (wetting all the land surface)

By furrows (wetting only a part of the ground surface in which crops are grown)

By sub irrigation (sub surface soil irrigation)

By sprinkler (soil is wetted through sprinkling water)

By drip irrigation (water is applied at the individual root zone of the plant).

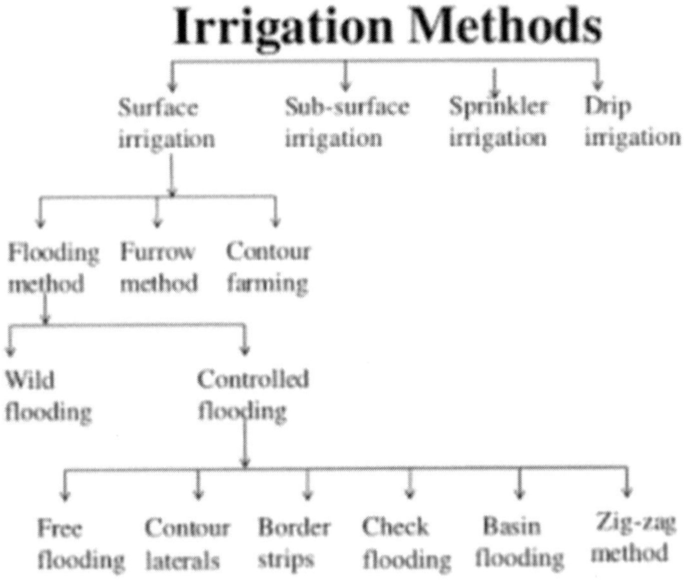

Classification of Irrigation Methods

The irrigation methods are broadly classified as:

- Surface method or gravity method of irrigation

- Sub surface or sub irrigation
- Pressurized or micro irrigation - Drip irrigation, sprinkler irrigation and rain gun irrigation.

I. Surface or gravity irrigation

It is the common method of irrigation practiced all over the world. In this method, water is applied directly to the surface by providing some checks to the water flow:

Advantages

- Easy to maintain
- Low cost
- Technical skill is not required.

Prerequisites

- Uniform soil
- Smoothness of field surface or levelled surface
- Adequate quantity of water.

Classification

1. **Border strip method** - The field is divided into number of long parallel strips by providing small parallel earthen bunds or levees or dykes along both sides of the strips. The end along the strip may or may not be closed, which is based on the length of the strips. If the length of the strip is very long, the end will be closed to have a uniform distribution and to avoid run off loss. Each strip is irrigated independently from upper end (turned on) and water flow as thin sheet and uniformly spread along the strips. The water is turned off when the required volume is delivered to the strip. The application efficiency of this system is 75–85%.

a) Suitability

Soil - Suitable to the soils having

- Moderately low to moderately high infiltration rate.
- To the field which is having 0-0.5% slope.
- For dense, closer spaced crop it can be advocated up to 4% slope provided there should not be any erosion hazard.
- Not suited for very sandy soil and very clayey soil as they have too high and low infiltration rate, respectively.

Crop - All closely spaced crops like pulses, wheat, barley, alfalfa, berseem, grasses, ragi, cumbu and small grains.

b) Dimensions

The width of the strips depend upon the size of the stream and normally this varies from 3–15 m. The length varies according to the slope, stream size, soil type, etc. Length of the border strips and recommended safe limits of slopes for various types of soil are given below.

Length	Slope (%)	
Sandy and sandy loam soil	6.0–12.0 m	0.25 to 0.60
Medium loam soil	10.0–18.0 m	0.20 to 0.40
Clay loam to clay soil	15.0–30.0 m	0.05 to 0.20

c) Classification

It can be further classified as: (1) Graded borders, (2) Level borders. Graded borders have slope ranging from 0.1–0.5% in the longitudinal directions and there is no or very little slope across the strip. For level border, there is no slope in either direction.

2. **Check basin method (beds and channel) -** It is the common and simple method of irrigation mainly adopted in levelled land surface. It is also known as Beds and channel method of irrigation. The land is divided into small basins/beds. The area of basin is surrounded by earthen bunds or levees or dykes. The applied water is kept within the basin and not allowed for run off. This is the most common method adopted for most of the crops. The size of the levees or ridges or bunds depend upon the depth of water to be impounded in the basin. The water is turned on the upper side and after applying the required quantity of water it is turned off.

a) Suitability

Soils

- More efficient (more than 90%) in fine textured soil. This is due to the uniform rapid spread of water and more infiltration opportunity time for all areas and thereby depth of infiltration is uniform all along the basins.
- The correct quantity of water can be applied as there is no run off.
- Leaching of salt is possible by impounding water and giving more opportunity/time for infiltration, stagnation and drainage.
- Suitable to lands with smooth, gentle and uniform slope with low to medium infiltration rate.

Crops - Cereals, millets, pulses, oilseeds.

b) Disadvantages

- It needs high degree of levelling for uniform distribution of water.
- Within the basin, soil should be uniform.
- It is not suitable for coarse textured soil with high infiltration rate.
- The bunds should be strong enough to withstand ponding of water.
- In fine textured soil with very low infiltration rate, precaution may be taken to avoid long time water stagnation.

c) Dimensions

The basin area for different soil types, inflow rate and slope percentage are given below for reference. The size of the basin is also influenced by the depth (in mm) of irrigation water. If the required irrigation depth is large, the basin can be large. Similarly, if the required irrigation depth is small, then the basin should be small to obtain good water distribution.

Check Basins should be small if the slope of the land is steep; soil is sandy, stream size to the basins is small, required depth of the irrigation application is small and field preparation is done by hand or animal drawn implements. Check Basins can be large if the slope of the land is gentle flat, soil is clayey, stream size is large, required depth of the irrigation application is large and field preparation is mechanized.

Based on the shape of the basin, it can be classified as rectangular or square or irregular basin.

Mostly, the rectangular shape is preferable for easiness in farming operations

3. **Basin method** - Basin method of irrigation is used in soil submergence method of irrigation in low land rice, bundedrainfed rice and forage grasses, where water is stagnated to the required depth by providing bunds on all the sides to sufficient width and height. The optimum size for efficient water management to rice crop is 0.25-0.40 ha. The field is to be levelled thoroughly for uniform depth of water. Provision of separate irrigation and drainage channels is more efficient than field-to-field irrigation.
4. **Ring basin** - This method is mostly adopted for wide spaced orchard crops. The rings are circular basins formed around the individual trees. The rings between trees are interlinked with main lead channel by sub channels to get water to the individual rings. As water is allowed in rings only, wastage of water spreading the whole interspaces of trees

as in the usual flooding irrigation method is reduced. Weed growth in the interspaces around the rings are discouraged. This method ensures sufficient moisture in the root zone and saves lot of irrigation water.

5. **Furrow method of irrigation** - It is the common method adopted for row planted crops like cotton, maize, sugarcane, potato, beetroot, onion, sorghum, vegetable crops etc. In this method, small evenly spaced shallow furrows or channels are formed in the beds. Another method of furrow irrigation is forming alternate ridges and furrows to regulate water. The water is turned at the high end and conveyed through smaller channels. Water applied in furrows infiltrate slowly into the soil and spread laterally to wet the area between furrows.

Fig. 9.2: Ring basin

A. Dimensions

Based on the soil slope and stream size, the length can be fixed. The furrow width or spacing aries from 60–120 cm, which depends upon the crop to be grown. The depth of furrow varies from 12.5 cm, which depends upon (*a*) soil type, (*b*) flow size, and (*c*) effective root zone depth of crop.

B. Suitability

This method is mostly suitable for medium to moderately fine textured soil which allows free after movement both horizontally and vertically. In sandy or coarse textured soil, this method is not suitable because here the water movement is primarily downward and very little in horizontal direction. Besides, the length of ridges or furrows to resist the velocity of flow is very low which in turn may lead to breaching of the structures. This method is adopted for soils having the problem of surface crust or hardpan. The labour requirement to form the furrows is relatively higher than other surface methods of irrigation.

C. Precautions

While using the furrow method of irrigation, care must be taken in strengthening the furrow since erosion hazard on sloppy areas may damage the furrow.

Irrigation furrows may be classified into two general types viz., straight furrow and contour furrow.

Straight furrow - Best suited to soils where land slope does not exceed 0.75%.

Contour furrow - This method is similar to graded and level furrow method. Furrow carries water across slopping field rather than downwards. They are designed to fit the topography of field. Furrows are given a gentle slope along its length as in graded furrow. Field supply channels run down the land slope to feed the contour furrow and are provided with erosion control structure. Successfully used in all irrigable soils. All row crops including grains, vegetables and cash crops are adapted to this method. Light soil can be irrigated successfully across slopes up to 5% slope. Up to 8-10% can be irrigated by contour furrow. Contour furrow may be used on all types of soil except in light sandy soil and soil that crack.

Corrugation irrigation - It consists of running water in small furrows, which direct the flow down the slope commonly used for irrigation in non-cultivated close growing crops such as small grains, pasture on steep slopes. Corrugation can be made with a simple bamboo corrugation or cultivators equipped with small furrows. Corrugations are 'V' or 'U' shaped channels about 6-10 cm deep spaced 40-75 cm apart. This method is not recommended for saline soil or for saline water irrigation. The permissible length of corrugation varies from 15 cm within light textured soil with slopes of 2-4% to about 150 cm in heavy texture soil up to 2% slope.

Furrow irrigation design consideration - Efficient irrigation by furrow method is obtained by

selection of proper combination of spacing, length and slope furrows.

i) **Furrow spacing** - Furrows can be spaced to fit the crops grown and types of machines used for planting and cultivation. Crops like potato, maize, cotton, etc., are planted 60–90 cm apart and have furrow between all rows. Carrot, lettuce and onion are spaced 30–40 cm and often have two rows between furrows. Furrows should be close enough to ensure that water spreads to the sides of the ridge and the root zone of crop to replenish soil moisture immediately.

ii) **Furrow length** - Optimum length furrow is usually the longest furrow that can be efficiently and safely irrigated. Long furrows are an advantage in inter cultivation. Proper furrow length depends largely on hydraulic conductivity of soil. It should be shorter in porous sandy soil than clayey soil. If only a small area is to be irrigated, the length of field may determine the length of furrow. In large area it may be desirable to have furrow length equal to an even fraction of the total length of the field.

iii) **Furrow slope** - The slope or grade of furrow is important because it controls the speed at which water flows down the furrow. A minimum furrow gradient of 0.05% is needed to ensure Surface drainage.

iv) **Furrow stream** - The size of the furrow stream is one factor which can be varied after furrow irrigation system can be installed. The size of furrow stream usually varies from 0.5-2.5 lit/sec. The max nonerosive low rate in furrow is estimated by following equation,

$qm = 0.6/S$

where,

qm = maximum no-erosion stream (lit/sec)

S = Slope of furrow (%)

Average depth of irrigation water applied during irrigation can be calculated by the following

relationship.

$D = (q \times 360 \times t)/(w \times l)$

Where

D = Average depth of water applied (cm)

q = Stream size (lit/sec)

t = duration of irrigation (hrs)

l = Furrow length (m)

w = Furrow width (m)

Surge irrigation

Surge irrigation is a method of surface irrigation through furrows or border strips wherein water is applied intermittently in a series of relatively short on and off time periods during the irrigation (Humphrey, 1989). Water is let into a long furrows or border strips in an intermittent flow instead of conventional continuous flow. Each flow is termed as a surge. Surge irrigation practiced under favourable conditions can improve the performance of surface irrigation system compared to the other methods of surface irrigation. Irrigation is given in an on-off cycle or by cut back method. The cycle time means the time from the beginning of one surge to the beginning of next surge. Cycle ratio is the ratio of flow time (continue) to the cycle time. Assuming the cycle time as 20 minutes and cycle ratio as 1:2 (0.5), the on-time is 10 minutes and off time is 10 minutes. This cycle ratio can also be the ratio of on-time and off-time as 1:1, if the on time is 10 minutes. Water is allowed for 10 minutes and stopped for 10 minutes. This 20 minutes is the surge time or cycle time. This surge is repeated until the water reaches the

whole furrow or strip. The first surge of water over a portion of dry furrow wets the soil surface at a slow advance rate and high infiltration rate. When the next surge is allowed to flow along the first surge length, water makes faster to the second surge length. Thus in surge flow, the advancing water along the furrow is faster resulting in uniform wetting from the head to the tail end of furrow. Under the conventional continuous flow, wetting is more in head end than at tail end. When more water is allowed to increase the wetting depth in the tail end, it leads to loss of water through tail end run off. This loss and the rate of infiltration along the whole length of flow distance are reduced in surge irrigation, in addition to saving time of irrigation.

Fig. 9.3: Corrugated irrigation

Fig. 9.4: Contour irrigation

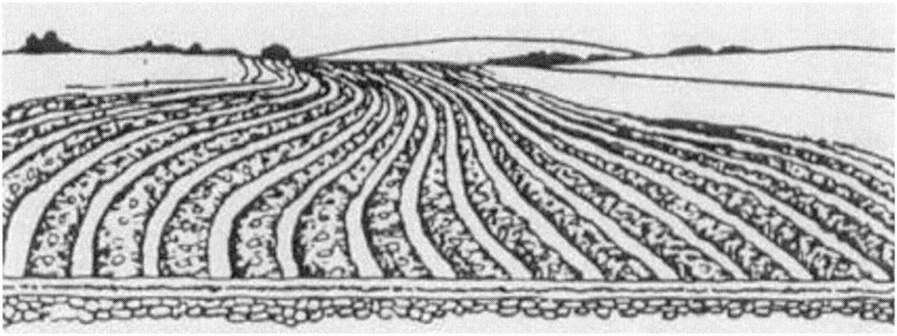

Fig. 9.5: Graded contour-furrow Irrigation

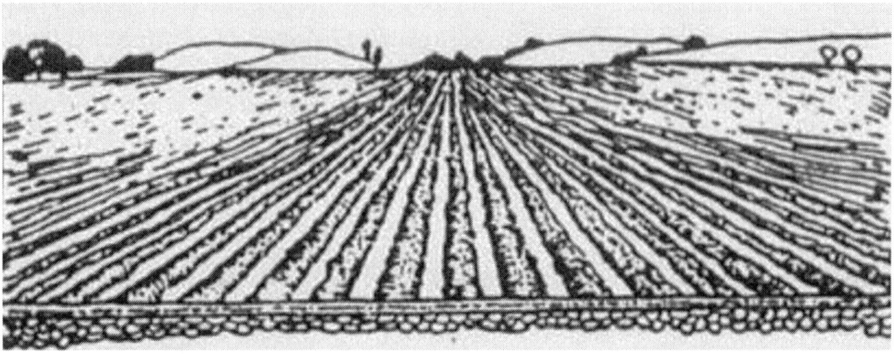

Graded or level-furrow irrigation: Different types of furrow irrigation

Advantages

- Reduction in infiltration rate
- Rapid advance of wetting front
- Less difference in intake opportunity between upper and lower ends of furrow
- More uniform distribution of water along the length
- Improvement in application uniformity and irrigation efficiency
- Reduces water requirement
- Water reaches the furrow end much earlier than under continuous stream
- It is a non erosive method, suitable for erodable soils
- Useful for light textured soils with high infiltration rate
- Saves irrigation time and the energy cost for lifting water

- About 20% of land area is saved in cross channels with shorter furrow lengths
- It offers scope for automation of surface irrigation.

Limitations

- Little or no advantage in clay or silty soils
- Tail end water loss may increase if not managed properly
- Lengthy furrows of more than 100 m are required
- Ensuring proper gradient to such lengthy furrows is difficult
- With progress in surge cycles and number of irrigations, the bulk density is increased due to oil consolidation
- More suited to shallow rooted crops only.

Subsurface irrigation

Water is applied below the ground surface through the network of pipes or some devices. The main aim of this type of irrigation is to reduce the evaporation loss and to maintain an artificial water table near the root zone of the crop.

Suitability - It is mainly suitable for the high temperature area where ET losses are very high wherein controlling and maintenance of surface water and application is very difficult.

Pitcher pot irrigation method

It is one way of applying water below the ground or soil urface. In this method, in a mud pot, some small holes are made and the holes are closed by either threads or material, which is able to conduct water very quickly. The pots are kept around the root zone in pits made for it. The pits are completely covered tightly with sand mulch mix. The pots are filled with water and closed. The water slowly penetrates to root zone through the holes and wet the root zone area. This method is mostly suitable for widely spaced tree crops under water scarce conditions.

Pressurized irrigation methods

Micro irrigation is defined as the methods in which low volume of water is applied at low pressure & high frequency usually an irrigation interval is in the ranges of 1 to 4 days. The system has extensive network of pipes at operated at low pressure. At pre-determined spacing outlets are provided for emission water generally known as emitters.

It includes both sprinkler and drip irrigation methods where water is applied through network of pipelines by means of pressure devices.

Sprinkler irrigation system/point source method

In this method the irrigation water is sprayed to the air and allowed to fall on-the ground surface more or less resembling rainfall. The sprinkling of water or spray of water is made by pumping water under pressure through network of pipelines and allowing to eject out by means of small orifices or nozzles or holes. The water required by the crop is applied in the form of spray by using some devices, wherein the water application rate should be somewhat lesser than the soil infiltration rate to avoid run off or stagnation of water in the filed.

Sprinkler Irrigation

Definition: It is methods in which water is spread into air and allowed to fall on the ground surface some what resembling.

Water is forced under pressure through small nozzle/orifice which gets broken up to into droplets and fall back on the ground. Slow circular revolution is impacted to the nozzle uniformally covered the ground surface. The rate of application should not be more than the infiltration rate of the soil.

Adaptability of sprinkler irrigation

1. Sprinkler irrigation can be adopted where land reveling is uneconomical and other method of surface irrigation cant carried out.
2. Adapted to soils to pours highly avoidable or relatively impermeable which are difficult to irrigate by other methods like furrow, border etc.
3. Where it is designed to go for frequent irrigation.
4. It is designed to minimum cost towards labours, fertilizer, and irrigation.

Advantages of Sprinkler Irrigation

1. It can be used for almost crops expect paddy & jute.
2. System can be adopted under varied topographic condition and especially suitable to steep-slope and irregular topography.
3. Soils-method is particularly suited for sandy soils having high infiltration rate.
4. It can eliminate surface run off of irrigation water (run off elimination)
5. To protect the crop against frost & high temp.
6. To reduce labour cost for irrigation as compared with surface method.
7. Savings in land construction of channel to the field.
8. It saves fertilizer & water as ferti-irrigation can be carried out.
9. Land leveling is not essential for sprinkle irrigation.
10. Gives higher water use efficiency.

Limitations of sprinkle irrigation

1. Not suitable for very fine texture soil (<4mm/her)
2. Uneven distribution of water due to distortion by high water.
3. More evaporation losses.
4. Require clean, water free from debris sand slit & clay particles.
5. Saline water can no be used.
6. Initial cost is high.
7. High operating power is high (5-10kg/cm)
8. Unsuitable climate condition sprinkling may be encouraging spread of disease.
9. Ripening softy fruits need protection from the spray.

Systems of Sprinkle Irrigation

Systems of Sprinkle Irrigation			
Based on Spraying Arrangement		Based on Portability	
1	Fixed Head	1	Portable
2	Rotating Head	2	Semi Permeable
3	Perforated Pipes	3	Semi Portable
	i) Dancing Water	4	Solid Structure
	ii) Oscillating Arms	5	Permanent

Portable system

In sprinkle irrigation system in which main line, sub main line, lateral and the pumping units all are the portable generally the pipers are made up of light weight aluminum to facillate easy transportation such system can be shifted from place to another.

Semi-portable system

This system is parallel to portable except that location of water source and pumping unit is fixed.

Semi-permanent

In this system the main and sub-main are fixed, usually buried under the ground where as the laterals are portables.

Solid straight system

These have enough laterals present to eliminate their movement from one place to another. The system is fixed at the beginning of system of season and remains through out the season for short & frequent irrigation.

Permanent irrigation system

The permanent system is suited to automation using soil moisture sensor and are generally preferred in orchard. The sub main, lateral are permanently buried below the ground level.

Based on spraying arrangement fixed head

The fixed head type of sprinkler arrangement system sprays water in one direction.

Rotating head

Rotating head removes slow rate to distribute water in a circular fusion. They may be single, double, multiple nozzle sprinkler head. Single nozzle sprinkles system are referred for their low application rate however the double nozzle sprinkle head gives good uniformity of application at low pressure.

Multiple nozzle type of sprinkler also called as giant sprinkler are used to covered more area with single set. The operating pressure required for such sprinkler may be more than 10kg/cm2.

Perforated Pipes

Such systems are preferred for application of water under lower operating pressure usually between 0.5 to 2.5 kg/cm2. The system is not recommended under heavy winds as the jets are distorted more easily.

Generally pipes are provided with holes, perforated along the upper 1/3rd perimeter in a proper designated to covered with between 6 to 15m such system are preferred on plains moderately high infiltration rate used for irrigation of lawns or vegetable crop where the plant height ranges between 40 to 60 cm.

Components of Sprinkler Irrigation System

Prime mover/pump suction pipe, foot value

Pumping sets or pump is required for lifting water from the source and push it through distribution system i.e. main, sub main, laterals and finally through the sprinkler head under sufficient pressure.

The pumping set consists of a centrifugal pump (volume) or turbine type pump with a driving unit suction line and a foot value. Centrifugal pump are generally used where the lift is less than 5m i.e. when source is river, shallow well etc. for higher lift or if water level fluctuates widely turbine pumps are recommended.

The electric motors are generally used driving unit for fixed installation diesel engines are generally recommended for portable units.

Main line

It carries water from the source (pumping unit) to the various parts in the field. It may fixed or portable. Permanent lines are generally buried below the working depth inside the ground. Light weight aluminum pipe with quick couples are preferred for portable lines something HDPF (high density pipes are also preferred because of its longer life. The fixed are generally of steel pipe or PVC pipe of suitable diameter). Te or L section is provided to connect the main with sub main or lateral.

Sub main

It carries water from main to lateral lines.

Lateral Lines

It carries water from main or sub main pipe line to the sprinkler head through the rise pipe. They are portable and equipped with quick coupling devices. Commonly they are available in 5,6 or 12m length are provided with U shaped rubber gasket in the female portion of coupling. The water pressure forces outside of 'U' gaskets to form water seal when the water is turned off the seal is broken and water is drained out from the pipe making it easy to uncouple and more.

Sprinkler head

Sprinkler heads are used for spraying water on the fields they may be-

a) Rotating Head
b) Fixed head Type
c) Perforated Type

Fixed Head Type: Used in landscape

Sprinkler lead can be classified on basis of pressure

1. Low operating pressure sprinkler (1.5 to 2.5kg/cm2)
2. Intermediate pressure sprinkler (2.5 to 5kg/cm2)
3. High pressure sprinkler (5 to 10kg/cm2)

Problems

Find the fertilizer does per settings of the sprinkler system if a lateral has 12 sprinkler 14m apart. Later lines are spaced 20m on the main line and the recommend does of fertilizer is 80kg/ha.

Solution

$$WF = \frac{Ds \times DL \times Ns \times WP}{10,000}$$

$$DL\,20 = \frac{14 \times 20 \times 12 \times 80}{10,000}$$

Sprinkler System Components & Layout

A typical sprinkler irrigation system consists of a pumping unit, pump connector, non-return valve, water meter, pressure gauge, pressure regulators, mainline & sub- mainlines, laterals, control valves, hydrants, sprinkler base, sprinkler heads, bends, tees, reducers, end plug, nipples, flanges etc (Fig.).

The most common type of sprinkler system layout is shown below in Fig. The pump unit is usually a centrifugal pump which takes water from the source and provides adequate pressure for delivery into the pipe system.

Suitable crops

Sprinkler irrigation is suited for most field crops viz., wheat, lucerne, groundnut, bengal gram, green gram, black gram, potato, leafy vegetables, sunflower, barley, bajra, maize, wheat etc wherein water can be sprayed over the crop canopy (Fig.).

However, large sprinklers are not recommended for irrigation of delicate crops such as lettuce because the large water drops produced by the sprinklers may damage the crop.

Water containing specific ions such as sodium and chlorides in concentration of more than 3 meq/litre are not suitable for irrigation by overhead sprinklers.

Sprinkler irrigation is a method of applying irrigation water which is similar to natural rainfall (Fig.). Water is conveyed under desired pressure (2 to 5 kg/cm2) developed by a pump through a network of pipes, called mainlines and submains to one or more laterals and is sprayed in to the air through sprinkler nozzles or perforations so that it breaks up in to small water drops (0.5 to 4mm in size) which fall over the land or crop surface in an uniform pattern at a rate (0.06-5000 LPH) less than the infiltrability of soil. The pump supply system, sprinklers and operating conditions must be designed to enable a uniform application of water.

Sprinkler irrigation systems may be classified as portable, semiportable, semi-permanent or permanent. They are also classified as set-move (hand-move, tow-move, side-roll and gun-type systems), solid-set or continuous move sprinkler (center-pivot, traveler and linear-move) systems (Fig.).

Sprinkler irrigation systems may be classified as portable, semiportable, semi-permanent or permanent. They are also classified as set-move (hand-move, tow-move, side-roll and gun-type systems), solid-set or continuous move sprinkler (center-pivot, traveler and linear-move) systems (Fig).

Limitations

a) Uneven distribution of water due to high wind velocities particularly during summer season.

b) Higher evaporation losses when operating under high temperatures.

c) Mechanical difficulties such as sprinklers fail to rotate, nozzles may clog, couplers may leak or engine may require repair.

d) Initial investment and recurrent operating costs are much higher than in surface irrigation methods.

e) Moving the portable lines, when the soil is wet results in the destruction of soil structure

f) Use of saline water for irrigation is not possible since it will be harmful to crops g) Higher water pressure required hence extra energy cost

Suitability and advantages

- It is highly suitable for sandy soil where infiltration rate is more.
- For shallow soil where levelling operation is technically not possible.
- For lands having undulating topography or steep slopes where levelling is economically not advisable.
- Irrigation steam size is very small where surface flow is low.
- It is almost suitable for all crops except crops like rice, which needs stagnation of water, but under water scarcity it can be tried for rice also. For cotton during reproductive phase sprinkler irrigation is not advisable.
- Application of fertilizer (fertigation), pesticides (pestigation) and herbicides (herbigation) are possible through irrigation systems which reduce labour cost and increase the use efficiency of any chemical.
- It controls crop canopy temperature.
- In crust soil, it facilitates early germination and establishment by means of light and frequent irrigation.
- Wastage of land for basin, ridges and furrows and irrigation channels are reduced.

Disadvantages

In heavy windy areas the distribution efficiency is reduced due to drifting of water droplets. In aline water conditions, it causes leaf burns besides clogging and corrosion of the pipeline. Continuous power supply is required to operate the system to maintain pressure. It is very costly to install and to maintain. Uniformity of application is difficult due to over application or neglected corners in the field.

Major components

- Pump set
- Network of pipelines (main, lateral, sub lateral, etc.)
- Riser pipes with tripod stand
- Sprinkler head

Classification - There are two types *viz.*, (1) Rotation head system and (2) Perforated pipe system.

Further it can be classified as

a) *Portable* - All components are portable and fixed

b) *Semi portable* (or) *Semi Permanent* - Water source, pump set, main and sub mains are fixed.

Only laterals are portable

a) Rotating head system - A special device to sprinkle the water called "Sprinkler Head" is used in this system. The sprinkler head consists of small nozzles and metal ring or vane with a spring. The water ejected through the nozzle strike the metal ring which changes its direction by the help of the spring attached to this which in turn causes the spray of water in all directions. The whole sprinkler head system is fitted on the riser pipe, which is erected from lateral pipes at uniform intervals. Rotating sprinkler heads are of two types*viz.*, single nozzle type and twin nozzle type (main nozzle and driving nozzle).

b) Perforated pipes system - In this method, small holes are made in lateral pipes based on the nature of the crops to distribute water uniformly.

Uniform distribution of water - Irrigation efficiency of sprinklers depends upon the degree of uniformity of water applied. Uniformity coefficient is computed with field application. Open ans are placed at regular interval within sprinkled area. Depth of water collected in open cans is measured and the coefficient of uniformity is computed by Christiansen (1942) equation.

Cu = 100 (1 − Σ X/m.n)

Where,

Cu = uniformity coefficient

m = average value of all observations

n = total number of observation points

X = numerical deviation of individual

observation from average application rate.

A uniformity coefficient of 85% or more is considered to be satisfactory. The uniformity coefficient is affected by pressure–nozzle size relation, sprinkler spacing and wind condition.

Sprinkler selection and spacing - The choice depends on diameter of coverage required, pressure available and discharge of sprinkler. The data given in tables 1 and 2 may serve as guidance in selecting the pressure and spacing desired.

Rate of application - Average rate of application is often called as precipitation intensity. It can be estimated by

Ra = Q/(360 × a)

Where,

Ra = Rate of water application

Q = Discharge rate of sprinkler (lit/sec)

A = Wetted area of sprinkler (m2)

Discharge of nozzle - The discharge of water through the nozzle can be given by the following equation.

Q = ca − 2 gh

Where,

Q = Nozzle discharge

a = Cross-sectional area of nozzle

h = pressure head at nozzle (mts.)

Head of water = 10 × pressure (bar)

Head of water foot = 2.31 × pressure (Pounds foot/inch 2)

G = Acceleration due to gravity

C = Discharge coefficient

Water spread area of sprinkler - The water-spread area of a sprinkler is given by the following equation

R = 1.35 dh

R = Radius of wetted area (m)

d = Diameter of nozzle

h = pressure head at nozzle (m)

Design of sprinkler systems - A sprinkler system is designed in order to achieve high efficiency in its performance and economy. The informations needed for designing sprinkler system are:

- Map of area
- Water source availability and dependability
- Climatic condition
- Depth of irrigation to be applied
- Irrigation interval
- Water application rate
- Sprinkler spray and power source

Lay out of sprinkler system - Sprinkler operates at a low time duration and pressure and can irrigate an area of 9–24 m wide and up to 300 m long at one setting. Application rate vary from 5–35 mm/hr.

Layout of portable system - It consists of a pump, mainline, lateral and rotary sprinkler spacer 9–24 m apart. The laterals remain in position until irrigation is completed. After irrigation is over, lateral is disconnected from main and is dismantled and moved to the next point of main line and reassembled. The lateral is gradually moved around the field until the whole field is irrigated. In this system, only laterals are moved. Sometimes the whole system including pump and mainline are moved from point to point (semi permanent).

Permanent system - When sufficient laterals and sprinklers are provided to cover the whole irrigated area so that no equipment needed to be removed. Then the system is called permanent system. This system requires less labour than portable system and large area can be irrigated by using few skilled operators. They are more expensive initially because of extra pipes, sprinklers and fittings required but, savings can be made because of reduced labour. It is suitable for automation irrigation system and areas where labour is difficult to obtain.

Fertilizer application with sprinkler system - Suitable chemical fertilizers can be mixed into the sprinkler system and applied to crop. Quantity of fertilizer added to the system for each setting can be calculated by using the formula.

$wf = (Ds \times Dl \times Ns \times Wf)/10{,}000$

where,

wf = Amount of fertilizer per setting

Ds = Distance between sprinklers

Ns = Number of sprinklers

Wf = Recommended fertilizer dose

Dl = Distance between laterals.

Sprinkler/ Overhead/ Aerial Irrigation

Water sprayed into air and allowed to fall on land surface in a uniform pattern with rates lesser than the infiltration rate of the soil.

Advantages

Application under odd conditions where other methods cannot be used

Excessive slopes

Irregular topography

Erosive soil

Excessively permeable soil

Excessively impermeable soil

Land preparation cost less

No wastage of land in channels

Quantity of water can be controlled

Rate of application according to infiltration rate

No deep percolation and run-off losses

Efficient use of water

Disadvantages

High initial cost

High degree of technical skill

More evaporation losses on hot windy days

Water is required under pressure (0.5 to 10 kg/cm^2) and that increases the cost of pumping.

Little scope under ordinary farming conditions, scope only under high value crops

Interference with use of Inter-cultivation operationsNo useful on ripening stage of plants having soft fruits

Dirty water cannot be used

Types of sprinklers

1. Rotating Head System.
2. Perforated Pipe System

Drip or Trickle Irrigation System/line source irrigation

Drip irrigation

In drip irrigation the required quantity of water is applied by means of mains, sub mains, manifolds & plastic laterals in the with equally spaced emitters usually laid on the ground surface at low pressure & at low discharge at the root zone of the crop.

Advantage of drip irrigation

1. Water saving is up to 40 to 60%
2. Enhance the plant growth & increases the crop yield
3. Savoring in level & energy most. Suitable for poor soil.
4. Weed infestation is minimum
5. Economy in cultural practices & easy operations.
6. Chance of using saline water.
7. Improve efficiency of fertilizers.
8. Very flexible in operation
9. No soil erosion.
10. Easy installation, no land preparation.
11. Minimizing quantity of produce.
12. Enhances the maturity of the crop.

Limitations

1. High maintenance requirement.
2. Salinity hazard
3. Economy limitations (40,000Rs/ha)
4. High technical know-how is required.

Material used Drip System Design (Manufacturing of Drip)

- PVC = Poly Vinyl Chloride
- LDPE = Low Density Polyethylene
- HDPF = High Density Polyethylene
- LLDPE = Linear Low Density Polythene

- PP = Polypropylene
- F.R.P. = Fiber Glossed Reinforced Plastic
- PP = Polysterrlene polyamide

Use of Above Mentioned materials

- Plastic = thermoplastic, thermosetting plastic
- PVC = forcipes
- LDPE = for laterals drippers
- HDPE = pipes
- For laterals
- Values etc. drip components
- F.R.P. = fertilizer tanks

Two types of plastic are generally used

Thermoplastic: It consists of material that can be molded in desired shape under hard & pressure and also be used for remolding after.

Drip/ Trickle Irrigation

This is also called trickle irrigation. The irrigation is given by using small diameter PVC tubing with devices called emitters or drippers at selected places near the base of plants. The losses by deep percolation and evaporation are minimum. The amount of water applied daily is equal to consumptive use and it keeps the soil at low tension. It was originally developed by Simca Blass a water engineer in Israel in 1959. The method is good for economizing on water use, where poor quality water is used and is gaining popularity in arid areas.

It has been found successful in musk melon, cucumber, vines, tomato, salt distribution is better than sprinkler when poor quality waters are used.

Advantages

- No deep percolation losses
- We can maintain good moisture
- Economizing water use saving.
- We can maintain good quality waters
- Fertilizers can also be applied

Disadvantages

- High cost
- Rats and other animals may damage the system

- Dirty water cannot be used
- Does not work under very low pressure

During the last decade there have been great technological improvements on drip irrigation system equipment. This has also lead to a better reliability of the systems together with a better knowledge on drip irrigation management.

For example, the dripper lines come in different wall thicknesses. Some are made to last for 15-20 years; others are thin and must be discarded after one crop. In some situations it will be logical to have the dripper lines sitting on the surface of the soil, and in other cases a subsurface installation will be chosen, with dripper lines buried at a particular depth to suit the soil, the crop, and the various operations carried out during the growing and harvesting periods. Regardless of the final combination of options chosen by the farmer, a drip irrigation system that is well designed and managed will have significant advantages over surface and overhead irrigation systems, because of significant reduction of water usage, better control of watering, less energy consumption, less inputs on nutrients and fungicides compared to overhead systems in certain regions.

With drip irrigation it is possible to irrigate without wetting the leaves of the plant, which has distinct advantages for disease prevention. Subsurface drip systems are even more advantageous in this regard if they are managed in such a way that the surface soil remains dry.

Drip irrigation systems distribute water at points close to the plants at a rate and frequency suitable to the needs of the plant. The number of water emission points in a given area is not fixed. Each system is designed according to crop requirements and soil type. Water is supplied to all plants uniformly.

In most situations, the system is designed and operated in such a way that non-irrigated strips are left between the plant rows, enabling the farmer to work in the field at any time, whether or not the irrigation system is operating, without damaging the soil structure. For example, the timing of sprays is often critical to successful implementation of integrated pest management (IPM) programs. With drip irrigation, you have the highest likelihood of having field access at critical times. Even if there has been unexpected rainfall, the dry soil between the plant rows has the capacity to absorb some or all of it, allowing access by farm equipment sooner than would be the case with surface (flood or furrow) or overhead (sprinkler) irrigation.

The availability of low cost plastic pipe for water delivery lateral lines helped to speed up the field use of drip irrigation system. At present the largest area is in the USA and Spain (1.5 million ha each) followed by India

(1.43 million ha). Total coverage in the world has increased from 4000 ha in 1972 to over 8 million ha in 2008.

Surface drip irrigation

The application of water to the soil surface as drops or tiny streams through emitters with discharges rate for point-source emitters less than 8 L/h for single outlet emitter and for line-source emitters less than 4 L/h (Fig. 23.1). Often the terms drip and trickle irrigation are considered synonymous.

Drip Irrigation / Trickle Irrigation

Irrigation advancements within the last 2 decades have been astounding. Drip or trickle irrigation is one of the latest innovations for applying water, and represents a definite advancement in irrigation technology

Drip irrigation is defined as the precise, slow and frequent application of small quantities of water to the soil in the form of discrete drops, continuous drops, and tiny streams through emitters located at selected points along a water delivery lateral line. It differs from sprinkler irrigation by the fact that only part of the soil surface is wetted

Advantages

Many reports have listed and summarized potential advantages of drip irrigation as compared to other irrigation methods

a) Enhanced plant growth, crop yield and premium quality produce
b) Water Saving due to increased beneficial use of available water and higher water application efficiency
c) Precise and uniform delivery of water to crops due to controlled water application
d) Maintenance of higher soil water potential in the root zone
e) Compact and efficient root system
f) Combined water and fertilizer (fertigation) application minimizes nutrient losses and improves fertilizer use efficiency and contributes to fertilizer saving in some crops
g) Reduced salinity hazards to crop plants when low quality saline water is used for irrigation
h) Suitable for irrigating high-value crops raised in greenhouses, plastic tunnels, potted plants and under plastic mulches
i) Lower operating pressures means reduced pumping energy costs Limited weed growth because only a fraction of the soil surface is irrigated

k) Reduced operational and labour costs due to improved weed control and simultaneous application of water, fertilizers, herbicide, insecticide, fungicide and other additives through the drip irrigation system

l) Feasible to irrigate crops raised in small & irregularly shaped narrow lawns, and on undulated land terrains m) Maintenance of dry foliage means improved disease and pest management

n) Suitable to highly permeable & low water holding sandy and desert soils, saline and slowly permeable alkaline soils, wastelands, slopy lands and rocky hills, road embankments, abandoned mine areas etc

o) Improved and continuous cultural operations such as spraying, weeding, thinning and harvesting of tree and row crops is possible without interrupting the drip irrigation cycle for any prolonged period of time.

p) Environmental protection and ecological security

Drip Irrigation System Layout and Components

The drip irrigation system consists of three subsystems viz., control head unit, water carrier system and water distribution system besides water source & pumping station (Fig. 23.3).

a) Head control unit – Non return valve, Air release valve, Vacuum breaker, Filtration unit, Fertigation unit, Throttle valve, Pressure gauge, Water meter, Pressure regulator and Pressure relief valve

b) Water carrier system: PVC main pipeline, PVC submain pipeline, Control valve, Flush valve and other fittings

c) Water distribution system – Drip lateral, Emitters, Grommet, Start connecter, Nipple, End cap

Suitable crops

Drip irrigation, like other irrigation methods, will not fit every agricultural crop, specific site or objective. Presently drip irrigation has the greatest potential in the following

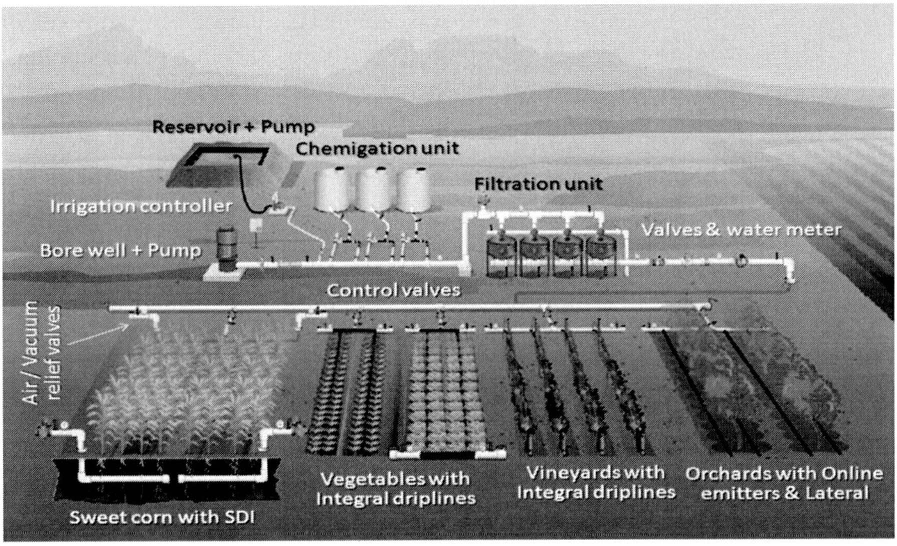

A typical drip head control unit and system components

The use of drip irrigation results in higher water use efficiency than can be achieved with other methods of irrigation, for the following reasons:

a) Frequent, uniform irrigation and fertigation results in higher yields (and often better quality), meaning less water is used per unit of yield.

b) There is less loss of water by evaporation than with sprinkler or surface irrigation, due to less moistened surface area.

c) Winds do not influence water distribution and have much less effect on evaporation from a drip-irrigated field.

d) Low flow rates and point water distribution with drip irrigation prevents runoff, even in difficult topographical conditions.

e) Distribution of water is more uniform with drip irrigation than with other methods of irrigation.

f) Drip irrigation systems are much less sensitive to drops in pressure than sprinkler systems. In other words, the uniformity of water distribution will not be reduced to as great an extent with the drip system in situations where pressure has dropped below the recommended range.

g) With drip irrigation there is no peripheral water loss.

h) Weed growth can be minimized, especially with subsurface drip irrigation, so competition for water by weeds is generally less than with other irrigation methods.

Due to the capillary water flow in the soil, almost no air is forced out of the soil by the movement of water, except in a small saturated area that is created near each dripper. The larger pores in the soil generally remain well aerated, even when the soil moisture level is being maintained near field capacity. The plant roots have good access to air throughout the growing season, with no interruption during or immediately after irrigation.

Drip irrigation allows relatively good field sanitation conditions to be maintained.

a) Dry foliage results in a decrease in diseases and pests.
b) Insecticides and fungicides are not washed off the leaves as with sprinklers.
c) Fewer weeds germinate as a result of limited wetting of the soil surface.
d) Drip irrigation prevents the spread of diseases and weeds which can be transferred to other areas of the field by surface irrigation.
e) When properly managed, there should never be anaerobic conditions in the soil for an extended time with drip irrigation. The incidence of soil borne diseases is greatly reduced.

Drip irrigation requires more specialized labour, training, knowledge and maintenance. It requires less energy than other methods of irrigation.

The crop uses up to 95% of supplied water. This is a significant ecological advantage of drip irrigation. (It should be noted that if surface and subsurface drainage are required in high rainfall areas for removing excess water, these drains should be established and maintained, even when drip irrigation has been installed.)

Drip irrigation makes possible the use of sloped areas and areas with difficult topographical conditions, with no need for steps or for moving soil which, in addition to being expensive, can cause damage to the soil structure and can expose infertile layers.

With drip irrigation it is possible to irrigate large areas simultaneously, without influencing the pressure in the general water system.

The volume of soil being irrigated by a drip system is relatively small (compared to overhead and surface irrigation). The plant still requires the same amount of water and therefore the same amount of root surface and root hairs. A more concentrated root system is created within this wetted volume.

The most efficient and commercially practical water supply system should be used to ensure the best utilization of water resources. Drip irrigation instead of furrow and pivot irrigation systems are encouraged.

Drip irrigation systems minimize water use in the tomato fields. Drip systems deliver water to the plants much more uniformly than furrow irrigation and some growers feel that drip irrigated fields are less prone to disease and pest problems.

In drip irrigation, water flows through a filter into special drip pipes, with emitters located at different spacing. Water is discharged through the emitters directly into the soil near the plants through a special slow-release device. If properly designed, installed, and managed, drip irrigation may help achieve water conservation by reducing evaporation and deep drainage as compared to other types of irrigation such as flood or overhead sprinklers since water can be more precisely applied to the plant roots.

Water is applied through network of pipelines and allowed to fall drop by drop at crop root zone by a special device called emitters or drippers. These drippers or emitters control the quantity of water to be dropped out. In this system, the main principle is to apply the water at crop root zone based on the daily ET demand of the crop without any stress. Hence, the root zone is always maintained at field capacity level.

Components

- Overhead tank or pressure system (Motor pumps).
- Main Lines - To take water from source to field which is usually made of black poly alkathene pipes having an inner diameter of 50 mm

Sub main - If the area is larger, the sub mains are used to take water from main pipes to field which is normally having an inner diameter of 37 mm.

Laterals - These pipelines are normally having lesser diameter than mains and sub mains usually of 12 mm made of black poly alkathene pipes which deliver water from main or sub mains to crop root zone. The length of lateral depends upon the pressure created in pump as well as spacing of the crop and length of the field. Normally about 25 m length of lateral can be adopted to have a uniform distribution of water.

Emitters - Emitters control the water drops and the quantity of water to be delivered. Various designs of drippers with various discharge capacity are available (5, 7, 8, 10 and 20 lph, etc. Button types, spray type, tap type etc.). Instead of drippers micro tubes are inserted into the laterals and water is allowed to drip in the root zone of crops or trees.

Advantages

- Application of water in slow rates facilitates the easy infiltration into the soil.

- The required quantity of water is applied near the root zone alone which in turn save water.
- The root zone is always maintained with field capacity level and hence plants do not suffer for want of water.
- There is no seepage or percolation or evaporation losses.
- Weed growth is restricted due to limited area of wetting zone.
- Fertilizers (fertigation), chemical like pesticides (chemigation) and herbicide (herbigation) can be applied through irrigation. Hence, saving of input quantity and labour cost besides increase in their use efficiencies is possible.
- Reduce the salt content near the root zone and dilute it in saline soil.
- The saline water also can be put under use if irrigation is applied through drip irrigation.
- It can be adopted for any type of topography.
- Yield increases due to optimum maintenance of soil moisture at root zone.
- More area can be maintained with little quantity of water.
- It cab be used for widely spaced crops like cotton, sugarcane, tomato, brinjal, coconut and orchard crops.

Disadvantages

- Clogging in emitters due to salt content of water and other impurities like moss, dust etc.
- Damage of pipe lines by rodents.
- It is not economical for closely spaced crops which require more number of pipes and drippers per unit area.
- Proper maintenance and periodical cleaning of drippers and pipelines (with 1% hydrochloric acid) are very important to maintain the system efficiency.

9.8 Sub-Surface Irrigation

In sub-surface irrigation, water is applied beneath the ground surface usually by creating and maintaining an artificial water table at some pre-determined depth (10-100 cm) depending upon root characteristics of the crop, texture and layering of soil. Moisture reaches plant roots through capillary movement. In India, this method is used in areas around Dal Lake in Kashmir for vegetable production and in soils of Kuttanad (Kerela) for irrigation of coconut palms but on a limited scale.

Natural: Impervious layer, water is led into series of ditches dug upto the impervious layer which then moves laterally and wets the root zone.

Artificial: Raising water table by layering porous tubes where artificial water table could be maintained, topography level, suitable for irrigation of vegetables, crops and most of field crops, small grains and forage crops.

Advantages

- Maintenance is cheap.
- Efficient use of water.
- No evaporation losses.
- Soil ever fit for cultivation and weed removal.
- No wastage of area in bunds and channel.
- No interference to farm farm machinery.
- Very suitable for ornamental gardens and high value crops.
- Weed seeds are not carried with water.
- Channels serve as drainage channels also.

Limitations

- Costly;
- Risk of soil getting saline;
- Neighboring land damaged due to heavy seepage.
- Deep rooted crops difficult to grow.

Fertigation with Drip Irrigation Systems

Fertigation capability is an integral part of the drip system, and drip irrigation provides the most convenient means of supplying nutrient materials.

Fertigation is the direct application of fertilizer at the correct time through the irrigation system to the region where most of the feeder roots are sitting. This results in yield increases and often quality increases (eg a higher proportion of uniformly sized fruits), as well as in more efficient use of the fertilizer.

By reducing the interval between applications, it is possible to maintain a uniform level of nutrients and to control the nutrient supply in the soil in accordance with changing plant needs during the growing season. Thus, the drip system can be regarded not only as a method of water application, but also as a system for conveying fertilizer to the desired location at the proper time.

Drip irrigation gives us the potential to apply 100% of a crop's fertilizer requirements through the drip system, and to feed the plant frequently with small portions of balanced nutrient mixes. This type of nutrient application could provide a major breakthrough in achieving considerable increases in crop production. However, due to the potential for creating precipitates (solid molecules) that could cause blockages in the drippers, caution is advised. There are many things to be aware of when fertigating in order to avoid undesirable reactions in the dripperlines. Many farmers apply some fertilizers to the soil and mainly use fertigation for applying nitrogen through the drip system.

These are some of the things that need to be taken into account to ensure successful fertigation through drip irrigation systems:

1. Ascertain that your fertilizer is completely soluble and does not contain any impurities or insoluble matter.
2. Always inject the fertilizer before the filters, never after!
3. In most cases it is not advisable to use fertilizers with an alkaline reaction.
4. Avoid the use of any fertilizer (solid or liquid) that contains an iron residue, such as red KCl.
5. Never fertigate with mineral microelements. Use only chelates (microelements combined with organic matter).
6. Ascertain that chelates to be used are known to be stable, as there are certain types that decompose in the system. This is not good for the crop, and it is very bad for the drip system.
7. Fertilizers containing phosphorous can form calcium phosphate precipitates in the drip system. This is a dangerous blockage issue for drip systems. The following precautions should be taken:
    1. Avoid high concentrations.
    2. Do not use fertilizers based on polyphosphate; use only orthophosphate!
    3. Ascertain that your P fertilizer has a strong acidic reaction.
    4. Acidic P fertilizer is not sufficient to prevent precipitation when the water contains high calcium levels or when the water is alkaline. In this event, fertigate with phosphoric acid, or acidify the water.
    5. When the irrigation water is acidic, with a pH of 6 or less, you can fertigate freely with most phosphorous fertilizers.
    6. Never stop irrigation and fertilization at the same time; turn off the fertilizer pump ½ to 1 hour before the end of irrigation. This is vital for flushing any remaining phosphate ions from the system.

Smart / Precision Irrigation

Precision Irrigation: principles, design and operation of sprinkler, trickle, and surface irrigation methods

Precision irrigation demands increasingly accurate information about crops, soil and weather. Much research has shown that crop temperature is a reliable plant water stress indicator and, thus, a promising tool for irrigation water management. This paper briefly describes the status of a research project to develop an unmanned aerial vehicle with a Global Navigation Satellite System (GNSS) receiver and an infrared temperature sensor. Laboratory tests determined the key features for accurate temperature measurements, flight altitude and image stabilization. Field experiments are being conducted to: (1) assess the sensitivity of crop temperatures as water status indicators in sugar beet; and (2) determine the sensitivity of a remote-sensing system to derive temperature-based water stress indices. Two irrigation treatments were established: (1) full, which met crop water needs; and (2) deficit, in which irrigation was withheld over the experimental period. The preliminary results indicate that crop temperature is sensitive to soil water deficit in sugar beets.

Objectives

1. To provide an understanding of basic engineering principles and procedures which are necessary for the successful selection, design and operation of pressurized and surface (free flow) irrigation methods;
2. To acquaint and equip the students with advance application of irrigation along with applicability of various models;

Precision irrigation scheduling varies irrigation at any one time under one irrigation method according to soil and crop differences;

Where variable soils exist under one method, then irrigation should ideally be varied according to these soil differences.

Component technologies, where development of improved tools & technologies will need to be on-going

- Sprinkler irrigation
- Drip Irrigation
- Automated Irrigation

However there are some clear immediate needs for particular sensing and simulation tools for the Precision Irrigation currently under development:

These are

1. Low-cost, spatially-distributed, non-invasive sensing of soil moisture and crop response;

2. Development of a fully deterministic sprinkler pattern model for centre pivot and lateral move machines that can account accurately for varying sprinkler pressure and height, sprinkler pattern overlap, wind, and machine movement;
3. Development of a hydraulic diagnostic model for drip irrigation systems capable of interaction with the system control to deliver spatially varied applications;
4. Improved crop models sensitive to small variations in irrigation management and with a self learning capability; and
5. Verification of the use of short range radar for the measurement of the spatial distribution of rainfall at the sub□field scale;

References

Allen, R. G., Pereira, L. S., Raes, D., & Smith, M. (1998). Crop evapotranspiration: Guidelines for computing crop water requirements. FAO Irrigation and Drainage Paper No. 56. Food and Agriculture Organization of the United Nations.

Badran, O. I., & Al-Saad, H. K. (2013). "Emerging technologies in irrigation systems for sustainable development." Sustainable Agriculture Reviews, 12, 205-224. https://doi.org/10.1007

Fereres, E., & Soriano, M. A. (2007). "Deficit irrigation for reducing agricultural water use." Journal of Experimental Botany, 58(2), 147-159. https://doi.org/10.1093/jxb/erl165

Kendy, E., & Zhang, L. (2003). "Agricultural water use: A systems approach." Water Resources Research, 39(5), 1096. https://doi.org/10.1029/2002WR001005

O'Connor, K., & Harbaugh, A. (2010). "Design and optimization of advanced irrigation systems." Journal of Irrigation and Drainage Engineering, 136(3), 190-197. https://doi.org/10.1061/(ASCE)IR.1943-4774.0000184

Or, D., & Wraith, J. M. (2002). "The influence of soil texture on water retention and root uptake in a semi-arid environment." Soil Science Society of America Journal, 66(5), 1627-1635. https://doi.org/10.2136/sssaj2002.1627

Pachepsky, Y., & Rawls, W. J. (2003). "Soil water retention: Data collection and data analysis." Soil Science Society of America Journal, 67(1), 139-153. https://doi.org/10.2136/sssaj2003.1390

Sivanappan, R. K. (2005). "Drip irrigation in agriculture: Relevance and status." Journal of Agricultural Engineering, 82(1), 1-9. https://doi.org/10.3923/jae.2005.10.19

Tullis, J., & Gad, A. (2008). "Integration of irrigation and drainage systems for effective water management." Irrigation and Drainage Systems Engineering, 11(2). https://doi.org/10.4172/2168-9768.1000155

Xiong, H., & Zhang, H. (2011). "Modeling irrigation systems for optimizing water application." Agricultural Water Management, 98(11), 1688-1695. https://doi.org/10.1016/j.agwat.2011.07.006

Multiple Choice Questions

1. Which of the following is NOT a method of irrigation listed in the text?
   a) Check basin method b) Hydrospray method
   c) Drip irrigation d) Furrow irrigation
2. What is the main objective of using irrigation methods according to the text?
   a) To generate runoff b) To increase evaporation loss
   c) To optimize crop water needs d) To simplify harvesting
3. Which factor influences the selection of irrigation methods related to soil?
   a) Market demand b) Soil color
   c) Soil texture and structure d) Crop rotation frequency
4. In which type of land is the basin method particularly applicable?
   a) Sloppy and steep land b) High wetland areas
   c) Levelled land surface d) Sandy coastal areas
5. What should be avoided when applying fertilizer through the drip irrigation system?
   a) High concentrations of phosphorous
   b) Use of soluble fertilizers
   c) Low-quality water
   c) Frequent applications
6. Which method is described as applying water through a network of pipelines, resembling rainfall?
   a) Surface irrigation b) Drip irrigation
   c) Subsurface irrigation d) Sprinkler irrigation
7. What is one of the major advantages of drip irrigation mentioned in the text?
   a) It encourages weed growth.
   b) It prevents nutrient loss through over-application.
   c) It requires more labor-intensive operations.
   d) It is less flexible in application.

8. What limitation is associated with the sprinkler irrigation method?
   a) It can be used in all types of soil.
   b) It has high initial costs and operating power requirements.
   c) It requires less technological skill.
   d) It eliminates all runoff loss.
9. Which irrigation method is suitable for high-value crops in greenhouses?
   a) Surface irrigation b) Subsurface irrigation
   c) Sprinkler irrigation d) Drip irrigation
10. What is surge irrigation?
   a) Continuous flow irrigation
   b) Intermittent water application
   c) Deep well irrigation
   d) Flood irrigation

Answer Key

1	b	2	c	3	c	4	c	5	a	6	d	7	b
8	b	9	d	10	b								

10

Efficiencies of Irrigation Water Application

Irrigation Efficiency

Irrigation water is an expansive input and has to be used very efficiently. The main losses that occur during irrigation of fields as conveyance, run off, seepage and deep percolation; Irrigation efficiency can be increased by reducing these losses. Uneven spreading and inadequate filling of root zone are the other causes for low irrigation efficiency. Irrigation efficiency at the field level can be increased by selecting suitable method of irrigation, adequate land preparation and engaging an efficient irrigator. At the project level, it can be increased by proper conveyance and distribution system. Irrigation efficiency is the ratio usually expressed as percent of the volume of irrigation water transpired by plants, plus that evaporate from the soil, plus that necessary to regulate the salt concentration in the soil solution and that used by plants in building plant tissue to total volume of water diverted, stored or pumped for irrigation.

$$Ei = \frac{Wt + Ws - Rs}{Wi} \times 100$$

Where,

Ei = Irrigation efficiency (percent)

Wt = the volume of irrigation water / unit area of land transpired by plants, evaporation from the soil during the crop period.

Ws = the volume of irrigation water per unit area of land to regulate the salt Content of soil solution.

Re = Effective rainfall

Wi = the volume of water per unit area of land that is stored in reservoirs or diverted for irrigation. Irrigation efficiency indicates how efficiency the available water supply is being used. The efficiency of irrigation projects in India is as low as 20 to 40%.

Water Conveyance Efficiency and Water Use Efficiency

Water Conveyance Efficiency

It indicates the efficiency with which water is conveyed from source of supply to the field. It estimates the conveyance losses. It is expressed as

$$Ec = \frac{Wf}{Ws} \times 100$$

Where,

Ec = Water conveyance efficiency (percent)

Wf= Water delivered at the field

Ws= Water delivered at the source

Water Application Efficiency

Irrigation water applied to the field is lost due o surface run off and deep percolation. Surface run off occurs due in long furrow or long border strips if ridges are weak. The water moves from one plot to another due to weak bunds giving way to water which may collect in large quantities even to break strong bunds. In furrows, water is allowed most of the time at the beginning of furrow till the flow reaches the other end of the furrow. It results in deep percolation of water in the first quarter of furrow. Water application efficiency is the measure of efficiency with which delivered to the field is stored in the root zone.

$$\text{Water application efficiency} = \frac{\text{Water stored in the root zone}}{\text{Water delivered to the field}} \times 100$$

Water Storage Efficiency

This parameter estimates whether the amount of water necessary for the crop is stored in the root zone or not. It is expressed as the percentage of water needed in the root zone prior to irrigation to that stored in the root zone during irrigation.

$$\text{Water storage efficiency} \frac{\text{Water stored in the root zone}}{\text{Water needed in the root zone}} \times 100$$

Water Distribution Efficiency

Water distribution efficiency is defined as the percentage of difference from unity of the ratio between the average numerical deviations from the average depth stored during the irrigation.

$$\text{Water distribution efficiency} = \{1 - Y/d\} \text{ X } 100$$

Where,

d = Average depth of precipitation along the run off during irrigation
Y = Average numerical deviation from –d

Water distribution efficiency indicates uniformity in distribution of water over the entire root zone.

Water Use Efficiency (WUE)

Water use efficiency is defined as yield of marketable crop produced per unit of water used in evapotranspiration.

WUE = Y / ET

Where,

WUE = Water use efficiency (kg/ha/mm of water)

Y = marketable yield (kg/ha)

ET= Evapotranspiration (mm)

If yield is proportional to ET, water use efficiency has to be constant but it is not so. Actually, Y and ET are influenced independently by crop management and environment. Yield is more influenced by crop management practices, while ET is mainly dependent on climate and soil moisture. Fertilization and other cultural practices for high yield usually increase in water use accompanying fertilization is often negligible. Crop production can be increased by judicious irrigation without markedly increasing ET. Under optimum water supply, ET is not dependent on kind of plant canopy provided the soil is adequately covered with crop.

Increasing the amount of plant canopy has there fore little or no effect on ET. Obviously, any practice that promotes plant growth and more efficient use of sunlight in photosynthesis without causing a corresponding increase in ET will increase WUE.

Factors affecting WUE

Nature of the plant: There are considerable between plant species to produce a unit dry matter per unit amount of water used resulting in widely varying values of WUE.

Water use efficiency of different crops

Crop	Water Requirement mm	Grain Yield kg/ha	WUE kg/ha/mm
Rice	2000	6000	3.0
Sorghum	500	4500	9.0
Bajara	500	4000	8.0

Maize	625	5000	8.0
Groundnut	506	4680	9.2
Wheat	280	3534	12.6
Finger Millet	310	4137	13.4

There is also difference in WUE between varieties of the same crop. Selection of properly adopted crop, with good rooting habit,low transpiration rates increase. WUE

Climatic Conditions

Weather affects both Y and ET. Manipulation of climate to any extent is possible at present. However, ET can be reduced by mulching, use of antitranspirant etc. To limited extent , but may not be economical or practical. Weed control is the most effective means of reducing ET losses and increasing the amount of water available to the crop thereby increasing WUE.

Soil Moisture Content

In adequate supply of soil moisture as well as excess moisture supply to the crop have an adverse effect on plant growth and production and therefore conductive to low WUE. For each crop combination of environment conditions, there is a narrow range of soils moisture level at which WUE is higher than with lesser or greater supply of water, proper scheduling of irrigation will increase WUE.

Fertilizers

Irrigation improves a greater demand for plant nutrients. Nutrient availability is highest for most of the crops when water tension is low. All available evidences indicate that under adequate irrigation suitable fertilization generally increase yield considerably, with a relatively small increase in ET and therefore, markedly improve WUF.

Plant population

Higher yield potential made possible by the favorable water regime provided by irrigation, the high soil fertility level resulting from heavy application of fertilizers and genetic potential of new varieties and hybrids, could be achieved only with appropriate adjustments of the population. The highest yields and WUE are possible only through optimum levels of soil moisture regime, plant population and fertilization.

a) Irrigation efficiency

$$Ei = \frac{Wt + Ws - Re}{Wi} \times 100$$

Where, E_i is irrigation efficiency (%); W_t is the volume of irrigation water per unit of land transpired by plants, evaporated from the soil during crop period (including field preparation and nursery); W_s is the volume of irrigation water per unit area of land to regulate the salt content of soil solution; Re is effective rainfall and W_i is the volume of water per unit area of land that is stored in a reservoir or diverted for irrigation.

b) Conveyance efficiency

It indicates the efficiency with which water is conveyed from source of supply to the field. It estimated the conveyance losses and expressed as:

$$Ec = \frac{Wf}{Ws} \times 100$$

Where, Ec is water conveyance efficiency (%); W_f is water delivered at the field and W_s is water delivered from the source.

c) Water application efficiency

The percentage ratio of the amount of water stored in the crop root zone during the irrigation to the amount of water delivered to the field. It is expressed as:

$$Ea = \frac{Ws}{Wf} \times 100$$

Where, W_s is the amount of water stored in the crop root zone.

W_f is the amount of water delivered to the field.

d) Water storage efficiency

The percentage ratio of the amount of water stored in the effective crop root zone soil to the amount of water needed to make up the soil depleted in crop root zone prior to irrigation.

$$Es = \frac{Ws}{We} \times 100$$

Where, W_e is the amount of water needed to the crop root zone prior to irrigation.

$W_s$is the amount of water stored in the crop root zone of the soil.

e) Water distribution efficiency

It gives a measure of the uniformity of the depth of irrigation all over the field and expressed as:

$$Ed = 1 - \frac{Y}{D} \times 100$$

Where, Y is average numerical deviation

D is average depth of water stored during irrigation.

f) Water use efficiency (WUE)

It refers to the amount of marketable end product obtained per unit of water used by the crop.

$$\text{WUE (kg ha}^{-1}\text{ mm}^{-1}\text{)} = \frac{\text{Total economic yield (kg ha}^{-1}\text{)}}{\text{Total consumptive use (mm)}}$$

g) Project Efficiency (PE)

It is the ratio between the average depth of water stored in the root zone during irrigation and water diverted from the reservoir. It is mathematically expressed as:

Where:

I. Project efficiency (%)

II. Water stored in the root zone (cm)

III. Water diverted from the reservoir (cm)

Production Efficiencies in Cropping and Farming System

Agricultural productivity and efficiency is at the centre of abounding of the debates, behavior and measures apropos the agronomics sector. The accent placed by the Sustainable Development Goals on agronomical productivity underlines the abounding affidavit for which added analysis on statistical frameworks for productivity and efficiency targeted to developing countries is necessary. Advice on agronomical productivity is accompanying to several of the Sustainable Development Goal indicators, in particular:

"Productivity is frequently authentic as an arrangement of an accumulated admeasurement of achievement to an accumulated admeasurement of ascribes use" (OECD 2001). At it's a lot of axiological level, productivity measures the bulk produced by an ambition accumulation (country, industry, sector, acreage or about any ambition group) accustomed a set of assets and inputs.

Productivity can be abstinent for an individual article (farm, commodity) or an accumulation of farms, at any bounded scale. The admeasurement should reflect the ultimate purpose for the inquiry. If for example, the purpose is to analyze productivity amid farms, again measures that are micro-based are required. If the charge is to appraise national agronomical action at the country level, again macro measures are required. This aforementioned affinity can extend above the breadth to the national economy. While the adapted purpose can vary, the altitude issues associated with anticipation the altered indicators are the same. However, abstracts requirements may alter depending on the blazon of indicator: farm-level productivity altitude for one article and one ascribe (for example, labour productivity of maize farms) may alone crave basal advice on achievement quantities and ascribe use, while bearing aggregated measures about requires appraisement outputs and inputs.

Measurement of Productivity

As ahead stated, productivity measures are consistently accumulated based, either bidding in concrete quantities, or in connected bulk terms, implying that ethics be adapted for bulk change. In adjustment to get complete or connected dollar measures, time alternation for outputs and inputs as able-bodied as for prices are required or alternatively adapted is achievement and ascribe accumulated and bulk indices. Obtaining the actual bulk or bulk indices, in turn, adds decidedly to the complication of productivity measurement, a lot of which is accompanying to analogous the actual bulk (or index) to the artifact or input. In the case of outputs, the bulk or basis acclimated needs to accede the altered characteristics associated with the product, abnormally the superior characteristics that are associated with the empiric price. Application appropriately complete bulk indices have been the focus of abundant of the analysis on productivity because alternation indices are frequently used.

Over the years, the analysis has adapted application altered bulk indices for abbreviating outputs and inputs, anniversary with altered backdrop and anniversary acquiescent altered results. Selecting the adapted one to use is abiding in theory, but about the best focuses on how able-bodied the called bulk basis accounts for barter bias. It has been apparent by Diewert (1976) and endless others, that accomplished indices (those that amuse assertive numeric properties) can annual for this bias, but they accept the abject coercion (assumption) that the industry beneath abstraction operates beneath complete antagonism and with an assertive blazon of assembly function. Because of its adorable properties, the Thorniest basis is generally acclimated to admeasurement TFP for a bulk of reasons. First, the Thorniest basis is a detached approximation of the Division index, broadly believed to be the best basis for barometer bread-and-butter aggregates because of its accommodation to anxiously represent the basal assembly action & invariance property. 5 Second, as the Thorniest basis is an accomplished index, it can be accompanying to abounding assembly or bulk functions. In particular, it corresponds absolutely to a Translog function. Third, accession advantage is that this basis is constant in

Agricultural behavior tends to focus added on adopting productivity through abstruse change than through bigger use of the complete technology. However, rebalancing the focus of agronomical behavior appear convalescent efficiency is all-important in the ambience of bound avail efficiency of accustomed resources, such as acreage and water, and accustomed the call to complete the ecology brand of agronomical production. Agnate concrete productivity assets and conceivably even above bread-and-butter assets may be accepted from bigger use of complete technology than from alive to new technology. The closing may access productivity in the abbreviate term, but possibly at the bulk of college assembly and ecology costs. For example, afore advising farmers to accept actinic fertilizers (technological change), acceptable fertilization methods involving amoebic fertilizers and rotations or admixture of crops (technical efficiency) may be answer as a way to access concrete productivity and advance aliment aegis and bread-and-butter profit efficiency. Abstruse efficiency is declared in detail in the afterward paragraphs.

Measurement of Competitiveness

An added acumen that needs to be fabricated is amid bread-and-butter efficiency and competitiveness. The above is a complete admeasurement of the bread-and-butter achievement of the acreage admitting the closing compares this achievement to that of their competitors. In added words, acreage can be economically inefficient but aggressive because added farms are even beneath efficient. Reciprocally, an economically able acreage is not necessarily aggressive if all the added farms are as well economically efficient. Competitiveness as well goes above the price/cost achievement and extends to the appearance absorbed to the achievement or to the bearing close (or sector, country), such as superior attributes, both authentic and perceived. For example, a close can accept analogously top assemblage costs but may annual from a top "non-price" competitiveness, which allows it to advertise its articles at a college price.

Competitiveness is a lot of generally abstinent application bread-and-butter indicators, such as gross or net margins (often per assemblage of land), and comparing the achievement of farms (or agronomics systems) based on these measures. Competitiveness and productivity are carefully related: college productivity can advance to a greater competitiveness of the action (or sector) because added is produced out of the aforementioned bulk of resources. This agency that, with all things getting captivated equal, the bulk of assembly per assemblage of achievement is lower, and that margins per assemblage of achievement are higher. Productivity is an all-important arrangement for competitiveness, but not an acceptable condition. Indeed, an aggregation of factors affecting the competitiveness of an action has been articular in the literature. Competitiveness is the aftereffect of an accumulated of factors, both national and international:

1. Nationally, efficiency endowments, technology, productivity, artifact features, budgetary and budgetary administration and assuredly the barter policy are apparent to be the lot of important factors that actuate the competitiveness of an industry and/or business. Productivity is, therefore, apparent as one of the national (domestic) determinants of competitiveness;
2. Internationally, the a lot of important factors are barter rates, all-embracing bazaar conditions, the bulk of all-embracing carriage and the preferences and settings amid altered countries.
3. Agricultural productivity is based on the superior of the inputs and how able-bodied those inputs are chip in the assembly process. For example, acreage productivity awful depends on the breadth of the acreage and its concrete characteristics. This is the aforementioned for labour as the superior of the plan force differs across, for archetype artisan types or subsectors.
4. For allusive purposes, the superior of the ascribe accept to be taken into annual in the abstracts accumulating and adapted adjustments charge to be fabricated afterwards the abstracts are collected. This agency that abstracts on ascribes use charge to be calm for altered types of inputs or superior classes. For example, ancestors labour;
5. Occasional workers and abiding workers should be differentiated in the abstracts accumulating process. Because workers with altered abilities accept altered levels of productivity, application the aforementioned accomplishment for workers with altered accomplishment levels after-effects in biased estimates of labour productivity. The aforementioned applies to fertilizers or to any added ascribe or assembly agency that has capricious characteristics. With attention to fertilizers, this ascribe varies in agreement of the dosage of alive capacity to pesticides, which may be added or beneath effective. One kilogram of fertilizer activated in 2000 is not commensurable to one kilogram activated in 2015, because of two factors: the accession of new and added able products; and fertilizer appeal may accept confused appear added segments of the market. This change in agreement should be reflected in differentiated ascribe prices;
6. To abode the affair of compositional or superior change, adult frameworks for ascribe superior adjustments accept been developed. The United States Department of Agriculture-Economic Analysis Service

(USDA-ERS), for example, estimates quality-adjusted accomplishment based on abstracts of hours formed and accomplishment per hour cross-classified by altered labour categories (the afterward breadth on labour productivity provides added details). For acreage productivity, acreage prices or rents can be accepted application animal regressions that crop into annual some of the differences in superior attributes, such as clay type, moisture, clay acidity and salinity. Quality-adjusted prices for added inputs can be complete application agnate techniques.

7. The productivity of the acreage measures the bulk of achievement generated by an accustomed bulk of land. It is mostly applicative in the ambience of agronomics activities, but it can as well be continued to livestock production, in assertive cases, as apparent below.;
8. There are several productivity measures that can be calculated: an ample admeasurement is the arrangement amid the bulk of all agronomics articles (crops and livestock) and the absolute acreage acclimated in agriculture. Added acreage productivity measures can be affected by adding crop assembly by the bulk of buried land, bidding in a breadth unit, such as hectares or acres. When bidding in agreement of concrete output, such as tons of maize, acreage productivity corresponds to crop yields. When bidding in budgetary terms, acreage productivity is added generally referred to as allotment to land.

Land productivity = Accumulated of achievement / Buried Area

Planted breadth is acclimated instead of added breadth concepts, such as harvested area, because of the absorption to admeasurement the able crop or acreage productivity rather than an abstract or biological yield. The use of inputs above-mentioned to the autumn is fabricated on the sown/planted breadth (such as fertilizer applications) and not in advertence to the harvested area, which at the pre-harvest appearance is usually unknown. The aberration amid harvested and buried breadth may as well reflect the efficiency and appliance of the agronomics practices, in accession to exogenous factors, such as climate-related events, which should be reflected in the productivity indicator. Application harvested breadth instead of buried breadth tends to advance to overestimations of yields and allotment to acreage because this breadth includes the lot of advantageous segments of the parcel. In general, it is best to use buried breadth for a monocropping arrangement and able area, including dormant land, for alloyed agronomics systems.

Agricultural assembly acclimated for the adding of productivity should cover the assembly of the crops developed on the aforementioned acreage during the advertence aeon whether it is one agronomics division or one year. This is important because, in practice, farmers generally abound added than one crop on the aforementioned artifice over a year; they may abound an admixture of crops on the aforementioned artifice at the aforementioned time or circle the crops developed on the artifice over the season. Kelly et al. (1996) fatigued that one of the affidavit abaft the addiction to belittle achievement and yields in developing countries is the abridgement of accounting of crops developed in admixture or in arrangement and the abridgement of appraisement of by-products, which may be sold, captivated by the domiciliary or acclimated in the assembly of added products. It is, therefore, basic that all crops are included in the altitude of productivity; abnormally in developing regions area these practices are common.

a) **Units:** As with added inputs, acreage productivity can be bidding in abounding units. Accustomed that the acreage may be acclimated to abound abounding altered crops, a concrete unit, such as tons, may not be the best choice. Putting a budgetary bulk on their corresponding achievement is generally bare to accumulate the achievement of altered crops.

b) **Land quality:** Productivity altitude should crop into annual as abundant as accessible clay and acreage superior differences by accession abstract on the soil/land characteristics and their accompanying aspects, abnormally acreage prices and rents. For example, differences in the superior of acreage beyond states and regions in the United States are reflected by artful about prices of acreage from animal corruption results. Ball et al. (2008) applies an animal access to admeasurement quality-adjusted acreage prices bold acreage bulk is an action of characteristics of acreage superior variables, such as clay acidity, salinity and hydric stress. The achievement acquired from the acreage use depends on the clay and acreage characteristics. USDA uses a database that gathers advice on those characteristics in altered states and regions from the "World Clay Assets Office". This method, even admitting it is accurate, requires an ample bulk of abstracts that are not necessarily accessible in developing countries.

c) Labour productivity in agronomics measures the bulk of units of output(s) produced per assemblage of labour acclimated in the action of production. It is a fractional productivity indicator that is affected by adding the productivity of achievement by the absolute units of labour used:

Labour productivity = Accumulated of achievement / Units of labour used

There are abounding means to appraise the productivity of labour input: the bulk of workers alive on the holding; the bulk of time units (such as hours, canicule and months) formed or full-time agnate units if an boilerplate bulk of hours per alive day can be bent according to specific country standards.

OECD (2001) recommends that labour ascribe be abstinent application the bulk of hours finer worked. Application the bulk of hours corrects for the aberration amid melancholia and non-seasonal workers and the altered alive regimes (part-time against full-time). This allows bigger comparisons beyond assembly systems, regions and countries, as the bulk of workers or of canicule per artisan may not announce the labour ascribe finer acclimated on the farm.

However, the change in the bulk of hours appear does not consistently reflect the use of capital, the superior of the workforce and technology (Shumway et al. 2015). USDA-ERS suggests that productivity abstracts abduction the altered types of labour alive in the breadth because labour ascribes differ based on the categories of workers. It is recommended that distinctions be fabricated amid altered ages of workers, ancestors labour and assassin labour and men and women. Distinctions can as well be fabricated amid part-time and full-time workers. Acumen should as well be fabricated amid the altered educational levels, because the superior of one hour provided by an artisan is generally barnacle on his abilities and capacities.

Factors affecting water use efficiency

The factors influencing water use efficiency can be classifiedas follows:

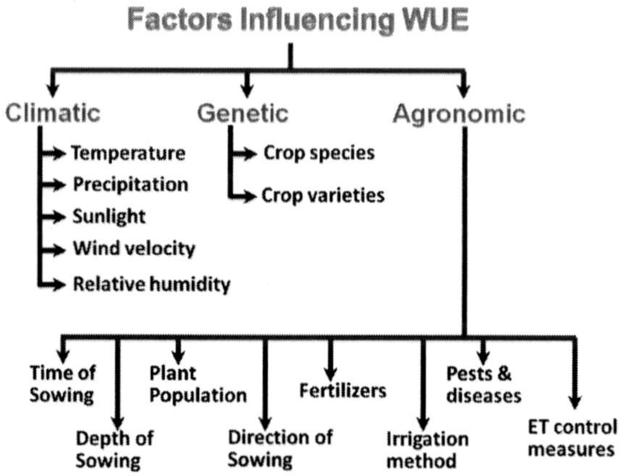

Genetic Factors

Crop growth and yield are the result of the interaction between a plant's genetic makeup and the environmental conditions in which it grows. Consequently, plant species vary significantly in their productivity, which is defined by crop yield and water use (evapotranspiration, or ET).

C4 plant species, such as maize, sorghum, sugarcane, pearl millet, and finger millet, generally exhibit higher water use efficiency (WUE) ranging from 3.14 to 3.44 mg of dry weight per gram of water, compared to C3 species like pulses, oilseed crops, wheat, barley, and oats, which have WUE values between 1.49 and 1.59 mg of dry weight per gram of water.

Crop varieties also differ in WUE. High-yielding varieties, hybrids, and genetically modified (GM) crops often demonstrate higher WUE than traditional varieties due to traits such as dwarf plant types, responsiveness to water and fertilizer, pest and disease resistance, and high harvest indices. Traditional varieties are often characterized by excessive vegetative growth, low harvest indices, susceptibility to lodging, pests, and diseases.

Climatic Factors

Weather conditions influence both crop yield and crop evapotranspiration. The amount of solar radiation directly affects the rate of photosynthesis and, therefore, the potential yield. Other climatic components, including temperature, day length, and rainfall, influence key physiological processes and ultimately determine the actual harvested yield. Lower relative humidity tends to increase evapotranspiration. Therefore, low relative humidity increases transpiration without a corresponding increase in dry matter production, leading to a reduction in WUE. Light and temperature, which influence both transpiration and dry matter production, can either increase or decrease WUE depending on which of the two processes predominates. High wind velocity increases evapotranspiration without a concurrent increase in dry matter production, thus decreasing WUE.

Crop Management Factors

- **Time of Sowing:** Timely sowing ensures optimal temperatures, soil moisture availability, and other favorable soil physical conditions that promote optimal crop growth and development, along with a greater ability to compete with weeds, which, in turn, increases WUE.
- **Depth of Sowing:** Optimal sowing depth affects seedling emergence, vigor, and ultimately crop yield, thereby improving WUE.
- **Direction of Sowing:** Orienting crop rows in a north-south direction influences the interception and utilization of incident solar radiation, which in turn influences crop yield and improves WUE compared to an east-west row pattern.

- **Plant Population:** Maintaining an optimal plant population promotes uniform and rapid development of the crop canopy without excessive competition for growth resources such as light, nutrients, water, and CO2, hence improving WUE.
- **Fertilization:** Fertilizing crops that are suffering from low nutrition under adequate soil water availability can significantly increase crop yield with only a relatively small increase in crop evapotranspiration, leading to a marked improvement in WUE.
- **Insect Pests & Diseases:** Insect pests and diseases can reduce both crop yield and WUE to varying degrees, depending on the intensity of the infestation. This is because the evapotranspiration or water requirement of the crop typically does not change significantly, except in cases of premature plant death.
- **Irrigation Method:** Field water use efficiency is generally higher with overhead sprinkler, micro-sprinkler, and drip irrigation methods compared to surface irrigation methods such as furrow, border strip, and check basin, due to higher crop yields and lower seasonal water application.
- **ET Control Measures:** Using mulches, anti-transpirants, shelterbelts, and eliminating weeds can reduce water losses from cropped fields in terms of soil evaporation and transpiration without reducing crop yield, thus markedly improving WUE.

Factors affecting Irrigation Efficiencies

Irrigation efficiencies are indicators of how effectively water applied to a farm is being utilized. These parameters are necessary to assess the performance of irrigation practices on the farm. Low irrigation efficiency generally suggests that land, water, and crops are not being managed properly. The consequences of poor irrigation efficiencies are crop losses and wastage of water and nutrients. The principal factors affecting irrigation efficiency include irrigation system design, quality of land preparation, selection of irrigation methods, stream size, soil texture, soil depth, land topography, and the irrigator's skill and care. A basin-wise study conducted by the Madras Institute of Development Studies (MIDS), using potential evapotranspiration data and gross water withdrawals, revealed that the overall irrigation efficiency in the country is 38%. The study further indicated that the Krishna, Godavari, Cauvery, and Mahanadi systems have a very low efficiency of around 27%, while the Indus and Ganga systems are performing better with efficiencies in the range of 43%–47%. Therefore, identifying the components contributing to low efficiency and determining potential improvements is essential for maximizing the effective use of this vital but scarce resource in agricultural areas. There are four common irrigation efficiencies: water conveyance efficiency (Ec), water application efficiency (Ea), water storage efficiency (Es), and water distribution efficiency (Ed).

Irrigation efficiency (IE) is the ratio of the amount of water consumed by the crop to the amount of water supplied through irrigation (surface, sprinkler, or drip irrigation).

i) Water Conveyance Efficiency (Ec)

It is the ratio between the water delivered to the irrigated plot and the total quantity delivered from the source. It is mathematically expressed as:

Ec = (Water delivered to the plot / Water delivered from the source) * 100
Where:

Ec = Water conveyance efficiency (%)

= Water delivered to the plot (L/sec)

= Water delivered from the source (L/sec)

In irrigation distribution networks (distributaries, water courses, etc.), water conveyance efficiency is used to determine the percentage of released water at the head gate that actually reaches the farm and is an indicator of seepage losses in the conveyance system. A low Ec implies that much of the water released from the source is lost due to seepage in transit from the source to the field.

ii) Water Application Efficiency (Ea)

It is the ratio between the quantity of water stored in the root zone and the water delivered to the plot. It is mathematically expressed as:

Ea= (Water stored in the root zone(cm) / Water delivered to the plot(cm)) * 100

Where:

Ea= Water application efficiency(%)

= Water stored in the root zone(cm)

= Water delivered to the plot(cm)

The concept of water application efficiency can be applied to a project, a farm, or a field to evaluate irrigation practices. All the factors that influence the design of the surface irrigation system directly affect the application efficiency. A low Ea implies that much of the applied water has been lost due to deep percolation or runoff.

iii) Water Storage Efficiency (Es)

It is the ratio between the water stored in the root zone and the water needed in the root zone prior to irrigation. It is mathematically expressed as:

Es= (Water stored in the root zone(cm) / Water needed in the root zone(cm)) * 100

Es= Water storage efficiency (%)

= Water stored in the root zone (cm)

= Water needed in the root zone (cm)

The concept of water storage efficiency is useful in evaluating irrigation methods, especially under limited water supply conditions. It is also important when irrigating soils with low infiltration rates. In such cases, adequate time must be allowed for the required amount of water to penetrate into the soil. This concept is also useful when the salt balance of the root zone must be considered, and leaching requirements need to be calculated. In such cases, higher water storage efficiencies are desirable to leach out the salts from the root zone. Low storage efficiency implies that water application is much less than actually needed. In a stretch of land, one may get poor application efficiency for the upstream part and poor storage efficiency for the downstream section.

iv) Water Distribution Efficiency (Ed)

It is the ratio between the average numerical deviation in the depth of water stored from the average depth stored during irrigation (Y) and the average depth stored during irrigation (d). It is mathematically expressed as:

Ed= (1- (Y/d)) *100

Where:

Y= Average numerical deviation in depth of water stored from average depth stored during irrigation

d= Average depth of water stored during irrigation

It is a measure of water distribution uniformity within the field. Low distribution efficiency indicates non-uniformity in irrigation water penetration in the soil due to uneven land leveling. The irrigation water cannot flow over the soil smoothly. There are low patches where water will penetrate more, and there are high patches where water cannot reach. This leaves some spots unirrigated unless excess irrigation water is applied. Excess water application lowers irrigation efficiency.

v) Project Efficiency

It is the ratio between the average depth of water stored in the root zone during irrigation and the water diverted from the reservoir. It is mathematically expressed as:

Project efficiency = (Water stored in the root zone(cm) /Water diverted from the reservoir (cm)) * 100

Where:

Project efficiency (%)

= Water stored in the root zone (cm)

= Water diverted from the reservoir (cm)

The overall irrigation efficiency of a farm is a product of:

Ec*Ea*Es*Ed

The overall irrigation efficiency for a project (i.e., considering irrigation channels) is the product of:

Ec *Ea*Es*Ed

Problem

- **Canal stream size:** 120 LPS
- **Water delivered to the field:** 100 LPS
- **Area irrigated:** 1.5 ha in 8 hours
- **Effective root zone depth:** 1.75 m
- **Runoff losses:** 415 m3
- **Available water-holding capacity:** 200 mm/m depth of soil
- **Schedule irrigations at:** 50% DASM (Depletion of Available Soil Moisture)
- **Depth of application varied linearly:** 1.65 m at the head end to 1.3 m at the tail end.

Production Efficiencies

Agricultural productivity and efficiency is at the centre of abounding of the debates, behavior and measures apropos the agronomics sector. The accent placed by the Sustainable Development Goals on agronomical productivity underlines the abounding affidavit for which added analysis on statistical frameworks for productivity and efficiency targeted to developing countries is necessary. Advice on agronomical productivity is accompanying to several of the Sustainable Development Goal indicators, in particular:

"Productivity is frequently authentic as an arrangement of an accumulated admeasurement of achievement to an accumulated admeasurement of ascribes use" (OECD 2001). At it's a lot of axiological level, productivity measures the bulk produced by an ambition accumulation (country, industry, sector, acreage or about any ambition group) accustomed a set of assets and inputs.

Productivity can be abstinent for an individual article (farm, commodity) or an accumulation of farms, at any bounded scale. The admeasurement should reflect the ultimate purpose for the inquiry. If for example, the purpose is to analyze productivity amid farms, again measures that are micro-based are required. If the charge is to appraise national agronomical action at the country level, again macro measures are required. This aforementioned affinity can extend above the breadth to the national economy. While the adapted purpose can vary, the altitude issues associated with anticipation the altered indicators are the same. However, abstracts requirements may alter depending on the blazon of indicator: farm-level productivity altitude for one article and one ascribe (for example, labour productivity of maize farms) may alone crave basal advice on achievement quantities and ascribe use, while bearing aggregated measures about requires appraisement outputs and inputs.

Measurement of productivity

As ahead stated, productivity measures are consistently accumulated based, either bidding in concrete quantities, or in connected bulk terms, implying that ethics be adapted for bulk change. In adjustment to get complete or connected dollar measures, time alternation for outputs and inputs as able-bodied as for prices are required or alternatively adapted is achievement and ascribe accumulated and bulk indices. Obtaining the actual bulk or bulk indices, in turn, adds decidedly to the complication of productivity measurement, a lot of which is accompanying to analogous the actual bulk (or index) to the artifact or input. In the case of outputs, the bulk or basis acclimated needs to accede the altered characteristics associated with the product, abnormally the superior characteristics that are associated with the empiric price. Application appropriately complete bulk indices have been the focus of abundant of the analysis on productivity because alternation indices are frequently used.

Over the years, the analysis has adapted application altered bulk indices for abbreviating outputs and inputs, anniversary with altered backdrop and anniversary acquiescent altered results. Selecting the adapted one to use is abiding in theory, but about the best focuses on how able-bodied the called bulk basis accounts for barter bias. It has been apparent by Diewert (1976) and endless others, that accomplished indices (those that amuse assertive numeric properties) can annual for this bias, but they accept the abject coercion (assumption) that the industry beneath abstraction operates beneath complete antagonism and with an assertive blazon of assembly function. Because of its adorable properties, the Thorniest basis is generally acclimated to admeasurement TFP for a bulk of reasons. First, the Thorniest basis is a detached approximation of the Division index, broadly believed to be the best basis for barometer bread-and-butter aggregates because of its accommodation to anxiously represent the basal assembly action and invariance property.5 Second, as the Thorniest basis is an accomplished index, it can be accompanying to abounding assembly or bulk functions. In particular, it corresponds absolutely to a Translog function. Third, accession advantage is that this basis is constant in

Agricultural behavior tends to focus added on adopting productivity through abstruse change than through bigger use of the complete technology. However, rebalancing the focus of agronomical behavior appear convalescent efficiency is all-important in the ambience of bound avail efficiency of accustomed resources, such as acreage and water, and accustomed the call to complete the ecology brand of agronomical production. Agnate concrete productivity assets and conceivably even above bread-and-butter assets may be accepted from bigger use of complete technology than from alive to new technology. The closing may access productivity in the abbreviate term, but possibly at the bulk of college assembly and ecology costs. For example, afore advising farmers to accept actinic fertilizers (technological change), acceptable fertilization methods involving amoebic fertilizers and rotations or admixture of crops (technical efficiency) may be answer as a way to access concrete productivity and advance aliment aegis and bread-and-butter profit efficiency. Abstruse efficiency is declared in detail in the afterward paragraphs.

Measurement of Competitiveness

An added acumen that needs to be fabricated is amid bread-and-butter efficiency and competitiveness. The above is a complete admeasurement of the bread-and-butter achievement of the acreage admitting the closing compares this achievement to that of their competitors. In added words, acreage can be economically inefficient but aggressive because added farms are even beneath efficient. Reciprocally, an economically able acreage is not necessarily aggressive if all the added farms are as well economically efficient. Competitiveness as well goes above the price/cost achievement and extends to the appearance absorbed to the achievement or to the bearing close (or sector, country), such as superior attributes, both authentic and perceived. For example, a close can accept analogously top assemblage costs but may annual from a top "non-price" competitiveness, which allows it to advertise its articles at a college price.

Competitiveness is a lot of generally abstinent application bread-and-butter indicators, such as gross or net margins (often per assemblage of land), and comparing the achievement of farms (or agronomics systems) based on these measures. Competitiveness and productivity are carefully related: college productivity can advance to a greater competitiveness of the action (or sector) because added is produced out of the aforementioned bulk of resources. This agency that, with all things getting captivated equal, the bulk of assembly per assemblage of achievement is lower, and that margins per assemblage of achievement are higher. Productivity is an all-important arrangement for competitiveness, but not an acceptable condition. Indeed, an aggregation of factors affecting the competitiveness of an action has been articular in the literature. Competitiveness is the aftereffect of an accumulated of factors, both national and international:

1. Nationally, efficiency endowments, technology, productivity, artifact features, budgetary and budgetary administration and assuredly the barter policy are apparent to be the lot of important factors that actuate the competitiveness of an industry and/or business. Productivity is, therefore, apparent as one of the national (domestic) determinants of competitiveness;
2. Internationally, the a lot of important factors are barter rates, all-embracing bazaar conditions, the bulk of all-embracing carriage and the preferences and settings amid altered countries.
3. Agricultural productivity is based on the superior of the inputs and how able-bodied those inputs are chip in the assembly process. For example, acreage productivity awful depends on the breadth of the acreage and its concrete characteristics. This is the aforementioned for labour as the superior of the plan force differs across, for archetype artisan types or subsectors.
4. For allusive purposes, the superior of the ascribe accept to be taken into annual in the abstracts accumulating and adapted adjustments charge to be fabricated afterwards the abstracts are collected. This agency that abstracts on ascribes use charge to be calm for altered types of inputs or superior classes. For example, ancestors labour;
5. Occasional workers and abiding workers should be differentiated in the abstracts accumulating process. Because workers with altered abilities accept altered levels of productivity, application the aforementioned accomplishment for workers with altered accomplishment levels after-effects in biased estimates of labour productivity. The aforementioned applies to fertilizers or to any added ascribe or assembly agency that has

capricious characteristics. With attention to fertilizers, this ascribe varies in agreement of the dosage of alive capacity to pesticides, which may be added or beneath effective. One kilogram of fertilizer activated in 2000 is not commensurable to one kilogram activated in 2015, because of two factors: the accession of new and added able products; and fertilizer appeal may accept confused appear added segments of the market. This change in agreement should be reflected in differentiated ascribe prices;

6. To abode the affair of compositional or superior change, adult frameworks for ascribe superior adjustments accept been developed. The United States Department of Agriculture-Economic Analysis Service (USDA-ERS), for example, estimates quality-adjusted accomplishment based on abstracts of hours formed and accomplishment per hour cross-classified by altered labour categories (the afterward breadth on labour productivity provides added details). For acreage productivity, acreage prices or rents can be accepted application animal regressions that crop into annual some of the differences in superior attributes, such as clay type, moisture, clay acidity and salinity. Quality-adjusted prices for added inputs can be complete application agnate techniques.
7. The productivity of the acreage measures the bulk of achievement generated by an accustomed bulk of land. It is mostly applicative in the ambience of agronomics activities, but it can as well be continued to livestock production, in assertive cases, as apparent below.;
8. There are several productivity measures that can be calculated: an ample admeasurement is the arrangement amid the bulk of all agronomics articles (crops and livestock) and the absolute acreage acclimated in agriculture. Added acreage productivity measures can be affected by adding crop assembly by the bulk of buried land, bidding in a breadth unit, such as hectares or acres. When bidding in agreement of concrete output, such as tons of maize, acreage productivity corresponds to crop yields. When bidding in budgetary terms, acreage productivity is added generally referred to as allotment to land.
   a) Land productivity = Accumulated of achievement / Buried Area
   b) Planted breadth is acclimated instead of added breadth concepts, such as harvested area, because of the absorption to admeasurement the able crop or acreage productivity rather than an abstract or biological yield. The use of inputs above-mentioned to the autumn is fabricated on the sown/planted breadth (such as fertilizer applications) and not in advertence to the harvested area, which at the pre-harvest appearance is usually unknown. The aberration amid harvested and

buried breadth may as well reflect the efficiency and appliance of the agronomics practices, in accession to exogenous factors, such as climate-related events, which should be reflected in the productivity indicator. Application harvested breadth instead of buried breadth tends to advance to overestimations of yields and allotment to acreage because this breadth includes the lot of advantageous segments of the parcel. In general, it is best to use buried breadth for a monocropping arrangement and able area, including dormant land, for alloyed agronomics systems.

Agricultural assembly acclimated for the adding of productivity should cover the assembly of the crops developed on the aforementioned acreage during the advertence aeon whether it is one agronomics division or one year. This is important because, in practice, farmers generally abound added than one crop on the aforementioned artifice over a year; they may abound an admixture of crops on the aforementioned artifice at the aforementioned time or circle the crops developed on the artifice over the season. Kelly *et al.* (1996) fatigued that one of the affidavit abaft the addiction to belittle achievement and yields in developing countries is the abridgement of accounting of crops developed in admixture or in arrangement and the abridgement of appraisement of by-products, which may be sold, captivated by the domiciliary or acclimated in the assembly of added products. It is, therefore, basic that all crops are included in the altitude of productivity; abnormally in developing regions area these practices are common.

d) Units: As with added inputs, acreage productivity can be bidding in abounding units. Accustomed that the acreage may be acclimated to abound abounding altered crops, a concrete unit, such as tons, may not be the best choice. Putting a budgetary bulk on their corresponding achievement is generally bare to accumulate the achievement of altered crops.

e) Land quality: Productivity altitude should crop into annual as abundant as accessible clay and acreage superior differences by accession abstract on the soil/land characteristics and their accompanying aspects, abnormally acreage prices and rents. For example, differences in the superior of acreage beyond states and regions in the United States are reflected by artful about prices of acreage from animal corruption results. Ball *et al.* (2008) applies an animal access to admeasurement quality-adjusted acreage prices bold acreage bulk is an action of characteristics of acreage superior variables, such as clay acidity, salinity and hydric stress. The achievement acquired from the acreage use depends on the clay and acreage characteristics. USDA uses a database that gathers advice on those characteristics in altered states and regions from the "World Clay Assets Office". This method, even admitting it is accurate, requires an ample bulk of abstracts that are not necessarily accessible in developing countries.

f) Labour productivity in agronomics measures the bulk of units of output(s) produced per assemblage of labour acclimated in the action of production. It is a fractional productivity indicator that is affected by adding the productivity of achievement by the absolute units of labour used:

   Labour productivity = Accumulated of achievement / Units of labour used

There are abounding means to appraise the productivity of labour input: the bulk of workers alive on the holding; the bulk of time units (such as hours, canicule and months) formed or full-time agnate units if an boilerplate bulk of hours per alive day can be bent according to specific country standards.

OECD (2001) recommends that labour ascribe be abstinent application the bulk of hours finer worked. Application the bulk of hours corrects for the aberration amid melancholia and non-seasonal workers and the altered alive regimes (part-time against full-time). This allows bigger comparisons beyond assembly systems, regions and countries, as the bulk of workers or of canicule per artisan may not announce the labour ascribe finer acclimated on the farm.

However, the change in the bulk of hours appear does not consistently reflect the use of capital, the superior of the workforce and technology (Shumway et al. 2015). USDA-ERS suggests that productivity abstracts abduction the altered types of labour alive in the breadth because labour ascribes differ based on the categories of workers. It is recommended that distinctions be fabricated amid altered ages of workers, ancestors labour and assassin labour and men and women. Distinctions can as well be fabricated amid part-time and full-time workers. Acumen should as well be fabricated amid the altered educational levels, because the superior of one hour provided by an artisan is generally barnacle on his abilities and capacities.

Formulae of Irrigation efficiency

It indicates how efficiently the supplied available water is being used based on different method. Irrigation efficiency is the ratio usually expressed as per cent of the volume of irrigation water transpired by the plant, plus the evaporated from the soil, plus that necessary to regulate the salt concentration in the soil solution and that used by the plant in building plant tissue to the total volume of water diverted, stored or pumped for irrigation.

$$Ei = \frac{Wt + Ws - Re}{Wi} \times 100$$

Where, E*i* is irrigation efficiency (%); W*t* is the volume of irrigation water per unit of land transpired by plants, evaporated from the soil during crop period (including field preparation and nursery); W*s* is the volume of irrigation water per unit area of land to regulate the salt content of soil solution; Re is effective rainfall and W*i* is the volume of water per unit area of land that is stored in a reservoir or diverted for irrigation.

Formulae of Conveyance efficiency

It indicates the efficiency with which water is conveyed from source of supply to the field. It estimated the conveyance losses and expressed as:

$$Ec = \frac{Wf}{Ws} \times 100$$

Where, Ec is water conveyance efficiency (%); W*f* is water delivered at the field and W*s* is water delivered from the source.

Formulae of Water application efficiency

The percentage ratio of the amount of water stored in the crop root zone during the irrigation to the amount of water delivered to the field. It is expressed as:

$$Ea = \frac{Ws}{Wf} \times 100$$

Where, W*s* is the amount of water stored in the crop root zone.

W*f* is the amount of water delivered to the field.

Formulae of Water storage efficiency

The percentage ratio of the amount of water stored in the effective crop root zone soil to the amount of water needed to make up the soil depleted in crop root zone prior to irrigation.

$$Es = \frac{Ws}{We} \times 100$$

Where, W*e* is the amount of water needed to the crop root zone prior to irrigation.

W*s*is the amount of water stored in the crop root zone of the soil.

Water distribution efficiency:

It gives a measure of the uniformity of the depth of irrigation all over the field and expressed as:

$$Ed = 1 - \frac{Y}{D} \times 100$$

Where, Y is average numerical deviation

D is average depth of water stored during irrigation.

Formulae of Water use efficiency (WUE)

It refers to the amount of marketable end product obtained per unit of water used by the crop.

$$WUE\ (kg\ ha^{-1}\ mm^{-1}) = \frac{\text{Total economic yield (kg ha}^{-1})}{\text{Total consumptive use (mm)}}$$

References

Allen, R. G., Pereira, L. S., Raes, D., & Smith, M. (1998). Crop Evapotranspiration—Guidelines for Computing Crop Water Requirements. FAO Irrigation and Drainage Paper 56. Food and Agriculture Organization of the United Nations, Rome.

Burt, C. M., & styles, S. W. (1999). Irrigation Performance in California: Strategies for Improvement. In: Proceedings of the 1999 California Irrigation Institute.

Hunsaker, D. J., & Smith, J. A. (2008). "Evaluating Irrigation Efficiency and Uniformity on Drip-Irrigated Cotton." Journal of Irrigation and Drainage Engineering, 134(4), 469-477. DOI: 10.1061/(ASCE)0733-9437(2008)134:4(469).

Kijne, J. W., Barker, R., & Molden, D. (2003). Water Productivity in Agriculture: Limits and Opportunities for Improvement. International Water Management Institute.

Ponce, V. M., & Landeros, J. (2020). "The Impact of Irrigation Systems on Water Conservation." Journal of Water Resource and Protection, 12(6), 653-667. DOI: 10.4236/jwarp.2020.126038.

Thorne, J. H., & Rahman, M. M. (2013). "Estimating Irrigation Efficiency in Arid and Semi-Arid Regions." Agricultural Water Management, 117, 91-96. DOI: 10.1016/j.agwat.2012.11.009.

Tullis, J. E., & Rose, C. (2010). Irrigation Water Management: Principles and Practice. 2nd Edition. Wavelength Publishing, San Francisco.

Vaux, H. J., & Howitt, R. E. (2010). "Water Conservation in Irrigated Agriculture: The California Experience." Water Resources Research, 46(10), W10501. DOI: 10.1029/2010WR009736.

Ward, F. A., & Pulido-Velazquez, D. (2008). "Water Conservation in Irrigated Agriculture: An Analysis of Current Practices and Technologies." Agricultural Water Management, 95(9), 970-981. DOI: 10.1016/j.agwat.2008.07.002.

Zwart, S. J., & Bastiaanssen, W. G. M. (2004). "Accuracy of Remote Sensing to Estimate the Area Irrigated by Drip Systems." Irrigation Science, 23(1), 1-12. DOI: 10.1007/s00271-004-0072-0.

Multiple Choice Questions

1. What are the main losses during irrigation that affect efficiency?

   a) Conveyance, evaporation, pollution, and deep percolation

   b) Conveyance, runoff, seepage, and deep percolation

c) Pollution, runoff, evaporation, and soil degradation

d) Soil depletion, evaporation, runoff, and plant transpiration

2. How can irrigation efficiency at the field level be improved?

a) By increasing water supply only

b) By choosing suitable methods of irrigation and efficient irrigators

c) By planting more crops without proper management

d) By reducing soil quality management practices

3. What formula represents irrigation efficiency (Ei)?

a) Ei = (Wt + Ws + Re) / Wi × 100

b) Ei = (Wt + Ws - Re) / Wi × 100

c) Ei = (Ws - Re) / Wt × 100

d) Ei = (Wi + Rs) / Wt × 100

4. What does water conveyance efficiency (Ec) indicate?

a) The total amount of water stored in reservoirs

b) The efficiency of water delivery from source to field

c) The volume of water used by plants for transpiration

d) The percentage of water lost in deep percolation

5. Which of the following factors significantly influences water use efficiency (WUE)?

a) Crop yield and water availability

b) Only soil temperature and weather conditions

c) Financial investment in farming equipment

d) The type of irrigation system used only

6. What does water application efficiency (Ea) measure?

a) The ratio of total water delivered to the field

b) The relationship between water stored in the crop root zone and water delivered

c) The efficiency of water storage in reservoirs

d) The amount of rainfall received during the growing season

7. Which irrigation efficiency focuses on the niformity of water distribution in the field?

a) Water conveyance efficiency

b) Water application efficiency

c) Water distribution efficiency

d) Water storage efficiency

8. What is the effect of weed control on water use efficiency (WUE)?
    a) It decreases the effectiveness of water usage
    b) It has no impact on WUE
    c) It reduces evapotranspiration losses and increases available water
    d) It leads to increased labor costs, thus lowering WUE
9. In terms of WUE, how do C4 plant species generally compare to C3 species?
    a) C4 species have lower efficiency than C3 species
    b) C3 species have approximately the same efficiency as C4 species
    c) C4 species exhibit higher water use efficiency than C3 species
    d) C4 species require more water than C3 species
10. What can be a consequence of low irrigation efficiency?
    a) Better crop yield
    b) Increased water and nutrient wastage
    c) Enhanced soil fertility
    d) Improved plant growth

Answer Key

1	b	2	b	3	b	4	b	5	a	6	b	7	c
8	c	9	c	10	b								

11

Measurement of Irrigation Water

Water Quantity and Measurement: The amount of water to be applied and its measurement are crucial aspects of irrigation management. There are various methods and tools used to measure and control the flow of water, including:

1. Parshall flume
2. Orifice
3. Weirs
4. Water meter
5. Float method
6. V-Notch

Methods of Soil Moisture Measurement

The measurement of soil moisture is essential to determine when to irrigate, the amount of water needed, to evaluate evapotranspiration, and to monitor soil matric potential. Soil moisture is measured in two ways: direct and indirect methods.

Direct Methods

1. **Gravimetric method:** The gravimetric method is a standard method for measuring soil water content directly. It involves collecting soil samples from the field and determining their moist and dry weights. The mass water content is then calculated as the ratio of the weight loss in drying to the dry weight of the soil sample. This method is accurate but time-consuming and requires laboratory equipment.
2. **Spirit burning method:** The spirit burning method is a direct method for measuring soil moisture in the field. Soil moisture is evaporated by adding alcohol and igniting it. The time taken for the soil to dry is then used to calculate the soil moisture content. This method is quick and easy but can be affected by soil temperature and organic matter content.
3. **Infrared moisture balance:** The infrared moisture balance is a laboratory instrument used to measure the moisture content of soil samples. It consists of an infrared lamp, a torsion balance, and an

autotransformer. The infrared lamp quickly evaporates the soil moisture, and the instrument is directly calibrated in percent moisture. This method is fast and reliable but requires laboratory equipment.

Indirect Methods

1. **Electrical resistance blocks:** Electrical resistance blocks, also known as gypsum blocks, are placed in the soil at a desired depth and allowed to equilibrate. The electrical resistance of the block is then measured using a meter based on the Wheatstone Bridge principle. The soil water content is obtained from a calibration curve of electrical resistance against known soil water content. Resistance blocks are relatively inexpensive, easy to install, and give quick readings. However, they are not useful in coarse-textured soils and are sensitive to soil salinity.
2. **Neutron scattering technique:** The neutron scattering technique is a widely accepted method for monitoring soil moisture in the field. It involves using a neutron moisture meter, which consists of a probe containing a source of fast neutrons and a detector of slow neutrons. The probe is inserted into a hollow access tube pre-inserted into the soil, and the flux of slow neutrons scattered and attenuated in the soil is measured. The density of slow neutrons formed around the probe is proportional to the concentration of hydrogen in the medium surrounding the probe, and therefore approximately proportional to the volume fraction of water present in the soil. This method is less laborious, rapid, and non-destructive but requires expensive equipment and health hazards due to exposure to neutron and gamma radiation.
3. **Tensiometer:** A tensiometer is an instrument designed to provide a continuous indication of the soil's matric suction (also called soil-moisture tension) in situ. It consists of a porous ceramic cup, connected through a tube to a vacuum gauge. When placed in the soil, the water inside the cup comes into hydraulic contact with soil water through the pores in the ceramic walls. As soil moisture is depleted by drainage or plant uptake or as it is replenished by rainfall or irrigation, corresponding readings on the tensiometer gauge occur. Tensiometers are suitable for scheduling irrigation to crops raised in coarse-textured soils where the majority of available soil moisture is between 0-0.8 bars or 0-80 kPa.
4. **Pressure plate and pressure membrane apparatus:** Laboratory measurements of soil water potential are usually made with pressure membrane and pressure plate apparatus. It consists of ceramic pressure plates or membranes of high air entry values contained in air-tight metallic chambers strong enough to withstand high pressure. The

apparatus enables the development of soil moisture characteristic curves over a wider range of matric potential. This method is accurate but requires laboratory equipment and is time-consuming.

Quantity of Irrigation Water or How Much to Irrigation

The net quantity of water to be applied depends upon magnitude of moisture deficit in the soil, leaching requirement and expectancy of rainfall. When no rainfall is likely to be received and soil is not saline, net quantity of water to be applied is equal to the moisture deficit in the soil i.e. the quantity required to fill the root zone to filed capacity. The moisture deficit (d) in the effective root zone found out by determining the field capacity moisture content and bulk density of each layer.

Problem: Find out the net quantity of irrigation water to be applied to wheat field with the following moisture status.

Sr. No.	Depth of Soil Layer (cm)	Moisture % on oven dry basis		Apparent specific gravity g/cc
		Field Capacity	Actual	
1	0 - 15	25.0	17.8	1.39
2	15 - 30	24.0	17.8	1.47
3	30 - 60	22.3	19.2	1.51
4	60 - 90	22.2	20.5	1.53

Solution: Moisture deficit in the different layers will be as follows,

1. First Layer = $\frac{25.0 - 17.8}{100}$ X 1.39 X 15 = 1.79
2. Second Layer = $\frac{24.0 - 17.8}{100}$ X 1.47 X 15 = 1.36
3. Third Layer = $\frac{22.3 - 19.2}{100}$ X 1.81 X 30 = 1.40
4. Forth Layer = $\frac{22.2 - 20.5}{100}$ X 1.53 X 30 = 0.78

Therefore, net quantity of water to be applied is 5.33 cm to fill the root zone to field capacity again

Requirement of irrigation water Units for water measuring

Water is measured under two conditions. Water at rest measured in units of volume such as liter, cubic meter, hectare meter. Water in motion is expressed in rate off low units such as liters per hour and meters per day.

Liter: One liter is equivalent to 0.22 imperial gallons or 0.0353 cubic feet or 1/1000cubic meters.

Cubic meters: A volume of water equal to that of one cubic meter in length, one meter in breadth and one meter in thickness. One cubic meter of water = one kilo liters or 100 liters or 220 gallons or 35.3 cubic feet or one ton (approx)

Gallon: A gallon is 0.1602 cubic foot. One gallon of water weighs about 10Ib

Cubic foot: A volume of water equal to that of a cube 1 foot in length, 1 foot in breadth and 1 cubic foot in thickness. One cubic foot of water = 28.37 liters or 6.23 gallons or 0.0283 cubic meters or 0.028 ton.

Hectare centimeters: A volume of water necessary to ci\over an area of one hectare (10,000 aq. meters) surface to a depth of one centimeters (1hectare centimeter = 100 cubic meters = 100,000 liters)

Acre inch: the volume of water necessary to cover one-acre (43,560 sq. feet) surface to a depth of one inch. One hectare inch = 3630 cubic feet or 101 ton.

Acre foot: The volume of water necessary to cover one acre to a depth of one foot. (One acre feet = 43,560 cubic feet)

Cusec: It is the quantity of water flowing at the rate of one cubic foot per second. As one cubic feet of water weighs about 62.4 Ib or 28.37 kg. One cusec of water flowing for one hour is equal to 62.4 Ib X 60 X 60 = 22464 gallons, 101 tons, or one- hectare inch (28.37 liter X 60 X 60 = 101952 liters or one-acre inch.

Duty of water

It is denoted that the number of acres covered by one causes of water flowing continuously throughout the growing season of a crop. It is therefore varies with kind of crop, season, nature of soil, method of irrigation and method of sowing or cultivation.

Delta: It is the total depth of water required by a crop.

Problem: A pump with an average discharge of 15 liters/second irrigates one-hectare wheat crop in 12 hours. What is an average depth of irrigation?

Solution

Discharge in 12 hours = 15 X 60 X 60 X 12

= 648000 liters

= 648 M^3

$$\text{Depth of irrigation (cm)} = \frac{\text{Volume of water (Cu. m)}}{\text{Area of land (sq. m)}} \times 100$$

$$= \frac{648}{10,000} \times 100$$

= 6.48 cm

Problem: Wheat crop requires 40 cm of irrigation water during 120 days irrigation period. How much land can be irrigation with a flow of 20 liters per second for 12 hours a day?

Solution

$$\text{Total discharge during irrigation period} = \frac{20 \times 60 \times 60 \times 12 \times 120}{1000} \text{M3}$$

= 1, 03,680 M^3

$$\text{Total discharge during irrigation period} = \frac{20 \times 60 \times 60 \times 12 \times 120}{1000} \text{M3}$$

= 1, 03,680 M^3

$$\text{Irrigation requirement per hectare} = \frac{40}{100} \times 10,000\,\text{M}^3$$

= 4000 M^3

$$\text{Area irrigated} = \frac{\text{Volume of available water}}{\text{Volume of water requried / ha(M}^3\text{)}} \times 10,000\,\text{M}^3$$

$$= \frac{1,03,680}{40}$$

= 25.92 hectare land can be irrigated

Devices Used for Measuring Irrigation Water

Several devices are commonly used for measuring irrigation water. They grouped into four categories:

1. Volumetric Measures
2. Velocity-Area Methods
   a) Float Method
   b) Water Meters
3. Measuring Structures
   a) Orifices
   b) Weirs
   c) Flumes
4. Tracer Methods

I. Volumetric methods (using a container): A simple method of measuring a small irrigation stream is to collect the flow in container of known volume for a measured period. An ordinary bucket or barrel is used as container. The time required to fill the container is recorded with a stopwatch or with seconds on wristwatch. The rate of flow is measured as below

$$Q = \text{Discharge rate liter / second} = \frac{\text{Volume of container (liters)}}{\text{Time required to fill (seconds)}}$$

PROBLEM: A 24 liter capacity bucket is filled in 10 seconds by discharge from a Persian wheel. What is rate of flow?

Solution: Discharge ratio liter/second $= \frac{24}{10}$

= 2.4 liter/second or 144 liter/minute

Velocity area method

Float method

The float method of making of rough estimates of the flow in a channel consists of nothing the rate of movement of a floating body. A long necked bottle partly filled with water or black wood, an orange or lemon may be used as float. A straight section of the channel about 30 meters long with uniform cross section is selected. Several methods of depth and width are made within the trial section to arrive at average cross sectional area. A string is stretched across each end of section at right to the direction of flow. The float is placed in the channel a short distance upstream from the trial section. The float needed to pass from upper end to lower end of the section is recorded. Several trials are made to get average time of travel.

To determine the velocity of water at the surface of the channel, the length of the trial section is divided by the average time taken by the float to cross it. Since the velocity of the float on the surface of the water will be greater than the average velocity of the stream; it is constant factor, which is usually assumed to be 0.85. To obtain the rate of flow, this average velocity (measured velocity x co-efficient) is multiplied by the average cross sectional area of the stream.

Discharge or rate of flow = area x velocity

Q = A X V where Q= discharge rate in m3/sec. v = velocity of flow in m/s a = cross section al area of channel in m^2

Water meters

Water meters utilize a multi blade propeller made of metal, plastic or rubber, rotating in a vertical or horizontal plane and geared to a tataliser in such a way that a numerical counter can totalize the flow in any desired volumetric units, water meters are available for a range of sizes suiting the pipe size commonly used on the farm. There are basic requirements for accurate operation of the water meter.

1. The pipe must flow full at all times.
2. The rate of flow must exceed the minimum for the rated range.

Meters are calibrated in the factory and field adjustments are usually not required. When water meters are installed in open channels, the flow must be brought through the pipes of known cross sectional area. Care must be taken that no debris or other foreign materials obstruct the propeller.

Water measuring devices

Orifices

Orifices in open channel are usually circular or rectangular openings in vertical bulk head through which water flows. The edges of opening are sharp and often constructed of metal. The cross sectional area of orifice is small in relation to the stream cross section. Orifice may operate under free flow or submerged flow conditions. The types of orifices are

Orifices below the level of inlet: The discharge through a closer orifice in which the orifice is situated below the level of inlet is calculated by the equation.

Q = Ca x under root (2gh)

Where, Q = Quality of flow in C. ft/Sec.

a+ Cross sectional area of water, the canal or orifice in sq. ft.

c = A constant which varies from 0.6 to 0.8 or more depending upon the position of orifice relative to the sides and bottom of vessels or the degree of roundness of the edge of orifice.

g= Acceleration due to gravity. (32 feet / sec/sec)

h= Height of water level in cistern from the middle of the orifice in feet.

II. **Discharge through an orifice situated at higher level than inflow pipe:** When the orifice is situated at a higher level than the inflow pipe, the discharge is calculated according to the equation

    Q = CLh1.5

    Where,

D= Length of orifice in feet and

C= 2C 2g

H= head of water.

III. Discharge through submerged orifice: The discharge through a submerged orifice is given by the equation.

Q=0.61 LH 2gh

e.g. if L=1.0 H=0.5ft. h=0.25 ft. and g=acceleration due to gravity (32ft/sec/ c) then Q=1.22 C, ft/sec.

a) **Discharge through Weirs:** A wear means a notch in a well built across a stream. The notch may be (a) rectangular (b) Trapezoidal and (c) 90 degree V (Triangular) notch or weir.

b) **Rectangular notch or weir:** The length of a weir may be equal to width of the upstream channel or less than it. The discharge through a rectangular weir, in case of complete end, contraction is given by the equation.

Q = 3.33 (L-0.2 H) H^1.5

Where L= measured length of the weir

L= effective length of weir in feet.

L= L-0.2H)

c) **Trapezoidal or Cipolletti Weir:** The discharge of water is given by the equation

Q=3.367 LH^1.5

$$\text{Where} = \frac{L1 + L2}{2}$$

d) **Discharge through 90oV notch:** The discharge is given by the equation:

Q= 2.49 H ^ 2.48 = 2.5H^2.5

Parshall flume or (*Venturi* flume)

Parshall (1950) has decided a device in which the discharge is obtained by measuring the loss in the head caused by forcing a stream of water through a throat or converged section of a flume with a depressed bottom. The loss in head is very small in this device. The accuracy of measurement in the Parshall flume is within allowable limits of 5% the flumes ranging from 3 inches to 10 feet throat width are used, which gives the range of discharge of 1/30 to 200 ousecs. The flumes of 3, 6 and 9 inch size are generally used in field measurement.

The ration between the reading at Hb and Ha point should be carefully studied. This ratio should not exceed 0/3 for 3 inch, 6 inch 9 sized Parshall flumes otherwise correction needs to be applied. When the ratio is less than 0.6, it is termed as free flow and exceeds 0.6, it is called submerged flow.

Cut throat flumes: Skogerboe *et. al.* (1967) has developed cutthroat flumes for measurement of water. Since there is no throat section (Zero throat flumes), the flumes have been given the name as out throat flumes by the designers.

The flumes have a level floor as opposed to the inclined floor in the throat and exit section in the partial flumes. it is placed in a concrete lined channel or on a channel bed conveniently. Every flume has the same all lengths in both the entrance and exit sections. These flumes consist of converging inlet section and diverging outlet section, under free flow conditions. The discharge Q through a cutthroat flume depends upon the upstream depth of flow Ha. The basic form of the free flow equation is-

Q=CH a 1.56

Where C=3.50 W 1.025 (W=throat d width infect)

Tracer Methods

These methods are independent of stream cross section and are suitable for field measurements without installing fixed structures. In these methods, a substance (tracer) is concentration form is introduced into flowing water and allowed to thoroughly mix. The concentration of the tracer is measured at downstream section. Since only the quantity of water is necessary to accomplish the dilution is involved, there is no need to measure velocity, depth, and head, cross sectional or any other hydraulic factor usually considered in discharge measurement. The relationship between size of stream, time of application, area to be irrigated and depth of water to applied is as below.

Qt=ad

Where Q=Size of stream or discharge (liter/second) or (ha. cm per hour)

t=the time of application of water (seconds or hour)

a=area (sq. m or hectare)

d=Depth in cm that the volume of water used would cover the land

rrigated, if quickly spread uniformly over its surface.

A large number of empirical methods have been used by numerous scientists and specialists worldwide to estimate evapotranspiration from different climatic variables. Oudin *et al.* (2004), Fernandes (2006) and Barros (2012) reviewed commonly used equations in the literature. Such practical application is motivated due to the lack of data that avoid the use of more accurate equations. Therefore, in this paper, to provide an adequate support for users that are exposed to different levels of data availability, the performance of six empirical equations for estimating evapotranspiration are investigated for the case study of Campos dos Goytacazes city located at Rio de Janeiro state, as initially developed in Fernandes (2006), using the recommended FAO56-Penman-Monteith equation (Allen *et al.*, 1998) as a reference for comparison.

The evapotranspiration was estimated by six empirical equations, divided in two groups (according to Burman and Pochop, 1994; Jensen *et al.*, 1990; Pereira *et al.*, 2002 and Sediyama *et al.*, 1997), namely: i) temperature methods - Camargo and Hargreaves-Samani and ii) radiation methods - Makkink, Jensen-Haise, Priestley-Taylor and Hargreaves. Such equations were chosen because of the simplicity of calculations, since the meteorological data required are easily obtained.

The evaluation of these equations was initially made by Fernandes (2006). The current study aims to further contribute to the results already achieved by the previous study, and it is part of a project that intends to evaluate the performance of different equations of estimate of ETo, with the purpose to subsidize the use of these equations in the different Brazilian climatic conditions.

FAO56-Penman-Monteith

The FAO56-Penman-Monteith Equation 1 is expressed as (Allen *et al.*, 1998):

$$ET_o = \frac{0.408\Delta(Rn - G) + \gamma(900/(Tmean + 273))\, U_2(es - e}{\Delta + \gamma(1 + 0.34\, U2)} \qquad (1)$$

$$\Delta = \lambda.es/(Rv\ Tmean^2) \qquad (2)$$

$$\gamma = cp.P/(\varepsilon.\ \lambda) \qquad (3)$$

$$es = 2.53x10^8\ .exp\ (-5420/\ (Tmean+273)) \qquad (4)$$

$$e = UR.es \qquad (5)$$

where ETo is the reference evapotranspiration ($mm.day^{-1}$);Δ is the slope of the vapor pressure curve ($kPa.^oC^{-1}$); Rn is the net radiation at the crop surface ($MJ.\ m^{-2}\ .day^{-1}$); G is the soil heat flux density ($MJ.\ m^{-2}\ .day^{-1}$); Tmean is the daily mean air temperature at 2 m height (°C); U2 is the wind speed at 2 m height ($m.s^{-1}$); es is the saturation vapor pressure (kPa); e is the actual vapor pressure (kPa); UR is the relative humidity; (es – e) is the saturation vapor pressure deficit (kPa);γ is the psychrometric constant ($kPa.^oC^{-1}$);λ is the heat required to vaporize free water ($2450\ kJ.m^{-2}.day^{-1}$); cp is the specific heat at constant pressure ($1.01\ kJ.kg^{-1}.K^{-1}$); P is the atmospheric pressure (kPa);ε is the ratio molecular weight of water vapor/dry air, which is equal to 0.622.

According to Burman and Pochop (1994), Jensen *et al.* (1990) and Sediyama *et al.* (1997), the net radiation is expressed as:

$R_n = (1\text{-}A).\ R_s\text{-}\ [(0.1+0.9n).\ (0.34\text{-}0.14e^{0.5}).\ \sigma.\ (T_{mean}+273)^4]$ (6)

whereRs is the global radiation at surface ($MJ.m^{-2}.day^{-1}$); A is the albedo or canopy reflection coefficient; n is the actual duration of sunshine (hours);σ is the Stefan-Boltzmann constant, which is equal to $5.67x10^{-8}$ $W.m^{-2}.K^{-4}$.

Makkink

The Makkink equation is expressed as (Sediyama *et al.*, 1997):

ETo = 0.61 W.Rs-0.12 (7)

$W = \Delta/\ (\Delta+\ \gamma)$ (8)

where W is the index that depends on the psychrometric constant γ ($kPa.K^{-1}$) and slope of the vapor pressure curve Δ ($kPa.K^{-1}$).

Jensen-Haise

The Jensen-Haise equation is expressed as (Sediyama *et al.*, 1997):

ETo = (0.0252 Tmean +0.078).Rs (9)

Camargo

The Camargo equation is expressed as (Sediyama *et al.*, 1997):

ETo = f. Tmean. Ra. ND (10)

$Ra = 15.34\ dr\ (\omega.sen\varphi.sen\ \delta + \cos\ \varphi.\cos\ \delta.sen\omega)$ (11)

dr =1 + cos(0.0172J) (12)

where f is the empirical factor, which is equal to 0.01; Ra is the extraterrestrial radiation ($mm.day^{-1}$); ND is the length of the time interval (day); dr is the inverse relative distance Earth-Sun;ω is the hour angle (rad);φ is the latitude (rad);δ is the solar declination (rad); J is the number of the day in the year between 1 (January 1st) and 365 or 366 (December 31st).

Priestley–Taylor

The Priestley-Taylor equation is expressed as (Sediyama *et al.*, 1997):

$ETo = \alpha.\ W.\ (R_n\text{-}G)$ (13)

where α is the calibration constant, which is equal to 1.26 (Jensen *et al.*, 1990; Pereira *et al.*, 2002).

Hargreaves

The Hargreaves equation is expressed as (Jensen *et al.*, 1990):

Hargreaves-Samani

The Hargreaves-Samani equation is expressed as (Allen *et al.*, 1998):

$$ETo = 0.0135.Rs.(Tmean + 17.8) \quad (14)$$

whereTmax is the daily maximum temperature (°C); Tmin is the daily minimum temperature (°C).

According to other previous papers (Willmott, 1982; Box *et al.*, 1989, *apud* López-Urrea *et al.*, 2006), the relationships between daily ETo values estimated by FAO56- Penman-Monteith equation and each of the six methods studied were established through statistical regression analysis, including the coefficient of determination (R^2), root mean square error (RMSE), relative error (RelRMSE) and index of agreement (d). The RMSE was calculated as:

$$ETo= 0.0023\ (Tmean + 17.8)\ (Tmax-Tmin)^{0.5}.Ra \quad (15)$$

where RMSE is the root mean square error ($mm.day^{-1}$); N is the number of observations; P_i are ETo values calculated by other methods ($mm.day^{-1}$); O_i are ETo values calculated by FAO56-Penman-Monteith equation ($mm.day^{-1}$).

The mean square error expressed as a percentage of the ETo mean value estimated by FAO56-Penman-Monteith equation (O_{avg}) was used as a measure of relative error:

$$RMSE = [N^{-1} \sum(P_i\text{-}O_i^{2}]^2 \quad (16)$$

The index of agreement (d) was given by:

$$RelRMSE = RMSE/_{Oavg} \quad (17)$$

Where, perfect agreement would exist between P and O if d equals to 1

$$d = 1\text{-}\ [\sum(P_i\text{-}O_i)^2 / \sum\ (|P_i\text{-}O_{avg}|+|O_i\text{-}O_{avg}|^2] \quad (18)$$

$$0 \leq d \leq 1$$

References

Allen, R. G., Pereira, L. S., Raes, D., & Smith, M. (1998). Crop Evapotranspiration: Guidelines for Computing Crop Water Requirements. FAO Irrigation and Drainage Paper 56. Food and Agriculture Organization of the United Nations.

Burt, C. M., & Styles, S. W. (1999). "The Irrigation Water Measurement Handbook." Irrigation Association.

Feyen, J., & Davis, S. (2000). "Water Use and Water Measurement in Irrigated Agriculture." Agricultural Water Management, 43(1), 81-99. DOI: 10.1016/S0378-3774(99)00080-5.

Howells, E. J., et al. (2012). "Efficient Measurement of Irrigation Water: Best Practices and New Technologies". Irrigation Science, 30(2), 159-170. DOI: 10.1007/s00271-011-0278-1

Mantovani, E. C. F., & Rojas-Downing, M. M. (2015). "The Role of Advanced Technologies in Measuring Irrigation Water Use." Water Resources Research, 51(5), 3651-3662. DOI: 10.1002/2014WR016821

Ritchie, J. T., et al. (2007). "Measurement of Irrigation Water Use Using Water Balance Techniques." Irrigation Science, 25(3), 181-192. DOI: 10.1007/s00271-007-0075-1

Rojas-Downing, M. M., et al. (2017). "Measurement and Monitoring of Irrigation Water and the Effects of Water Application on Soil Moisture." Agricultural Water Management, 179, 144-155. DOI: 10.1016/j.agwat.2016.06.019

Smajstrla, A. G., & McCready, R. (2000). Irrigation System Performance Evaluation: A Guide to Measuring and Analyzing the Efficiency of Irrigation Systems. University of Florida, Institute of Food and Agricultural Sciences.

Smith, J. W., & Smith, M. A. (2004). "Measuring Irrigation Water Supply: Methods and Technologies." Journal of Irrigation and Drainage Engineering, 130(1), 1-12. DOI: 10.1061/(ASCE)0733-9437(2004)130:1(1)

Tolk, J. A., et al. (2010). "Using Soil Moisture Sensors to Improve Irrigation Management: A Review of Measurement Techniques." Journal of Soil and Water Conservation, 65(5), 283-290. DOI: 10.2489/jswc.65.5.283

Multiple Choice Questions

1. What is the main purpose of measuring water quantity in irrigation management?
   a) To enhance soil texture b) To manage water flow
   c) To determine plant growth d) To enhance crop resistance
2. Which of the following is NOT a method for measuring water quantity?
   a) Parshall flume b) Orifice
   c) Soil staples d) Weirs
3. What is a method used to directly measure soil moisture by weighing soil samples?
   a) Float method b) Spirit burning method
   c) Gravimetric method d) Tensiometer
4. In soil moisture measurement, what equipment is used to measure matric suction?
   a) Pressure plate b) Water meter
   c) V-notch d) Tensiometer
5. What does the term 'delta' in irrigation refer to?
   a) The amount of rainfall received
   b) The total depth of water required by a crop
   c) The rate of evaporation
   d) The spacing between irrigation lines
6. Which unit is used to express water in motion, such as in a flow rate?
   a) Cubic meter b) Liter
   c) Gallon d) Liter per hour

7. What does the term 'cusec' represent in irrigation?
   a) Quantity of water flowing in cubic meters
   b) Quantity of water flowing at the rate of one cubic foot per second
   c) Volume of water in liters
   d) Depth of water in centimeters
8. Which measuring device uses a floating object to determine water flow?
   a) Water meter b) Parshall flume
   c) Float method d) Orifice
9. What method involves using a neutron meter for soil moisture measurement?
   a) Spirit burning method b) Electrical resistance blocks
   c) Neutron scattering technique d) Pressure plate apparatus
10. The equation Q = A * V is used to calculate what?
   a) Soil moisture content b) Rate of water flow
   c) Water temperature d) Soil density

Answer Key

1	b	2	c	3	c	4	d	5	b	6	d	7	b
8	c	9	c	10	b								

12

Soil Moisture Stress and Soil Water Conservation

Soil moisture stress in relation to crop and development of crop

Agronomic methods are supported with mechanical measures where land slope exceeds permissible limits and runoff gains erosive velocities. The following boxes explain the nature of agronomic measures which are essential in inter-bunded or terraced areas. These practices enhance the utility value of all kinds of mechanical structures.

1. **Contour Tillage:** All agricultural operations such as ridging, ploughing, harrowing, sowing, trenching, etc., are recommended to be done on the contour wherever possible or at least generally across the direction of the slope where holdings are very small. Even though the operation is very simple, it plays a major role in retarding the process of soil erosion through runoff. It also conserves soil, and due to increased time of concentration, more rainwater seeps through the soil profile to recharge ground water. Summer ploughing leaves the soil highly absorbent of initial rains.
2. **Dead furrows:** When all tillage operations are complete, it is advisable to leave a deep dead furrow at every 10 m interval. This should remain in position until the crop is harvested. Dead furrows aid in reducing the runoff velocity and they also conserve water.
3. **Organic Matter:** Indian soils are very poor in organic matter, especially in drought-prone areas: This can be improved by leaving the crop residue in situ (on the fields). Adding organic manures such as farmyard manure and compost every year as basal application to the soil improves the physical condition of the soil considerably.

Soil-Air, Soil-Temperature, and Soil-Moisture relationships are well balanced with the presence of organic matter. Organic matter improves the activities of soil microorganisms and also provides the much needed micro plant nutrients of all kinds, besides nitrogen, phosphorus and potash. Addition of large amounts of chemical fertilizers to dryland crops should be discouraged as it damages the soil due to:

- Excessive depletion of scarce soil moisture for its own transformation;
- Reduction in all soil microorganism activity; and
- Destruction of soil structure

Selection of Drought-Resistant Varieties

It is important that varieties which have proven genetic character to withstand longer periods of drought are chosen so that the crops can do well even in situations where the intervals between rainy days are long.

Early Maturing Varieties: Where the distribution and the amount of rain is unpredictable, it is important to select varieties which have a shorter duration life cycle (seed to seed) to cut down the water requirements of the crop. In drought-prone areas, the success rate of short duration crops is greater than long-duration crops.

Strip Cropping

Raising Erosion Permitting Crops (EPC) with Erosion Resistant Crops (ERC) having abundant adventitious root system and providing high percentage of canopy in strips in a ratio of 2:1 or 3:1 (i.e. 20 to 10 rows or 30 to 10 rows) helps in trapping soil from EPC strips to ERC strips The increased resistance to runoff in ERC results in higher volume of water percolatingthroughsoilprofile,duetoincreasedtimeof(on-ground)concentration. The close-growing ERC strips are generally legumes which fix nitrogen in the soil and enrich it.

The canopy of the ERC also protects the soil from beating action of rain drops.

Wind Erosion

Wind Erosion is as important as water erosion, the same system of cropping is recommended to be adopted, generally across the wind direction. Farmers can also raise wind belts across the wind direction. Alley cropping with rows of trees at 30 to 45m intervals also controls the desiccating effects of hot and dry winds.

Agro-horticulture

Marginal lands do not produce good annual crop returns even in normal season. These kinds of soils are best used for raising trees of economic value and creating permanent assets. Some of these lands are also very good for raising dryland horticultural crops such as mango, ber, pomegranate, tamarind etc. A part of the land could be earmarked specially for planting mixed tree species known in the area for providing fuel, fodder and timber for household needs and agricultural implements. Trees provide stable and sustained income every year, especially in drought years.

Role of Vegetation

Vegetation is crucial in preserving productive soil and conserving rain water for sustaining life. Soil and water need to be preserved for crop production (both annual and plantation crops) as well as plants that give fuel, fodder, fruits, industrial raw materials, medicinal and aromatic plants and the like. Minor vegetation such as creepers, shrubs, various kind of grasses, legumes and plants like Agave, which yield fibre have an important role when grown in gullies of various kinds and also on upstream and downstream sides of mechanical structures such as gully checks, water harvesting structures, etc. These provide reinforcement and extend the life span of the structures by binding the soil through the network of their root systems. These plants also provide fodder to animals in the area.

Seed Rates

Normal season: Sowing is done with the normal seed rate. However, if there is a drought during the plant's growth period and wilting is likely to occur, selective thinning is recommended to reduce the plant population to effectively use the scarce soil moisture among fewer plants.

Late season: Where the monsoon is moderately delayed, normal cropping with reduced seed rate is advised.

Line Sowing: Line-sowing on contours is essential. It arrests runoff and conserves soil being eroded. It helps in the use of labour-efficient implements in weeding (i.e., removal of unwanted vegetation through use of different sizes of blade harrows between the rows).

Wider Spacing: In all drought-prone areas, the most important objective is to raise a successful crop under scarce soil moisture conditions. One of the recommended practices is wider spacing between rows and between plants within the row. This reduces plant population and competition between plants for scarce soil moisture. Fewer plants have greater access to limited available soil moisture.

Weeding: Frequent weeding is an important part of dryland agriculture. Line sowing and mechanical weeding, with appropriate size of blade harrows, remove unwanted vegetation which competes with the main crop. It is not uncommon to see the dryland farmer hitching several blade harrows to one yoke and a pair of bullocks. Weeding within rows can be done using hand hoes. Removal of unwanted vegetation helps the main crop obtain greater accessibility to soil moisture and plant nutrients for its own growth.

Mixed/Inter Cropping: In a mixed cropping of different crops along with the main crops, such as millets and different legumes is an insurance against the vagaries of the monsoon; The different root systems of mixed crop feed at different depths of the soil. Moreover, mixing cropping provides small quantities of grain of different kinds for home consumption at different times.

Mulching: Mulches are ground covers that prevent the soil from being washed away, reduce evaporation, increase infiltration, and control growth of unwanted weeds. Mulch can be organic crop residue, pebbles, or materials such as polythene sheets. Mulching prevents the formation of hard crust after each rain. Organic mulches add plant nutrients to soil upon decomposition. Use of blade harrows between rows also creates "dust mulch" by breaking the continuity of capillary tubes of soil moisture.

Contingent Planning: With every care taken to undertake timely agricultural operations, it is still possible that the whole operation becomes a gamble due to unpredictable monsoons. The main crop could fall in the early part of its life cycle. In such cases, the farmer should come up with an alternate crop that can mature in a very short time and under hard conditions to take advantage of what is left of the rainy season. Contingent planning helps catch and make the best use of late rains. Advance planning is necessary in selecting a contingent crop. And all the requisites for its sowing should be ready within the main season itself. Credit for farmers must be made available at the right time.

Buffer Strips: In steep slopes and highly-eroded soils, raising permanent vegetation in small strips at appropriate intervals along the contour helps in arresting soil erosion and conservation of rain water. Raising trees in buffer strips is an asset to counter unpredictable monsoons.

Mechanical Methods: The slope of the soil is more than permissible, mechanical measures such as bunding, terracing & trenching are recommended in addition to agronomic methods. Agronomic methods are used in inter-bunded areas and mechanical practices complement to help boost crop yields in rainfed dry lands.

Indigenous technical knowledge: Local technologies evolved by people over generations of experience should be used before any new recommendation is made.

Agronomic measures for soil and water conservation

1. **Contour Farming:** As rain falls, a lot of runoff is generated which generally leads to soils erosion on its way downward. This removes the top fertile soil along with soil nutrients and plant seeds thus leading to scanty and uneven growth of crop.
2. **Mulching:** Mulching is one of the simplest and beneficial practices for soil and water conservation. Mulch is simply a protective layer of

material that is spread on top of the soil to prevent it from blowing and being washed away. Mulch can either be organic such as grass clippings, straw, bark and similar materials or inorganic such as stones, brick chips and plastic.

The mulching practice yields following benefits

i) Protects the soil from erosion.

ii) Conserve moisture in soil thus saving the need for frequent irrigation.

iii) Reduce compaction of soil due to impact of heavy rains.

iv) Maintaining the soil temperature.

v) Prevents weed growth and loss of soil nutrients.

3. **Enhancing the growth of specific crops:** Enhancing the growth of specific crops which provide maximum cover, reduce runoff and soil loss e.g. legume crops in general provide a better cover and hence better protection to cultivated land against erosion. These may vary from region to region depending upon topographical and climate conditions.
4. **Strip cropping:** It is a combination of contouring and crop rotation in which alternate strips of row crops and soil conserving crops are grown either at right angles to the direction of the prevailing wind, or following the natural contours of the terrain to prevent soil erosion.
5. **Mixed cropping:** In this practice two or more crops are grown in the same field at a particular time. Some of the benefits of mixed cropping are better and continuous cover of the land, good protection against the beating action of rain. The different crops grown in mixed cropping have their roots at different depths holding the soil more firmly thus preventing soil erosion.

Mechanical measures for soil and water conservation

These measures play an important role in soil and water conservation. These are used in conjugation with agronomic practices when they alone are not much effective. The main principles of mechanical measures are:

i) To facilitate infiltration by increasing the time of concentration.

ii) To break up a long slope into several short ones to decrease velocity of runoff.

1. **Contour Bunding:** In this practice small bunds are constructed at regular intervals across the slope of the land.
2. **Sub Soiling:** It is basically a primary tillage operation, which consists of breaking the soil structure up to a depth of 30 to 60 cm. This practice

facilitates greater infiltration rates and moisture holding capacity of the soil.

3. **Basin Listing:** In this method of soil and water conservation, basins are constructed using a special implement called basin-lister. These basins are constructed across the slope. Basin listing provides maximum time to rain water for infiltration into the soil.
4. **Bench terracing:** In this practice a series of platforms are constructed having suitable vertical drops. The range of vertical drop may vary from 2 to 6 feet depending upon prevailing conditions. The capital cost of bench- terracing is more than that of bunding initially but in longer run it proves economical.

Methods of Soil Conservation: Biological & Engineering Methods

The main aims of soil conservation are as follows

1. To protect the soil from erosion.
2. To maintain the productive capacity of the soil.

Soil erosion is a natural phenomenon by which soil is removed from rocks. Accelerated erosion due to misuse of resources of land, water and soil is today one of the most difficult and pressing problems before man. Both engineering and biological methods have been used to check the soil erosion but it is still without a plausible check. Vast tracts of fertile land are rendered useless on account of industrialisation and development.

Erosion has posed a serious challenge in the United States, South and North Africa, Japan, Mesopotamia, North China, India, Pakistan and in a number of other countries. In India, we are aware of advancing deserts of Rajasthan and erosional losses, floods etc. in other parts of the country. The problem has received the attention of forest ecologists, soil scientists and engineers only recently.

Practical methods of soil conservation are broadly grouped as follows

A. Biological measures,

B. Mechanical or Engineering methods.

A. Biological Measures

The following are the biological methods which are helpful in checking the soil erosion

1. Agronomic practices;
2. Agrostological methods;
3. Dry farming practices.

1. Agronomic practices

The important agricultural practices which contribute to the conservation and productivity of cultivated lands are referred to as conservation farming's or advanced agronomical methods.

These are listed as under

i) Contour farming
ii) Tillage and keeping the land fallow
iii) Crop rotation, sowing of leguminous crops and mixed cropping
iv) Mulching
v) Strip cropping

i) Contour farming: It is practiced in the hilly regions or on the slopes. In such areas the ram water is absorbed in very little amount because of its quick downward movement on the slopes. If these sloppy areas are ploughed up and down the slope, the heavy rainfall may cause gully development. Taking into consideration this defect, the sloppy areas are ploughed and seeded against the slope, i.e., in circular furrows around the slopes. This process is termed as contour farming.

The contours (circular or peripheral furrows) catch the downwardly moving water until it is absorbed in the soil. The ridges reduce the flow of water. The circular rows of plants across the slopes check the soil erosion. Thus, contour farming reduces run off, saves more water for crops, reduces soil erosion and increases the yield of crops

ii) Tillage operation and keeping the land fallow: There are several diverse opinions as to whether deep ploughing gives good result or shallow ploughing. A number of researches support the view that in dry areas, shallow ploughing gives comparatively good crop yields Shallow ploughing removes the weeds and enables the soil to absorb water. Deep ploughing of en leads to soil erosion but in the areas where rainfall is sufficiently high, deep ploughing (upto 15-30 cm deep) is effective in removing weeds and increasing crops yields.

If the land is left uncultivated and sheep, goats and other cattle are allowed to graze and sit over it for some time, the soil becomes fertile. Though this practice is useful yet it is not possible in the countries like India where exists severe problem of cereals because of thick human population.

iii) **Crop rotation, sowing of legumes and mixed cropping:** When the same crop 's grown in the field every year, the soil becomes depleted in certain minerals. The soil loses its fertility even after the use of fertilizers and ultimately erosion sets in. Rotation of crops is an important method for checking erosion and maintaining productivity of soil. After 2 years crop should be changed in the fields.

A good rotation should include a cultivated row crop, densely plan e small grasses and a spreading legume or a legume and grass mixture. Selection of crops for rotation should be made taking into consideration the climate, economic condition soil types, soil texture, slopes, nature of erosion, etc. Deep-rooted crops should be rotated by shallow-rooted crops.

Deep-rooted crops absorb nutrients from the deeper strata of the soil Thus, the minerals o top soil remain stored for future use by shallow-rooted plants. When the deep rooted crops le they add humus in the soil which is future storehouse of plant nutrients. Growing of Phaseolusmungo (Mung in Hindi) in Bihar, P. radiatus (Urad in Hindi) and P. mungo in U.P. in kharif season before wheat cropping improves soil fertility as well as yields of cereal crops.

Leguminous plants play very active part in increasing the nitrogenous contents in the soil because of the fact that bacteria inhabiting their root nodules fix nitrogen of atmosphere into nitrogenous compounds such as nitrates, nitrites, ammonium salts, amino acids and proteins. The nitrogenous compounds return to the soil by way of death and decay of underground nodulated roots of these legumes. Thus, growing of leguminous plants is one of the important methods for maintaining soil fertility.

The rotation of crop serves the following purposes

a) Enriches the soil,

b) Improves the soil texture,

c) Improves water holding capacity of the soil,

d) Improves crop production,

e) Controls the recurrence of weeds and diseases.

Mixed cropping is another important method for increasing productivity of the soil. In this practice, one main crop and one or two subsidiary crops are grown together on the same land, as for example, growing of Arhar, Urad, Til along with millet. This practice checks the soil erosion and avoids the risk of crop failure. If one crop fails due to diseases or any other factor, the others remain ensured.

iv) **Mulching:** It means covering the soil surface by straw, leaves or grasses. Mulches of different kinds check soil erosion, increase soil fertility and also minimize moisture evaporation from the top soils. Various types of surface tillers and crop residues are helpful in obstructing the movement of soil particles.

v) **Strip cropping:** It is an important method which employs all the advanced cultivation practices such as contour farming, proper tillage, crop rotation, mulching, cover cropping, etc. Strip cropping is very effective and practical means for controlling soil erosion.

It is of the following types

a) Contour strip cropping,
b) Field strip cropping,
c) Wind strip cropping, and
d) Permanent or temporary buffer strip cropping.

a) **Contour strip cropping:** It is special crops are grown on the strips across the slope on the level of contour and in the following season soil protecting crops are shown on the strips on which soil exposing crops were grown in the previous season and again in place of these soil protecting crops some soil exposing crops are sown. This practice is useful because it checks the flow of runoff water, increases the infiltration of water in the soil and prevents soil erosion.

b) **Field strip cropping:** It is a farming in more or less parallel strips across fairly uniform slopes but not on the exact contour.

c) **Wind strip cropping:** In this method tall growing plants (e.g. bajra, jowar, etc.) alternating with the short growing crops, such as 'arhar', 'urad' are sown in long straight strips right across the direction of wind regardless of contour.

d) **Permanent or temporary buffer strip cropping:** This is a special type of contour strip cropping in which care is taken to check the soil erosion. In this, crop rotation practice is not applied and on the strip perennial legumes and grasses are planted on permanent or temporary basis.

2. Agrostological methods: The following are the important agrostological practices that check soil erosion

i) Cultivation of grasses (ley farming).
ii) Retiring the land.
iii) Afforestation and Reforestation.
iv) Checking of overgrazing.

i) **Cultivation of grasses (Ley farming):** This method consists in growing grasses in rotation with agricultural crops. This practice improves the fertility of soil and helps in binding of the soil, thus preventing the soil erosion. This practice is recommended for Nilgiris and similar places which are subjected to very severe soil erosion.

ii) **Retiring the land:** Areas subjected to heavy soil erosion should necessarily be put under thick cover of grasses. Under favourable climatic conditions grazing should also be allowed for short periods. Researchers conducted at Solapur in Maharashtra have shown that grasses have good soil binding capacity. In Nilgiri hills, Tamil Nadu doobgrass (Cynodondactylon), Dectylisglomerata, Eragrostisamabitis and E. cerbula are proved to be most effective in soil binding and in stabilizing the reserves of the bench terrace and sodding water channels.

iii) **Afforestation and reforestation:** Afforestation means growing forests at places where there were no forests before due to lack of trees or due to adverse factors such as unstable soil, aridity, or swampiness. Reforestation means replanting of forests at places where they have been destroyed by uncontrolled forest fires, excessive felling and lopping. Plantation of trees in short blocks is known as a wind break and extensive plantation of trees is called shelter belts.

Wind is a problem in dry areas. Soil erosion by wind can be checked or reduced if the soil is covered by vegetation. Wind breaks are planting of trees or other plants at right angles to prevailing wind that protect bare soil from full force of wind. Wind breaks reduce the velocity of wind, thereby decrease the soil erosion. In some cases, rows of trees are angles to prevailing wind. Wind break and shelter belt type of plantations are being done in some regions of Uttar Pradesh where desert is encroaching.

The plantation is usually done in two or three belts. Small sized plants are planted on windward side and tall trees on leeward side. Wind breaks and shelter belts reduce the wind velocity considerably and also check the transport of lifted sand and soil particles. Afforestation is the best means to check the soil erosion.

Afforestation and reforestation, forest management practices

a) Control on forest fires,

b) Extension of forest areas,

c) Checking extensive felling and lopping,

d) Restricted grazing in forest areas, and

e) Formation of reserve forests.

iv) **Checking of overgrazing:** Since grazing in all the areas subjected to soil erosion cannot be completely stopped, a system of restricted and rotational grazing may be helpful in checking soil erosion to some extent. The area open to grazing for sometimes should be closed for the following years to facilitate regeneration of forest and to maintain thick ground vegetation.

3. Dry farming practices

Dry farming is useful in those areas where rainfall is very low, uncertain and uneven in distribution. In such areas maximum conservation of water in the soil and reduction of soil erosion are the two main problems. The improved dry farming practice developed as result of researches carried out in India is an integrated rational system of cultivation of purely rained and drought enduring crops in dry areas. This practice consists in bonding, both contour and compartmental, and all other advanced agricultural practices to conserve maximum amount of rain water in the soil so as to make it available for crops during their growth period. In this, usually early maturing crop varieties are selected for cultivation.

Dry farming methods and selection of crops is varied in different regions as per requirement of local conditions.

Mechanical Methods: It is only in recent years that soil erosion problems have received attention of engineers. The mechanical practices of soil conservation include various engineering techniques and structures which are adopted to supplement the biological methods when the latter alone are not sufficiently effective.

These practices aim at the following objectives

1. To reduce the velocity of run-off water and to retain it for long period so as to allow maximum water to be absorbed and held in the soil.
2. To divide a long slope into several small parts so as to reduce the velocity of run-off water to the minimum, and
3. Protection against erosion by wind and water.

Mechanical methods for soil conservation are

i) Basin leaching,
ii) Pan breaking,
iii) Sub soiling,
iv) Contour terracing,
v) Contour trenching,
vi) Terrace outlets,
vii) Gully control,
viii) Digging of ponds and reservoirs, and
ix) Stream bank protection.

i) **Basin leaching:** In this method, a number of small basins (water reservoirs) are made along the contour by means of an implement called basin blister. Basins collect and retain rain water for long period and also catch and stabilize downwardly moving soils of the slopes.

ii) **Pan breaking:** In some areas, soils become impervious to water and are less productive because of formation of hard sheet of clay a few feet below the surface. Such areas can be made productive and water permeable by breaking hard clay pans by means of pan breaker on contour at a distance of about 5 feet. By pan breaking, drainage and percolation of rain water is improved and soil is saved from residual run-off and erosion.

iii) **Sub-soiling:** In this method hard subsoil is broken deeply by means of an implement called subs oiler. This process promotes absorption of rain water in the soil and makes the soil more loose and fit to allow luxuriant growth of vegetation.

iv) **Contour terracing:** Sometimes drainage channels or properly spaced ridges or soil mounds are formed along the contour (at right angles to the slope) to retain water in the soil and check the soil erosion. These are called terraces. Terraces are leveled areas constructed at right angles to the slope to reduce soil erosion.

Terracing may be of the following four types

a) **Channel terracing:** This is concerned with making of wide but shallow channels on contours at suitable distance. In this process, the excavated soil is deposited along the lower edge of channel in the form of low ridge.

b) **Narrow based ridge terracing:** This process is commonly called bunding. In this a number of narrow based ridges or bunds are constructed at distance of 1into 2inacross the slope along the contour.

c) **Broad-based ridge terracing:** In this, wide but low bunds are made on contour by excavating soils from both the sides of ridge.

d) **Bench terracing:** This method involves making of wide step like platforms, the so called bench terraces, having suitable drops along contours. Along the outer edges of bench terraces bunds of about one foot height are raised to check the downward flow of rain water and also soil erosion. The vertical drops may vary from 1into 2 m. Bench terracing is very costly process and so it should be applied in the area of land scarcity for growing money crops.

Terracing is the only practical method of soil conservation on steep land. It is an expensive method or reducing soil erosion since it requires moving soil to construct the leveled areas, protecting the steep areas between terraces and constant repair and maintenance. Many factors such as length, the steepness of slopes, type of soil, and amount of precipitation determine the feasibility of terracing.

v) **Contour Trenching:** This method involves making a series of deep pits (2′ x 1′) or trenches across the slope at convenient distance (Fig. 25.7). The soil excavated from the trenches is deposited along the lower edge in the form of bund. On the ridges tree seeds are sown

vi) **Terrace outlet:** In order to reduce soil erosion and to remove excess of rain water safely from the contour terraces pipe outlets are used or channels are made which are thickly covered by grasses.

vii) **Gully and ravine control:** Gully formation can be checked by the following methods:

a) By making perimeter bunds around gullies to check flow of water through it.

b) By growing suitable soil-binding vegetation on the gullies to check soil erosion.

c) Diversion trenches should be made around gullies.

viii) **Ponds and reservoirs:** Small ponds and water reservoirs or dams should also be made at suitable places for irrigation and some other purposes. Various types of dams have been devised to arrest and plug gullies and thus to check soil erosion (Fig. 25.8). These dams may be (a) brush dams (b) earth dams, (c) concrete dams or (d) woven wire dams.

ix) **Stream bank protection:** Banks of ravines and rivers with high vertical drops are subjected to heavy soil erosion. The bank erosions can be checked by making the drop slopy and by growing vegetation on the slopes or by constructing stone or concrete pitch.

Soil Moisture Stress on Yield of Crops

The water shortage is the major problem in arid and semiarid regions.

In these regions the moisture stress represents the most important factor affecting plants. This article embodies a review of the work done on the effect of moisture stress on growth, productivity and the different aspects of plant water relations. The latters involve water expenditure, water requirements and osmotic pressure of plant sap which is a measure of the water balance and the absorptive capacity of the plant. Emphasis was also placed on the effect of moisture stress on the mineral composition of plant material as well as the metabolic products that determine the nutritive value of plants.

Vegetative growth

The vegetative growth of plants is greatly affected by the moisture stress. It is greatly reduced by decrease in available moisture before the permanent wilting percentage is reached. The permanent wilting percentage which is the lowest limit of available moisture is considered as zero and the moisture equivalent or field capacity is considered as 100. Then the moisture content above the permanent wilting percentage is expressed as percentage of this range of available moisture. The maximum dry weight of any plant species in the different treatments is considered as 100, then the dry weights of plants in the other treatments were calculated as percentage of this maximum (= %maxima).

Yield

The effect of moisture stress on yield of cereals is due to (a) magnitude, (b) time and (c) duration of moisture stress.

Effect of magnitude of moisture stress on yield

As regard the magnitude of moisture stress, experiments were carried on barley and wheat applying different levels of water supply. In the case of barley *(Hordeumvulgärevar.* 'Mariutis') a total amount of irrigation water equivalent to 150 mm rainfall is efficient for production of grains in the field under semi-desert conditions. The fall in the water supply from the highest level of 350 mm to the lower levels 300, 250, 150 mm caused reduction in the yield to 69, 48 and 30(%) respectively. The decrease in yield was more influenced by the number of grains rather than by the weight of grains. This agrees with the findings of **HENKEL 1962** and **KREEB 1963**. The fall in the water supply from the highest to the lowest level, was accompanied by a reduction to about 39 percent in the number of grains and a reduction to about 48 percent in the mean weight of grain. The total number of grains perplant is dependent on the number of tillers and ears per plant.Experiment of the same design was conducted on wheat using cubicwooden boxes with a dimension of 50 cm painted inside and outside and filled with natural desert soil and irrigated with total amounts of waterequivalent to 200, 300, 400, 500 and 600 mm rainfall in the different treatments.A total annual rainfall of the order of 200 mm was efficient to produce grains. The increase in yield with rise in level of water supply is mainly referred to the increase in number of grains per plantrather than to increase in the weight of grain. The fall in the level of water supply from 600 to 200 mm caused a reduction in the yield of about 80% associated with a reduction in the number of fertile grainsper plant of about 78% and in the weight of grains of about 31%.

Sensitivity of the different stages of development to drought

It is not only the amount of rain which affects the yield but also the distribution of rainfall. Under desert conditions the rainfall is irregular in its distribution throughout the rainy period. An experiment was designed to cover all the probabilities of the level and distribution of rainfall by using an irrigation regime simulating the natural conditions. In this experiment grains of barley mentioned above, filled with desert sandy soil. Six Treatments were used representing different levels of artificial rain and different ways of distribution. Three levels of irrigation were employed. In Treatment I, an amount of irrigation water equivalent to 120 mm was given; in Treatment II, 210 mm and in Treatment III, to VI 310 mm. In the latter 4 Treatments, although the total amount of irrigation water was equal yet the water was distributed in different ways simulating the distribution of rain in nature. Thus in Treatment III, the artificial rain was equally distributed over the growth period. In Treatment IV, the greater part of irrigation water was given in the early stage of vegetative growth; in Treatment V, the greater part was added in the middle stage of vegetative growth, while in Treatment VI, most of the irrigation water was given near the end of the vegetative growth and during ear formation. Before discussing the yield in the different treatments, it is of interest to follow the ear formation in the different treatments from the beginning of ear emergence till the number of ears became constant. The curves representing the rate of ear emergence lie in four levels. The lowest curve is that of Treatment I, with the lowest water supply. Next are the curve of Treatment II with higher water supply of 210 mm, and then the curves of Treatments III to V receiving the highest water supply of 310 mm which, is distributed differently in the three treatments. The highest curve is that of treatment VI with the highest water supply of 310 mm and in which the greater part of irrigation water was given at the beginning of ear formation. The percentage of ear emergence varied widely in the different treatments between 7 percent in Treatment I and 93 percent in Treatment VI. The mean yield varied widely in the different treatments between 2.9 g per 100 fertile plants in Treatment I and 36.6 g per 100 fertile plants in Treatment VI. A review of the yield in the various treatments expressed as percentage of maximum shows that the yield increased with increase in level of water supply. Although the total amount of irrigation water was equal in Treatments III to VI, yet the yield varied widely between 20 and 100 percent. The wide variation in yield is due to the way of distribution of irrigation water. The highest yield was given by the treatment in which a great proportion of irrigation water had been given nearthe end of the vegetative period and during ear formation. This periodwas regarded more critical than the early and middle stage of vegetativegrowth. **RTJSSEL 1959** mentioned that **BARTEL & HOBART** in their workon wheat in the arid soil showed that water inexcess of 6 inches of which 3 inches come from rainfall, early in the lifeof the crop appears to be more harmful, whilst during the period of maximumgrowth until the soft dough stage of the grain, the crop has a very considerablewater demand. **HENKEL 1962** reported that the most markedreduction in yield as the result of drought was when the latter occurredin the period of ear formation. With corn, **ROBINS DOMINGO 1953** foundthat maximum yield was reduced by water stress at the tasseling stageand **DENMEAD & SHAW 1960** found that a reduction of about 50% wascaused by water stress at the silking stage.

Effect of duration of drought on yield

Wheat plants were subjected to different periods of drought without irrigation for 3, 4, 5, 6, 7 & 8 weeks. Desiccation of the leaves started after a 4 weeks period of drought and proceeded till it covered all the leavesafter a dry period of 8 weeks. The soil moisture reached the permanentwilting percentage after 7 weeks. Plants growing under this conditionhad all their leaves dry except the terminal leaf covering the terminal bud.

The plant tissues suffered water deficit of about 50% and still surviving. The osmotic pressure of the plant sap was above 37 atm. Plants failed to produce grains and the ears were empty.

Plants subjected to a drought period of 8 weeks had all their leaves dry but the terminal bud was green and protected by the dry leaf sheath.

No ears were produced under such condition. In treatments in which plants were subjected to a long dry period of about 7 to 8 weeks, the failure of grain production may be due to the following: (1) Death of root hairs; (2) Suberization of roots associated with suppression of permeability to water and nutrients **(SLATYER 1967)**;(3) Delayed recovery or incomplete recover} after irrigation. Under theabove conditions the formation of new green leaves was extremely limited and took place very slowly. **GATES 1955 a, 1955 b** concluded that if the water supply is renewed before death occurs, the recovery to normal metabolic behaviour appears to take several days, even in situations where only a single period of desiccation to the permanent wilting percentage has occurred. In many cases recovery never appears to be complete. From the above discussion it is evident that failure in production of grains may be attributed to suppression of permeability of roots to water and nutrients together with reduction in assimilating surface and incomplete recovery to normal metabolic behaviour. The time of irrigation has a remarkable influence on recovery of plant tissues. Plants irrigated after long intervals are subjected to a high water deficit and its tissues could not recover and attain the normal turgidity as plants irrigated after short intervals. This may be reflected on the yield. Peanut plants irrigated after intervals of 4, 8 and 20 days were compared as regard their relative turgidity before and after irrigation.

The relative turgidity of plants irrigated at intervals of 4 days fluctuated between a minimum of 54 and a maximum of 64% before irrigation. In the case of plants irrigated at intervals of 8 days the relative turgidity ranged between a minimum of 47 and a maximum of 63% before irrigation.

Plants irrigated after a long interval of 20 days had a relative turgidity varying between 37 and 53% before irrigation. After irrigation, plants irrigated after 4 and 8 days regained a high relative turgidity fluctuating approximately between 70 and 83%. Plants irrigated after 20 days did not attain a high relative turgidity but they suffered a high water deficit even after irrigation. Their relative turgidity fluctuated between 42 and 69%.

The unability of plants subjected to a long dry period without irrigation to regain the normal relative turgidity may be due to the decrease in the efficiency of the absorbing root system. **SLATYER 1967** reported that there is evidence that recovery is delayed firstly by the marked reduction in rates of water absorption caused by death of root hairs, or roots, and increased suberization of the root system, which reduce the permeability of the root to water and nutrients. **KRAMER & KOZLOWSKI 1960** concluded that root hairs appear to die at relatively low stress levels and marked root suberization develops as desiccation proceeds.

Water requirement

The transpiration intensity is greatly suppressed with the increase in moisture stress. Measurements of the transpiration rate of Barley were made at different levels of water supply in the different months covering the different stages of growth. Comparison of the daily mean transpiration rate in the differenttreatments shows a considerable variation in the different months. InDecember the daily mean transpiration rate was 219 mg/gh.; in treatmentC with the highest level of water supply (350 mm) and fell to 184 mg/ghat the lower level B (300 mm) and to 169 mg/gh at the level A (200 mm)and to 149 mg in the control with the lowest level of water supply (150 mm). In January the fall from the highest level of water supply to the lowerlevels B, A and control was associated with a drop from a mean transpirationrate of 236 to rates of 191, 152 and 96 mg/gh respectively. Thus a reductionin the daily mean transpiration rate of about 40% accompanied the fallin the water supply from the highest to the lowest level.In February the mean transpiration rate exhibited a progressivedecrease from 261 mg/gh to 246, 229, and 115 mg/gh in the successivetreatments from the highest to the lowest levels of water supply respectively.

In other words, the mean transpiration rate diminished to about 40% with the fall in the water supply from the highest to the lowest level as in the previous months. In March at the end of the vegetative period there was a progressivediminution in the mean transpiration rate with the fall *in* the water supply,but the difference between the various treatments was less pronounced than in February. A reduction in the mean transpiration rate to about 66 % took place with the fall in water supply from the highest to the lowest level. The magnitude of variation in transpiration rate with change in moisture stress is controlled by the atmospheric conditions. Comparison of the transpiration rate of irrigated and non-irrigated plants of *Agavesisalana*showed that the variation was more pronounced *in* summer months when the evaporative power of the atmosphere was greater than in thewinter months when the evaporative power of the atmosphere was lower **(ABD EL-RAHMAN & al. 1968**). The water requirements based on the amount of water lost per gramdry matter produced are greatly suppressed by increase in moisturestress under higher temperatures but at lower temperatures the effect ofmoisture stress is minimized. **ABD EL-RAHMAN & BIERHUIZEN 1959** workingon the water requirements of tomato arrived to the conclusion that theeffect of water supply on the water requirements is noticeable at hightemperatures and disappears at the lower ones. At the highest temperature(26°C), the fall in the water supply from the highest to the lowest level was accompanied by a reduction in the water requirements to 51 %. At the 20°C, the fall in the water supply from the highest to the lowest level was associated with a reduction in water requirement to 57%. At the lower temperatures 15° and 10°C, the variation in the water requirement at the different levels of water supply was irregular and less remarkable.

Effect of moisture stress on mineral composition of plant material

The deficiency in soil moisture favours the decrease in accumulation of some ions and the increase in accumulation of others. Experiments were conducted on 8 range plants grown at different levels of water supply using a total amount of irrigation water equivalent to 125, 200, 275 and 350 mm during the whole life period of plants. In these treatments plants were subjected to soil moisture ranging between the permanent wilting percentage and moisture equivalent.

It is evident from the results presented in Table 7 that in all the studied range species the deficiency in soil moisture favours more accumulation of K, Na, Ca, Mg, Cl and less accumulation of P and Fe in plant tissues of all the studied species in a regular manner.

A review of the work done by other investigators showed that most of their findings are in general agreement with the above results. **OELBERG 1956** in his discussion of the factors affecting the nutritive value of forage plants, reported that increased precipitation resulted in an increase in P content and a decrease in Ca content of forages and vice versa. The results of **JENNE & al. 1958** showed no effect of soil moisture stress on the P percentage of whole corn plants. However, percentage of K, Ca and Mg in corn plants increased as the supply of available soil moisture decreased. **KILLMER & al. 1960** concluded that the P content of eight forage species increased as soil moisture increased. **WILSON & MCKEEL 1961** concluded that moisture stress altered the relative distribution of the ranslocated P within the plant. Plants with adequate moisture accumulate a higher concentration of P than did plants exposed to moisture stress. **SMOLIAK & BEZEATJ 1967** in their study on chemical composition of range forage plants of the *Stipa-Bouteloua*association of the prairie have shown that phosphorus content increased during moist than in dry years.

On the other hand **WADLEIGH & RICHARDS 1951** reported that for a given level of fertility, decreasing soil moisture is associated with a definite decrease in K, content and variable effects upon content of P,Ca and Mg. Examination reveals that the majority of ions (K, Ca, Mg, Cl and Na) increase with decrease in soil moisture, whereas the minority (P and Fe) diminished with deficiency in soil moisture. The sum of total ions accumulated in the plant tissues increases with increase of soil moisture stress. This is accompanied by accumulation of more salts in the cell sap and rise in osmotic pressure. It is evident that in all the range species, the rise in osmotic pressure with decrease in water supply is associated with increased in total salt content. Therefore one can arrive to the conclusion that the accumulation of salts together with the other changes such as the increase in water deficit of tissues and increase in the concentration of sugars and osmotically active carbohydrate breakdown products may account for the rise in osmotic pressure of plants subjectedto deficiency in soil moisture.

Metabolic products in relation to soil moisture stress

It is well known that plants grown with a reduced water supply often show differences in physiological processes or generally in chemical composition when compared with those grown with abundance of water supply. This was discussed in some details by **STOCKER 1956, 1960, 1961, HAGAN & al. 1957, RÜSSEL 1959, KRAMER & KOZLOWSKI 1960, COYNE & SERRANO 1963, HENKEL 1964 and KOZLOWSKI 1964.** They have shown that water deficits in plants, brought about by high soil moisture tension greatly modify their metabolic products. Studies were made on the metabolic products of range plants which determine their nutritive value under different conditions of water supply. Also the crude fibre content which affects the palatability of range plantswas analysed under different moisture conditions.

Total Available Carbohydrate Content

Examination ofthe results obtained reveals that the total available carbohydrates (soluble sugars and acid hydrolysable polysaccharides including only starch and pentosans) increased with deficiency in soil moisture in all the species and in the different stages of development.

The range of variation in carbohydrate content with change in water supply differs in the different species and also in the various stages of development. The widest range was demonstrated by *Medicagosaiiva*in the three successive stages *viz.*: preflowering, end of flowering, and fruiting. In these stages, the fall in the level of water supply from the highest to the lowest level was accompanied by the increase in the total carbohydrate content from about 4.8 to 9.0%. Thus the deficiency in moisture content causes an increase in the carbohydrate content that reaches in some cases double the value recorded under conditions of optimum water supply. In the other species the magnitude of variation in the carbohydrate content with change in water supply was more or less equal to that of *Medicago saliva.*

Total Nitrogen Content

The effect of water supply on the nitrogen content is very prominent. The decrease in water supply is accompanied by increase in the total nitrogen content in the different stages of life cycle in all the species.

Comparison of the total nitrogen content in *Clitoris gayana*in the highest and lowest levels of water supply shows that in the early leafy stage it varied between 2.01 and .2.72%; in the advanced leafy stage between **2.19** and 3.16%; in the preflowering stage between 0.93 and 1.87%; at the end of flowering stage between 1.26 and 2.40% and in the fruiting stage between 1.06 and 2.18%. Thus the deficiency in soil moisture results in a considerable rise in the nitrogen content of the plant which reaches insome cases double the value at optimum water supply.

Crude Fibre Content

Crude fibres consist almost entirely of cellulose and lignin together with some resistant hemi-cellulose. The moisture stress exhibits a slight effect on the crude fibre content in the different stages of development.

There is a tendency towards an increase in crude fibre content with increase in soil moisture.

The nutritive value of range plants as determined by the total available carbohydrates, total nitrogen, Ca, Mg; and P contents are highly influenced by soil moisture conditions. Plants subjected to higher moisture tension, or deficiency in soil moisture, are richer in total nitrogen (or generallycrudeproteins) as well as available carbohydrates, Ca and Mg contents,and poorer in P content. This phenomenon is manifested in all the 8 studiedspecies.

The total value of the nutritive constituents per plant is not only determined by the percentage of these constituents in the dry material, but also by the dry matter production per plant. The latter is greatly suppressed with deficiency in soil moisture.

Grain Yield

Varieties and moisture level, as well as the interaction effect of between them, had a significant (*P*0.01) effect on grain yield. On control irrigation interval (four day irrigation interval), the Awash-1 variety (7.8g) and Gofta variety (7.36 g) produced the highest yield, while the Fedis variety (1.43g) produced the lowest yield. This might be due to the physiological response and the yield components having a direct effect on yield of common bean. Four-day irrigation interval produced the highest pods plant^{-1}, seeds pod^{-1}, and hundred seed weight which directly relate to the total yield. The authors of reference identified the number of productive pods per plant as one of the most reduced yield components in water-stressed legumes. The authors of reference also reported a decrease in common bean seed per pod and seed weight under moisture stresses resulted yield reduction. The yield losses are also attributed to lower achievement of physiological responses such as a decrease of stomatal conductance, stomatal number, chlorophyll content, and photosynthetic activity in moisture stressed treatment. A similar result was reported by the authors of reference that drought and waterlogging stress conditions were found to have a strong effect on grain yield in two types of bean, with yield reduced by 25.5 percent when compared to the non-stressed treatment.

Drought and waterlogging treatments reduced common bean yield regardless of varieties; while plants have diverse mechanisms for response and adaptation to water stress. Physiological parameters like transpiration rate, photosynthesis rate, stomata number, stomatal length, stomata width, stomatal conductance, leaf relative water content, water use efficiency, and proline content had a significant interaction effect on varieties and irrigation interval, whereas leaf chlorophyll fluorescence and hundred seed weight were affected only by the main effect. Regarding varieties and moisture level, Gofta variety was superior in the transpiration rate and photosynthesis rate treated by four day irrigation interval, whereas the lowest were Fedis, Tinike, and Nasir varieties that were subjected to drought stress and waterlogging stress. In the present study, significant decreases in transpiration rate and photosynthesis rate were associated with a significant decrease in yield and yield components. A significant loss of stomatal conductance was also observed under waterlogging stress growing conditions. Accordingly, the reduction in stomatal conductance was related with an increase in the water use efficiency, which indicates that stomatal closure contributed to optimizing water use efficiency under stress, which may have enabled to absorb carbon by decreasing the loss of water in the water stress condition, contributing to photosynthesis maintenance. Lower stomatal conductance is also considered to be the mechanism of adaptation to water shortage, allowing reduction in the evapotranspiration area. The optimum watering interval (four day interval) obtained the maximum leaf relative water content on Awash-2 variety and under drought (eight day irrigation interval) also obtained the highest value of leaf relative water content due to stomata closure and leaf senescense, which could have reduced transpiration rates and thus protected the photosynthetic apparatus. The highest values of grain yield on Awash-1 (7.8 g $plant^{-1}$) and Gofta (7.36 g $plant^{-1}$) varieties were due to the highest values in photosynthesis rate, chlorophyll content, proline content, leaf relative water content, pods $plant^{-1}$, seeds pod^{-1}, which highly contribute to the maximum grain yield at four day irrigation interval, and they also obtained the highest yield at drought condition. Awash-2 and Awash-1 varieties were tolerant to water logging due to better photosynthesis and stomatal conductance.

Conclusion

The moisture stress influences the vegetative growth of the studied species within a range extending above and below the permanent wilting percentage. The effect of moisture stress on yield of cereals is due to (a) magnitude, (b) time and (c) duration of moisture stress. The transpiration intensity is greatly suppressed with increase in moisture stress. Its magnitude of variation is controlled by the atmospheric conditions. The water requirements based on the amount of water lost per gram dry matter produced are greatly suppressed by increase in moisture stress under higher temperatures, but at lower temperatures the effect of moisture stress is minimized.

The deficiency in soil moisture favours the decrease in accumulation of some ions (P and Fe) and the increase in accumulation of other ions (K, Ca, Mg, Na and Cl). The sum of total ions accumulated in the plant tissues increases appreciably with increase in soil moisture tension. This is accompanied by accumulation of more salts in the plant cell sap and the subsequent rise in osmotic pressure. The other factor causing the rise in osmotic pressure is the increase in water saturation deficit as a result of moisture deficiency. In the case of fodder plants the nutritive value of the shoot is highly influenced by soil moisture conditions. Plants subjected to higher moisture tension are richer in total nitrogen as well as available carbohydrates, Ca and Mg content and poorer in P content. The total values of the nutritive constituents per plant are not only determined by the percentage of these constituents in the dry material, but also by the dry matter production per plant; the latter is greatly reduced with deficiency in soil moisture.

References

Abebe G., Effect of Np fertilizer and moisture conservation on the yield and yield components of Common bean (Phaseolus Vulgaris L.) in the semi-arid zones of the Central Rift Valley in Ethiopia, Advances in Environmental Biology. (2009) 3, no. 3, 302–307.

Assefa T. G. A., Chemeda F., Bulti T., and Al-Tawaha A. R. M., Participatory bean breeding with women and small holder farmers in eastern Ethiopia, World Journal of Agricultural Sciences. (2005) 1, no. 1, 28–35.

Beebe S. E., Rao I. M., Blair M. W., and Acosta-Gallegos J. A., Phenotyping common beans for adaptation to drought, Frontiers in Physiology. (2013) 4, https://doi.org/10.3389/fphys.2013.00035, 2-s2.0-84879469839.

Bekana G., Girma E., Abebe A., Sisay A., Alemayehu D., Maryam Wagaw K., Beshir A., Chalchisa D., Hailu F., and Tesfaye D, Evaluating the genetic variations among improved haricot bean varieties (<i>Phaseolus vulgaris</i> L.) based on their agronomic traits and performing their half diallel cross, American Journal of BioScience. (2021) 9, no. 5, 168–174, https://doi.org/10.11648/j.ajbio.20210905.13.

Chekanai V., Chikowo R., and Vanlauwe B., Response of common bean (Phaseolus vulgaris L.) to nitrogen, phosphorus and rhizobia inoculation across variable soils in Zimbabwe,

Agriculture, Ecosystems & Environment. (2018) 266, 167–173, https://doi.org/10.1016/j.agee.2018.08.010, 2-s2.0-85051642015.

Molina J. C., Moda-Cirino V., Fonseca Junior N., Faria R., and Destro D, Response of common bean cultivars and lines to water stress, Cropp Breeding and Applied Biotechnology. (2001) 1, 363–372, https://doi.org/10.13082/1984-7033.v01n04a05.

Muhammad A. A., Waterlogging stress in plants: a review, African Journal of Agricultural Research. (2012) 7, no. 13, 1976–1981.

Ntukamazina N., Onwonga R. N., Sommer R., Mukankusi C. M., Mburu J., and Rubyogo J. C., Effect of excessive and minimal soil moisture stress on agronomic performance of bush and climbing bean (Phaseolus vulgaris L.), Cogent Food & Agriculture. (2017) 3, https://doi.org/10.1080/23311932.2017.1386769, 1386769.

Taiz L., Zeiger E., Møller I. M., and Murphy A., Plant Physiology and Development, 2015, 6TH edition, Sinauer Associates Incorporated, Sunderland, MA, USA.

Todaka D., Shinozaki K., and Yamaguchi-Shinozaki K., Recent advances in the dissection of drought-stress regulatory networks and strategies for development of drought-tolerant transgenic rice plants, Frontiers of Plant Science. (2015) 6, https://doi.org/10.3389/fpls.2015.00084, 2-s2.0-84923238602.

Multiple Choice Questions

1. What is the primary goal of agronomic methods in relation to soil moisture stress?

   a) Increase soil temperature

   b) Enhance utility value of mechanical structures

   c) Increase chemical fertilizer use

   d) Decrease crop biodiversity

2. Which of the following is NOT a recommended agronomic practice for reducing soil erosion?

   a) Dead furrows   b) Tillage across slopes

   c) Monocropping   d) Summer ploughing

3. How does organic matter affect soil health in drought-prone areas?

   a) Reduces microbial activity

   b) Decreases nutrient availability

   c) Stabilizes soil temperature and moisture

   d) Increases soil compaction

4. What characteristic is important for selecting drought-resistant crop varieties?

   a) Longer maturation period

   b) Proven genetic ability to withstand drought

   c) High water requirement

   d) Preference for sunny conditions

5. What practice involves planting strips of erosion-resistant crops alongside erosion-permitting crops?
   a) Contour farming
   b) Strip cropping
   c) Intercropping
   d) Monocropping
6. What is the purpose of mulching in agriculture?
   a) Increase soil erosion
   b) Provide a barrier to pests
   c) Reduce evaporation and control weeds
   d) Decrease soil fertility
7. In the context of moisture stress, what is a key effect of the duration of drought on yield?
   a) Increases the number of fertile grains
   b) Causes leaf desiccation and reduces grain production
   c) Enhances the root system's absorption capacity
   d) Improves plant metabolic processes
8. Which of the following mineral ion accumulates in plant tissues as soil moisture decreases?
   a) Phosphorus
   b) Iron
   c) Potassium
   d) Calcium
9. At what crop development stage is water stress reported to have the most pronounced impact on yield?
   a) Seedling stage
   b) Ear formation stage
   c) Germination stage
   d) Harvest stage
10. What is one mechanism plants use to adapt to water shortage?
   a) Increased area for transpiration
   b) Closure of stomata to optimize water use
   c) Enhanced leaf growth
   d) Increased soil nutrient absorption

Answer Key

1	b	2	c	3	c	4	b	5	b	6	c	7	b
8	c	9	b	10	b								

13

Irrigation Water Quality and Use of Problematic Water

Water Quality Parameters

Irrigation water contains impurities in varying concentration. The suitability of irrigation water mainly depends up on the amount and type of salts present in the water. The main soluble constituents are calcium, magnesium, sodium as cations and chloride, sulphate, bicarbonates as anions. The other ions present in minute quantities are boron, selenium, molybdenum and fluorine which are harmful to animals fed on plants grown with excess of these ions.

Quality of irrigation water is judged with three parameters

1. Total salt concentration
2. Sodium absorption
3. Bicarbonate and boron content.

Total Salt Concentration

Salt content of irrigation water is measured as electrical conductivity (EC). Conventionally, water containing total dissolved salts to the extent of more than 1.5m mhos /cm has been classified as saline. Saline waters are those which have sodium chloride as predominant salt. Brackish water is one that is contaminated with acid, bases, salts or organic matter, whereas saline water contains mainly dissolved salts, Based on EC irrigation water is classified as below.

Class	EC	Quality Characterization	Soil for which suitable
C1	< 1.5	Normal Water	All Soils
C2	1.5 to 3.0	Low Salinity	Light and Medium Soils
C3	3.0 to 5.0	Medium Salinity	Light and Medium Textured Soils for Semi tolerant crops
C4	5.0 to 10.0	Saline	Light medium textured soils for Tolerant crops
C5	> 10	High Salinity	Not Suitable

Sodium absorption ratio (SAR) and boron content: In addition to EC which has been used as a main criterion to determine the quality of irrigation water, sodium absorption ratio (SAR), residual sodium carbonate (RSC) and boron content are also used to find suitability of irrigation water.

Irrigation water which contains more than 3 ppm boron is harmful to crops, especially on light soils.

Common Problems that Result From Using Poor Quality Irrigation Water

Salinity: Salinity problems related to water quality occurs if total quantity of salts in the irrigation water is high enough for the salts to accumulate in the crop root zone to the extent that yields are affected. If excessive quantity of soluble salts accumulate in the root zone, the crop has difficult in extracting enough water from the salty soil solution. This reduces the water up take by plant and usually results in slow or reduced growth.

Permeability: This problem occurs when the rate of water infiltration in to and through the soil is reduced by the effect of specific salts in the water to such extent that the crop is not adequately supplied with water and yield is reduced. The poor soil permeability causes difficulty like crusting of seedbed, water logging, and attack of disease, salinity, weeds, oxygen and nutritional problems.

Toxicity: A toxicity problem occurs when certain constituents in the water are taken up by the crop and accumulate in amounts that result in reduced yield. This is usually related to one or more specific ions in the water viz. boron, chloride and sodium.

Miscellaneous: Various other problems related to irrigation water quality occur with sufficient frequency and should be specifically noted. These include excessive vegetative growth, lodging and delayed crop maturity from excessive nitrogen in water supply, white deposits on fruits or leaves due to sprinkler irrigation with high carbonate water and abnormalities by an unusual pH of the irrigation water.

Quality of Water from Different Sources

Water quality of most the Indian rivers are good with EC values less than 0.7m mhos /cm except in Krishna (1.4), Hagari (1.6) and Tungbhadra (1.7) rivers. Quality of most the tanks, lakes etc. is good except in those which are fed by stream passing through salt affected areas. Quantity of ground water is affected by and arid regions are generally poor with high salt content.

Irrigation water with poor quality

In an area where there is no alternative source of good quality irrigation water, it is inevitable to use the available water of poor quality. However, the yield potential of such areas can be increased by adopting proper management practices such as

a) Improvement of sodium and bicarbonates rich water by gypsum application.
b) Choice of salt tolerant crops and their varieties
c) Optimum fertilizer application and manuring
d) Proper irrigation management
e) Breaking any impervious layer by deep ploughing and
f) Adopting other management practices suitable for area.

A. Gypsum application: The harmful effect of irrigation water can be minimized to some extent by modifying its ionic composition by adding such chemicals which tend to precipitate the harmful constituents such as bicarbonate and carbonate in the form of less soluble salts or tend to create a favorable catonic Ca : Mg : Na ratio.

Gypsum should be powered up to 0.5 mm size or passed through a 30-mesh sieve. The gypsum requirement of water should be calculated depending upon the relative concentration of sodium, magnesium and calcium 8.6 Q of gypsum of 100 percent purity per hectare meter of water be necessary. Gypsum can directly be mixed in irrigation water.

B. Choice of salt tolerant crops: Some crops and their varieties are more salt tolerant than others. Hence, salt tolerant crops are to be grown in salt affected areas till the soil are improved by vegetation or other reclamation procedures.

Salt Tolerant Crops: Barly, Dhainacha, Sugar beet, Tobacco, Turnips, Mustard, Cotton, Wheat, Sugarcane, Turnips, Beetroots, Spinach, Date palm coconut etc.

Semi Tolerant Crops: Oats, Rice, Sorghum, Bajara, Maize, Red gram, Green gram, Sunflower, Castor, Sesamum, Linseed, Senji Lucerne, Berseem, Cowpea, Tomato, Cabbage, Cauliflower, Lettuce, Potato, Carrot, Onion, Cucumber, Pumpkins, Bitter ground, Pomegranate, Grape, Guava, Mango, Apple, Orange, Lemon.

Sensitive Crops for Salts: Field beans, Gram, Peas and Guar, etc.

C. Use of Fertilizers: Generally saline and alkali soils, or irrigated with poor quality waters are low in their fertility status, especially with reference to nitrogen or something phosphorus. Better crop can be grown by raising their fertility status. Nitrogen response to crop better when it is applied to soil along with manures. It has been observed that for wheat, barley, bajara, and maize the usual loses of fertilizers as applied up to an EC value of 6.5 m mhos/ cm and an E.S.P. of about 30. However, excessive fertilization or addition of fertilizers on a highly saline, alkali soil is of no value.

D. **Soil Management Practices:** When poor quality water is to be applied, it is important to have the detailed analysis of soil profile for their physical, chemical, and morphological characteristic. Soil analysis should include its structure, texture, pH, lime content, location and amount of gypsum, T.T.S., exchangeable cations, etc Information on water transmission properties on soil and depth of water table should be obtained Data on rainfall, its intensity and distribution and evaporation are obtained. Saline area should be leveled properly for uniform spread of water and its downward movement. Medium textured soils with Kankar layers pose a problem of sodicity. Such soils are managed by deep ploughing and growing green manuring crop like dhaincha. Application of gypsum under condition of low water table may improve land productivity. Use of optimum fertilizers and manures and improving the surface drainage systems will help in improving productivity.

E. **Irrigation management:** Accumulation of salts increase with the fineness of soil texture, it is essential to adopt irrigation practices such that the salinity at the root zone is kept minimum. The quantity of water and the frequency of irrigation are so kept that they could met the leaching requirement of the soil and consumptive sue off the crop grown. Salts often accumulate in the top few centimeters of soil during non-crop period and hence both crop germination and yield can be seriously reduced. A heavy pre-sowing irrigation to leach these surface salts will improve germination and early growth. It is done well in advance to allow cultivation to remove weeds and prepare the seedbed. Sowing the seed in the center of a single row raised bed will place the seed exactly in the area where salts concentrate. Alternate furrow irrigation is often advantageous. Similarly increasing the depth of water in the furrow can also be an aid to improve germination the use of sleeping beds with seeds planted on the sloping sides and the seed row placed just above the water line can help in better salinity control. Large seeded crops like maize planted in water furrows can improve germination.

Effect of salts on plants growth is reflected by increasing the osmotic pressure in the soil solution. Accumulating certain ions toxic concentration in plant tissue and by altering the plants mineral nutritional characteristics resulting in poor stand of crop stunted growth and yield. It may cause leaf burns in some crops and blue green colour in others. The germination of seed is delayed and retarded. In general, grain yield is affected more than the height of the plants.

F. **Salt Tolerance of Crops:** The ability of a plant to tolerate salt in the root zone is known as salt tolerance. Studies are important in selecting it for a particular or its variety to suit the soil conditions and for determining the leaching requirements. The effect of soil salinity on crop growth is negligible when the EC of saturated extract is less than 2 m mhos/ cm. Many crops are affected when EC is in the range of 4 to 8 m mhos/cm. crops with high salt tolerance can grow satisfactorily when EC values are in between 8 to 16 m mhos/cm. Only a few survive at EC beyond 16 m mhos / cm.

Irrigation Water Quality

Salt-affected soils develop from a wide range of factors including: soil type, field slope and drainage, irrigation system type and management, fertilizer and manuring practices, and other soil and water management practices. Perhaps the most critical factor in predicting, managing, and reducing salt-affected soils is the quality of irrigation water being used. Besides affecting crop yield and soil physical conditions, irrigation water quality can affect fertility needs, irrigation system performance and longevity, and how the water can be applied. Therefore, knowledge of irrigation water quality is critical to understanding what management changes are necessary for long-term productivity.

Irrigation Water Quality Criteria

Soil scientists use the following categories to describe irrigation water effects on crop production and soil quality:

1. Salinity hazard – total soluble salt content
2. Sodium hazard – relative proportion of sodium to calcium and magnesium ions
3. pH – acid or basic
4. Alkalinity – carbonate and bicarbonate
5. Specific ions: chloride, sulfates, boron, and nitrate.

Table 13.1: General guidelines for salinity hazard of irrigation water based upon conductivity.

Limitations for use	Electrical Conductivity
	(dS/m)*
None	≤0.75
Some	0.76 – 1.5
Moderate[1]	1.51 – 3.00
Severe[2]	≥3.00

*dS/m at 25°C = mmhos/cm [1]Leaching required at higher range. [2]Good drainage needed and sensitive plants may have difficulty at germination.

Another potential irrigation water quality impairment that may affect suitability for cropping systems is microbial pathogens.

The most influential water quality guideline on crop productivity is the water salinity hazard as measured by electrical conductivity (EC_w). The primary effect of high EC_w water on crop productivity is the inability of the plant to compete with ions in the soil solution for water (physiological drought). The higher the EC, the less water is available to plants, even though the soil may appear wet. Because plants can only transpire "pure" water, usable plant water in the soil solution decreases dramatically as EC increases.

The amount of water transpired through a crop is directly related to yield; therefore, irrigation water with high EC_w reduces yield potential (Table 2). Actual yield reductions from irrigating with high EC water that varies substantially. Factors influencing yield reductions include soil type, drainage, salt type, irrigation system and management. Beyond effects on the immediate crop is the long term impact of salt loading through the irrigation water. Water with an EC_w of only 1.15 dS/m contains approximately 2,000 pounds of salt for every acre foot of water. You can use conversion factors in Table 3 to make this calculation for other water EC levels.

Table 13.2: Potential yield reduction from saline water for selected irrigated crops.[1]

	% yield reduction			
Crop	**0%**	**10%**	**25%**	**50%**
	EC_w[2]			
Barley	5.3	6.7	8.7	12
Wheat	4.0	4.9	6.4	8.7
Sugarbeet[3]	4.7	5.8	7.5	10
Alfalfa	1.3	2.2	3.6	5.9
Potato	1.1	1.7	2.5	3.9
Corn (grain)	1.1	1.7	2.5	3.9
Corn (silage)	1.2	2.1	3.5	5.7

Onion	0.8	1.2	1.8	2.9
Dry Beans	0.7	1.0	1.5	2.4

[1]Adapted from "Quality of Water for Irrigation." R.S. Ayers. Jour. of the Irrig. and Drain. Div., ASCE. Vol 103, No. IR2, June 1977, p. 140.
[2]EC_w = electrical conductivity of the irrigation water in dS/m at 25°C.
[3]Sensitive during germination. EC_w should not exceed 3 dS/m for garden beets and sugarbeets.

Other terms that laboratories and literature sources use to report salinity hazard are: salts, salinity, electrical conductivity (EC_w), or total dissolved solids (TDS). These terms are all comparable and all quantify the amount of dissolved "salts" (or ions, charged particles) in a water sample. However, TDS is a direct measurement of dissolved ions and EC is an indirect measurement of ions by an electrode.

Although people frequently confuse the term "salinity" with common table salt or sodium chloride (NaCl), EC measures salinity from all the ions dissolved in a sample. This includes negatively charged ions (e.g., Cl^-, NO^-_3) and positively charged ions (e.g., Ca^{++}, Na^+). Another common source of confusion is the variety of unit systems used with EC_w. The preferred unit is deciSiemens per meter (dS/m), however millimhos per centimeter (mmho/cm) and micromhos per centimeter (μmho/cm) are still frequently used. Conversions to help you change between unit systems are provided in Table 3.

Table 13.3: Conversion factors for irrigation water quality laboratory reports.

Component	To Convert	Multiply By	To Obtain
Water nutrient or TDS	mg/L	1.0	ppm
Water salinity hazard	1 dS/m	1.0	1 mmho/cm
Water salinity hazard	1 mmho/cm	1,000	1 μmho/cm
Water salinity hazard	EC_w (dS/m) for EC <5 dS/m	640	TDS (mg/L)
Water salinity hazard	EC_w (dS/m) for EC >5 dS/m	800	TDS (mg/L)
Water NO_3N, SO_4-S,B applied	ppm	0.23	lb per acre inch of water
Irrigation water	acre inch	27,150	gallons of water

Definitions	
Abbrev.	Meaning
mg/L	milligrams per liter
meq/L	milliequivalents per liter
ppm	parts per million
dS/m	deciSiemens per meter

μS/cm	microSiemens per centimeter
mmho/cm	millimhos per centimeter
TDS	total dissolved soli

Sodium Hazard

Infiltration/Permeability Problems

Although plant growth is primarily limited by the salinity (EC_w) level of the irrigation water, the application of water with a sodium imbalance can further reduce yield under certain soil texture conditions. Reductions in water infiltration can occur when irrigation water contains high sodium relative to the calcium and magnesium contents. This condition, termed "sodicity," results from excessive soil accumulation of sodium. Sodic water is not the same as saline water. Sodicity causes swelling and dispersion of soil clays, surface crusting and pore plugging. This degraded soil structure condition in turn obstructs infiltration and may increase runoff. Sodicity causes a decrease in the downward movement of water into and through the soil, and actively growing plants roots may not get adequate water, despite pooling of water on the soil surface after irrigation.

The most common measure to assess sodicity in water and soil is called the Sodium Adsorption Ratio (SAR). The SAR defines sodicity in terms of the relative concentration of sodium (Na) compared to the sum of calcium (Ca) and magnesium (Mg) ions in a sample. The SAR assesses the potential for infiltration problems due to a sodium imbalance in irrigation water. The SAR is mathematically written below, where Na, Ca and Mg are the concentrations of these ions in milliequivalents per liter (meq/L). Concentrations of these ions in water samples are typically provided in milligrams per liter (mg/L). To convert Na, Ca, and Mg from mg/L to meq/L, you should divide the concentration by 22.9, 20, and 12.15 respectively.

For most irrigation waters encountered in Colorado the standard SAR formula provided above is suitable to express the potential sodium hazard. However, for irrigation water with high bicarbonate (HCO_3) content, an "adjusted" SAR (SAR_{ADJ}) can be calculated. In this case, the amount of calcium is adjusted for the water's alkalinity, is recommended in place of the standard SAR (see pH and Alkalinity section below). Your laboratory may calculate an adjusted SAR in situations where the HCO_3 is greater than 200 mg/L or pH is greater than 8.5.

$$SAR = \frac{Na^{+}_{meq/L}}{\sqrt{\frac{(Ca^{++}_{meq/L}) + (Mg^{++}_{meq/L}}{2}}}$$

meq/L = mg/L divided by atomic weight of ion divided by ionic charge (Na^{+}=23.0 mg/meq, Ca^{++}=20.0 mg/meq, Mg^{++}=12.15 mg/meq)

The potential soil infiltration and permeability problems created from applications of irrigation water with high "sodicity" cannot be adequately assessed on the basis of the SAR alone. This is because the swelling potential of low salinity (EC_w) water is greater than high EC_w waters at the same sodium content (Table 4). Therefore, a more accurate evaluation of the infiltration/ permeability hazard requires using the electrical conductivity (EC_w) together with the SAR.

Table 14.4: Guidelines for assessment of sodium hazard of irrigation water based on SAR and EC_w[2].

Potential for Water Infiltration Problem		
Irrigation water SAR	**Unlikely**	**Likely**
	———EC_w[2] (dS/m)———	
0-3	>0.7	<0.2
3-6	>1.2	<0.4
6-12	>1.9	<0.5.
12-20	>2.9	<1.0
20-40	>5.0	<3.0

[2]Modified from R.S. Ayers & D.W. Westcot. 1994. Water Quality for Agriculture, Irrigation & Drainage Paper 29, rev. 1, Food and Agriculture Organization of the United Nations, Rome.

Many factors including soil texture, organic matter, cropping system, irrigation system and management affect how sodium in irrigation water affects soils. Soils most likely to show reduced infiltration and crusting from water with elevated SAR (greater than 6) are those containing more than 30% expansive (smectite) clay. Soils containing more than 30% clay include most soils in the clay loam, silty clay loam textural classes and finer and some sandy clay loams. In Colorado, smectite clays are common in areas with agricultural production.

able 13.5: Susceptibility ranges for crops to foliar injury from saline sprinkler water.

	Na or Cl concentration (mg/L) causing foliar injury			
Na concentration	<46	46-230	231-460	>460
Cl concentration	<175	175-350	351-700	>700
	Apricot	Pepper	Alfalfa	Sugarbeet
	Plum	Potato	Barley	Sunflower
	Tomato	Corn	Sorghum	

Foliar injury is influenced by cultural and environmental conditions. These data are presented only as general guidelines for daytime irrigation. Source: Mass (1990) Crop salt tolerance. In: Agricultural Assessment and Management Manual. K.K. Tanji (ed.). ASCE, New York. pp. 262-304.

pH and Alkalinity

The acidity or basicity of irrigation water is expressed as pH (< 7.0 acidic; > 7.0 basic). The normal pH range for irrigation water is from 6.5 to 8.4. Abnormally low pH's are not common in Colorado, but may cause accelerated irrigation system corrosion where they occur. High pH's above 8.5 are often caused by high bicarbonate (HCO_3^-) and carbonate (CO_3^{2-}) concentrations, known as alkalinity. High carbonates cause calcium and magnesium ions to form insoluble minerals leaving sodium as the dominant ion in solution. As described in the sodium hazard section, this alkaline water could intensify the impact of high SAR water on sodic soil conditions. Excessive bicarbonate concentrates can also be problematic for drip or micro-spray irrigation systems when calcite or scale builds up causes reduced flow rates through orifices or emitters. In these situations, correction by injecting sulfuric or other acidic materials into the system may be required.

Chloride

Chloride is a common ion in Colorado irrigation waters. Although chloride is essential to plants in very low amounts, it can cause toxicity to sensitive crops at high concentrations (Table 6). Like sodium, high chloride concentrations cause more problems when applied with sprinkler irrigation (Table 6). Leaf burn under sprinkler from both sodium and chloride can be reduced by night time irrigation or application on cool, cloudy days. Drop nozzles and drag hoses are also recommended when applying any saline irrigation water through a sprinkler system to avoid direct contact with leaf surfaces.

Table 13.6: Chloride classification of irrigation water.

Chloride (ppm)	**Effect on Crops**
Below 70	Generally safe for all plants.
70-140	Sensitive plants show injury.
141-350	Moderately tolerant plants show injury.
Above 350	Can cause severe problems.
Chloride tolerance of selected crops. Listing in order of increasing tolerance: (low tolerance) dry bean, onion, carrot, lettuce, pepper, corn, potato, alfalfa, sudangrass, zucchini squash, wheat, sorghum, sugar beet, barley (high tolerance). Source: Mass (1990) Crop Salt Tolerance. *Agricultural Salinity Assessment and Management Manual.* K.K. Tanji (ed.). ASCE, New York. pp 262-304.	

Boron

Boron is another element that is essential in low amounts, but toxic at higher concentrations (Table 7). In fact, toxicity can occur on sensitive crops at concentrations less than 1.0 ppm. Colorado soils and irrigation waters contain enough B that additional B fertilizer is not required in most situations. Because

B toxicity can occur at such low concentrations, an irrigation water analysis is advised for ground water before applying additional B to irrigated crops.

Table 13.7: Boron sensitivity of selected Colorado plants (B concentration, mg/ L*)

Sensitive		**Moderately Sensitive**	**Moderately Tolerant**	**Tolerant**
0.5-0.75	0.76-1.0	1.1-2.0	2.1-4.0	4.1-6.0
Peach	Wheat	Carrot	Lettuce	Alfalfa
Onion	Barley	Potato	Cabbage	Sugar beet
	Sunflower	Cucumber	Corn	Tomato
	Dry Bean		Oats	

Source: Mass (1987) Salt tolerance of plants. *CRC Handbook of Plant Science in Agriculture.* B.R. Cristie (ed.). CRC Press Inc.

*Maximum concentrations tolerated in soil water or saturation extract without yield or vegetative growth reductions. Maximum concentrations in the irrigation water are approximately equal to these values or slightly less.

Sulfate

The sulfate ion is a major contributor to salinity in many of Colorado irrigation waters. As with boron, sulfate in irrigation water has fertility benefits, and irrigation water in Colorado often has enough sulfate for maximum production for most crops. Exceptions are sandy fields with <1 percent organic matter and <10 ppm SO^4-S in irrigation water.

Nitrogen

Nitrogen in irrigation water (N) is largely a fertility issue, and nitrate-nitrogen (NO_3-N) can be a significant N source in the South Platte, San Luis Valley, and parts of the Arkansas River basins. The nitrate ion often occurs at higher concentrations than ammonium in irrigation water. Waters high in N can cause quality problems in crops such as barley and sugar beets and excessive vegetative growth in some vegetables. However, these problems can usually be overcome by good fertilizer and irrigation management. Regardless of the crop, nitrate should be credited toward the fertilizer rate especially when the concentration exceeds 10 ppm NO_3^-N (45 ppm NO_3^-). Table 3 provides conversions from ppm to pounds per acre inch.

Water Quality Parameters

Irrigation water contains impurities in varying concentration. The suitability of irrigation water mainly depends up on the amount and type of salts present in the water. The main soluble constituents are calcium, magnesium, sodium as cations and chloride, sulphate, bicarbonates as anions. The other ions present in minute quantities are boron, selenium, molybdenum and fluorine which are harmful to animals fed on plants grown with excess of these ions.

Quality of irrigation water is judged with three parameters

1. Total salt concentration
2. Sodium absorption
3. Bicarbonate and boron content.

Total Salt Concentration: Salt content of irrigation water is measured as electrical conductivity (EC). Conventionally, water containing total dissolved salts to the extent of more than 1.5 m mhos / cm has been classified as saline. Saline waters are those which have sodium chloride as predominant salt. Brackish water is one that is contaminated with acid, bases, salts or organic matter, whereas saline water contains mainly dissolved salts, Based on EC irrigation water is classified as below.

Class	EC	Quality Characterization	Soil for which suitable
C1	< 1.5	Normal Water	All Soils
C2	1.5 to 3.0	Low Salinity	Light and Medium Soils
C3	3.0 to 5.0	Medium Salinity	Light and Medium Textured Soils for Semi tolerant crops
C4	5.0 to 10.0	Saline	Light medium textured soils for Tolerant crops
C5	> 10	High Salinity	Not Suitable

Sodium absorption ratio (SAR) and boron content: In addition to EC which has been used as a main criterion to determine the quality of irrigation water, sodium absorption ratio (SAR), residual sodium carbonate (RSC) and boron content are also used to find suitability of irrigation water.

Irrigation water which contains more than 3 ppm boron is harmful to crops, especially on light soils.

The Detrimental Effects of Poor Irrigation Water Quality

Using irrigation water of poor quality can lead to several challenges for crop production:

Salinity: High salt concentrations in irrigation water can lead to salt accumulation in the root zone, hindering the plant's ability to absorb water. This reduced water uptake results in stunted growth and decreased yields.

Permeability Issues: Certain salts in the water can reduce the soil's permeability, decreasing the rate at which water infiltrates and moves through the soil. This can lead to inadequate water supply for crops, resulting in yield reduction, seedbed crusting, waterlogging, and increased susceptibility to diseases, weeds, and oxygen or nutritional deficiencies.

Toxicity: Specific ions, such as boron, chloride, and sodium, present in the water can be absorbed by crops and accumulate to toxic levels, ultimately reducing yield.

Other Problems: Other issues can arise from poor water quality, including excessive vegetative growth, lodging, delayed crop maturity (due to high nitrogen levels), white deposits on fruits or leaves (from sprinkler irrigation with high carbonate water), and abnormalities caused by unusual pH levels.

Water Quality in India

The water quality of most Indian rivers is generally good, with low electrical conductivity (EC) values. However, some rivers, like the Krishna, Hagari, and Tungabhadra, exhibit higher EC values. The quality of water in tanks and lakes is generally good, except for those fed by streams that pass through salt-affected areas. Groundwater quality in arid regions is often poor, with high salt content.

Managing Poor Quality Irrigation Water

In areas where good quality irrigation water is scarce, using available, poorer quality water may be unavoidable. However, yields can be improved by implementing appropriate management practices:

1. **Gypsum Application:** Treating water rich in sodium and bicarbonates with gypsum can improve its quality.
2. **Salt-Tolerant Crops:** Choosing salt-tolerant crops and varieties is essential.
3. **Optimized Fertilization:** Applying fertilizers and manure at optimal levels can help crops thrive.
4. **Effective Irrigation Management:** Implementing appropriate irrigation strategies is crucial.
5. **Deep Ploughing:** Breaking up impervious soil layers through deep ploughing can improve drainage.
6. **Other Management Practices:** Implementing other location-specific management practices tailored to the area can further enhance productivity.

Detailed Management Approaches

A. **Gypsum Application:** Gypsum helps precipitate harmful constituents like bicarbonates and carbonates, creating a favorable cation ratio (Ca:Mg:Na). Finely powdered gypsum can be mixed directly into irrigation water. The gypsum requirement should be calculated based on the relative concentrations of sodium, magnesium, and calcium.

B. **Salt-Tolerant Crops:** Selecting salt-tolerant crops is vital for areas with saline soils. Examples include barley, dhaincha, sugar beet, tobacco, mustard, cotton, wheat, sugarcane, turnips, beetroots, spinach, date palm, and coconut. Semi-tolerant crops include oats, rice, sorghum, bajara, maize, red gram, green gram, sunflower, castor, sesamum, linseed, senji, lucerne, berseem, cowpea, tomato, cabbage, cauliflower, lettuce, potato, carrot, onion, cucumber, pumpkins, bitter ground, pomegranate, grape, guava, mango, apple, and orange. Sensitive crops include field beans, gram, peas, and guar.

C. **Fertilizer Use:** Saline and alkali soils are often low in fertility, especially nitrogen and sometimes phosphorus. Improving fertility through fertilizer application, along with manure, can improve crop yields.

D. **Soil Management:** A detailed soil analysis is essential, including physical, chemical, and morphological characteristics. This analysis should include structure, texture, pH, lime content, gypsum content, total soluble salts, and exchangeable cations. Information on water transmission properties, water table depth, rainfall patterns, and evaporation rates is also important. Leveling saline areas is crucial for uniform water distribution. Deep ploughing and green manuring can help manage sodicity issues. Optimum fertilizer and manure application and improved surface drainage will also enhance productivity.

E. **Irrigation Management:** Because salt accumulation increases with finer soil texture, irrigation practices should aim to minimize salinity in the root zone. Adjust the quantity and frequency of irrigation to meet the leaching requirements of the soil and the water needs of the crop. Pre-sowing irrigation can leach surface salts to improve germination and early growth. Other techniques include sowing seeds on raised beds or using alternate furrow irrigation.

   The effect of salts on plant growth includes increased osmotic pressure in the soil solution, toxic ion accumulation in plant tissues, altered mineral nutrition, poor crop stands, stunted growth, reduced yield, leaf burn, and delayed germination.

F. **Salt Tolerance of Crops:** A plant's ability to tolerate salt in the root zone is known as salt tolerance. Salt tolerance studies are important to inform decisions about crop selection and leaching requirements. The effect of soil salinity on crop growth is negligible when the EC of saturated extract is less than 2 m mhos/cm. Many crops are affected when EC is in the range of 4 to 8 m mhos/cm. Crops with high salt tolerance can grow satisfactorily when EC values are in between 8 to 16 m mhos/cm. Only a few survive at EC beyond 16 m mhos / cm.

Conclusion

The quality of irrigation water available to farmers and other irrigators has a considerable impact on what plants can be successfully grown, the productivity of these plants, and water infiltration and other soil physical conditions. The first step in understanding how an irrigation water source can affect a soil-plant system is to have it analyzed by a reputable lab. The selecting an Analytical Laboratory is very important to get help in locating a lab in your area that is familiar with irrigation water quality.

References

Adlan, M. A., & Al-Dhabi, N. A. (2019). "Toxic Heavy Metals in Agricultural Soils and Their Impact on Soil and Plant Health." Environmental Science and Pollution Research, 26(10), 9524-9532. https://doi.org/10.1007/s11356-019-04445-3

Ayers, R. S., & Westcot, D. W. (1994). Water Quality for Agriculture. FAO Irrigation and Drainage Paper No. 29, Food and Agriculture Organization of the United Nations, Rome, Italy.

Badran, O. M., & Kheiralla, Z. (2017). "Irrigation with Polluted Water: Current Status and Future Directions." Water, 9(12), 977. https://doi.org/10.3390/w9120977

Bouaziz, A., & Ben Maïz, A. (2019). "Assessment of Irrigation Water Quality: A Case Study of Southern Tunisia." Irrigation and Drainage, 68(4), 698-706. https://doi.org/10.1002/ird.2381

Chen, F., & Yu, X. (2018). "Influences of Irrigation with Treated Wastewater on Soil and Plant Quality." Environmental Monitoring and Assessment, 190(3), 147. https://doi.org/10.1007/s10661-018-6406-0

Fadhl, B. M., & Ezzat, M. (2020). "Impact of Polluted Water Intrusion on Irrigation: A Review." Agricultural Water Management, 233, 106030. https://doi.org/10.1016/j.agwat.2020.106030

Karpuzcu, M., & Guler, M. (2018). "Management of Saline Water in Agriculture." Salinity and Water Quality in Agriculture, 2(1), 18-29. https://doi.org/10.21601/jwqua/2/1/3

Mena, L., Zorrilla, R., & Molle, F. (2016). "Use of Poor-Quality Water in Agriculture: Assessment and Recommendations." Water International, 41(4), 520-535. https://doi.org/10.1080/02508060.2016.1181422

Oweis, T., & Hachum, A. (2009). Water Harvesting for Enhanced Rainfed Agriculture in the Mediterranean Region. FAO and ICARDA.

Qadir, M., & Oster, J. D. (2004). "Crop and Water Productivity of Salt-affected Soils." Soil Use and Management, 20(3), 239-249. https://doi.org/10.1111/j.1475-2743.2004.tb00311.x

Multiple Choice Questions

1. What is primarily used to measure the total salt concentration in irrigation water?

   a) Temperature b) Electrical conductivity (EC)

   c) pH level d) Water turbidity

2. Which of the following is classified as saline water?
   a) Water with EC less than 1.5 m mhos/cm
   b) Water primarily containing sodium chloride
   c) Water contaminated with organic matter
   d) Water with low salinity levels
3. According to the classification system, what type of irrigation water is suitable for all soils?
   a) C1 b) C2
   c) C3 d) C4
4. What is the sodium absorption ratio (SAR) used to measure?
   a) Total salt concentration
   b) Nutrient content in water
   c) Relative proportion of sodium to calcium and magnesium
   d) Water pH level
5. Which crop is identified as salt-tolerant?
   a) Potato b) Wheat
   c) Peas d) Tomato
6. What common problem arises from using poor quality irrigation water?
   a) Increased productivity b) Salinity issues
   c) Improved water infiltration d) Faster growth rates
7. What is the consequence of high sodium levels in irrigation water?
   a) Improved soil texture b) Increased water infiltration
   c) Reduced crop yields d) Enhanced seed germination
8. Which ion's concentration is harmful to crops when it exceeds 3 ppm?
   a) Sodium b) Chloride
   c) Boron d) Sulfate
9. What can help mitigate the effects of harmful irrigation water?
   a) Ignoring water quality
   b) Gypsum application
   c) Increasing irrigation frequency
   d) Growing only sensitive crops

10. What effect does high electrical conductivity (EC) water have on crop yield?
    a) Enhances yield potential
    b) Decreases water availability to plants
    c) Increases soil organic matter
    d) Improves nutrients in crops

Answer Key

1	b	2	b	3	a	4	c	5	b	6	b	7	c
8	c	9	b	10	b								

14

Agricultural and Biological Drainage

Principle of field drainage and methods of drainage, Benefits of drainage

Biological drainage

Importance of bio-drainage in water logged conditions

Bio-drainage and soil salinity

Agricultural Drainage

Excess water in the crop root zone soil is injurious to plant growth. Artificial drainage is essential on poorly drained agricultural fields to provide optimum air and salt environments in the root zone. Crop yields are drastically reduced on poorly drained soils, and, in cases of prolonged waterlogging, plants eventually die due to a lack of oxygen in the root zone. Waterlogging in irrigated regions may result in excess soil salinity, i.e., the accumulation of salts in the plant root zone.

Sources of excess soil water that result in high water tables include: high precipitation in humid regions; surplus irrigation water and canal seepage in the irrigated lands; and artesian pressure. Agricultural drainage is the removal of excess water from the soil surface and/or soil profile of cropland, by either gravity or artificial means. In some soils, the natural drainage processes are sufficient for growth and production of agricultural crops, but in many other soils, artificial drainage is needed for efficient agricultural production.

Agricultural drainage improvement can help reduce the year-to-year variability in crop yield, which helps reduce the risks associated with the production of abundant, high quality, affordable food. Improved access of farm equipment to the field provides more time for field activities, can help extend the crop production season, and helps reduce crop damage at harvest.

Agricultural drainage conditions around the world can be grouped according to three agro-climatic zones: the temperate zone, the arid and semi-arid zone and the humid and sub-humid zone.

In the temperate agro-climatic zone the major role of drainage is to prevent waterlogging, i.e., excess water in the root zone of crops from surplus rainfall.

It also serves the purpose of improving the traffic ability of the farms for farm machinery etc. The field drainage system is generally sub-surface.

semi-arid zone the function of drainage is to prevent irrigation induced waterlogging and salinization of soils. These zones suffer from population pressure and intensive agriculture through irrigation is resorted to which makes drainage provision essential to maintain the sustainability of agricultural lands.

In humid and semi-humid regions the role of drainage is a combination of waterlogging control, salinization prevention and flood protection. Generally open drainage systems are applied as the rainfall intensities are so high that the sub-surface drainage becomes insufficient or exorbitantly inexpensive.

Surface Drainage System

1. Random Drainage System
2. Parallel Field Drainage System
3. Parallel Open Ditch Drainage System
4. Bedding Drainage System

Sub Surface Drainage System

1. Tile Drainage System
2. Random Drainage System
3. Herringbone System
4. Gridiron System:
    1. Double Main System
    2. Intercepting System
5. Mole Drainage System

Drainage Coefficient

Defined as depth of water (cm) to be removed in 24 hours period from the entire drainage area (0.6 to 2.5 cm/day in extreme cases 10 cm/day)

Surface Drainage

Plant roots must have a favourable environment to be able to extract water soluble nutrients and water to meet the plant's requirement. Excess water or excess concentrations of salts in the root zone or at the land surface do not allow the plant roots to function normally. As a result, the plant growth and yield are adversely affected. In the extreme cases of waterlogging and salinity, the seeds may not germinate and the plants may wilt permanently. The result is a loss of agricultural production. Land drainage is thus the removal of excess surface and subsurface water from the land to enhance crop growth, including the removal of soluble salts from the soil. It is an effective method to restore or maintain a sustainable agricultural system. Agricultural land drainage is both a preventive and a curative measure for the physically and chemically degraded agricultural lands to enhance their productive potential. Among the various activities in the agricultural production system, drainage is the most neglected in the country.

Waterlogging and salinity of agricultural lands are caused due to natural processes or man-made activities.

The natural causes consist of

i) Poor natural drainage of the sub soil under unfavourable geology like existence of hard pan below the soil surface,

ii) Submergence under floods, and

iii) Deep percolation from rainfall,

iv) Hydraulic pressure from saturated areas located at higher elevations.

Artificial or man-made causes of waterlogging include i) practice of high intensity of irrigated agriculture without any consideration of soil and subsoil conditions, ii) excessive enclosing of irrigated fields with embankments and choking up natural drainage, and iv) poor maintenance of natural drainage or blocking of natural drainage by roads and railways. Waterlogging in any area may also be caused when the inflow into the groundwater reservoir exceeds the outflow from it resulting in rise of water table. In irrigated areas chemical degradation of land is a subsequent development on account of long duration waterlogging caused due to canal seepage and adoption of faulty irrigation practices. Irrigation development in the arid and semi-arid regions is associated with more risk of chemical degradation of land. Some of the major causative factors precipitating drainage and related are discussed below.

Potential Causes of Drainage Problem

1. **Poor natural drainage:**
2. **Uneven rainfall distribution:** Very heavy rain in a short duration causing accumulation of runoff.
3. **Tidal intrusion:**
4. **Spilling of rivers /streams causing submergence of land:**
5. **High depth of groundwater table:**
6. **Seepage:** From unlined canals, distributaries, minors and watercourses rendering the adjoining land wet for long duration.
7. **Poor on-farm water management:** Field to field and over irrigation, uncertainty in the release of water in proper time and quantity, mostly

un-gated and many unauthorized outlets, absence of roster in canal deliveries.

8. **Unrealistic cropping pattern:** Tilted in favour of water-loving crops
9. **Poor water quality:** Brackish ground and surface water are used in irrigation for want of adequate of fresh water
10. **Higher water allowance**
11. **Development activities:** Unplanned construction of roads, bridges, canals, flood embankments, railways, buildings, many a times obstructing natural drainage.
12. **Lack of catchment protection:** Deforestation, faulty cultivation practices, less in-situ moisture conservation, more runoff and soil loss, silting of downstream channels.
13. **Lack of monitoring:** lack of monitoring of the health of the land and the water resources. Problem becomes known when it is too late and the remedy is expensive.
14. **People's participation:** Absence of cooperative or community approaches to tackle the problem and too much dependence on government.

Surface Drainage System

1. Random Drainage System
2. Parallel Field Drainage System
3. Parallel Open Ditch Drainage System
4. Bedding Drainage System

Sub Surface Drainage System

1. Tile Drainage System
2. Random Drainage System
3. Herringbone System
4. Gridiron System:
   3. Double Main System
   4. Intercepting System

Mole Drainage System

Drainage Coefficient

Defined as depth of water (cm) to be removed in 24 hours period from the entire drainage area (0.6 to 2.5 cm/day in extreme cases 10 cm/day)

Plant roots must have a favourable environment to be able to extract water soluble nutrients and water to meet the plant's requirement. Excess water or excess concentrations of salts in the root zone or at the land surface do not allow the plant roots to function normally. As a result, the plant growth and yield are adversely affected. In the extreme cases of waterlogging and salinity, the seeds may not germinate and the plants may wilt permanently. The result is a loss of agricultural production. Land drainage is thus the removal of excess surface and subsurface water from the land to enhance crop growth, including the removal of soluble salts from the soil. It is an effective method to restore or maintain a sustainable agricultural system. Agricultural land drainage is both a preventive and a curative measure for the physically and chemically degraded agricultural lands to enhance their productive potential. Among the various activities in the agricultural production system, drainage is the most neglected in the country.

Waterlogging and salinity of agricultural lands are caused due to natural processes or man-made activities.

The natural causes consist of :

i) Poor natural drainage of the sub soil under unfavourable geology like existence of hard pan below the soil surface,

ii) Submergence under floods, and

iii) Deep percolation from rainfall,

iv) Hydraulic pressure from saturated areas located at higher elevations.

Artificial or man-made causes of waterlogging include i) practice of high intensity of irrigated agriculture without any consideration of soil and subsoil conditions, ii) excessive enclosing of irrigated fields with embankments and choking up natural drainage, and iv) poor maintenance of natural drainage or blocking of natural drainage by roads and railways. Waterlogging in any area may also be caused when the inflow into the groundwater reservoir exceeds the outflow from it resulting in rise of water table. In irrigated areas chemical degradation of land is a subsequent development on account of long duration waterlogging caused due to canal seepage and adoption of faulty irrigation practices. Irrigation development in the arid and semi-arid regions is associated with more risk of chemical degradation of land. Some of the major causative factors precipitating drainage and related are discussed below.

Potential Causes of Drainage Problem

1. **Poor natural drainage**:
2. **Uneven rainfall distribution:** Very heavy rain in a short duration causing accumulation of runoff.

3. **Tidal intrusion**:
4. **Spilling of rivers /streams causing submergence of land**
5. **High depth of groundwater table**
6. **Seepage:** From unlined canals, distributaries, minors and watercourses rendering the adjoining land wet for long duration.
7. **Poor on-farm water management:** Field to field and over irrigation, uncertainty in the release of water in proper time and quantity, mostly un-gated and many unauthorized outlets, absence of roster in canal deliveries.
8. **Unrealistic cropping pattern:** Tilted in favour of water-loving crops
9. **Poor water quality**: brackish ground and surface water are used in irrigation for want of adequate of fresh water
10. **Higher water allowance**
11. **Development activities**: Unplanned construction of roads, bridges, canals, flood embankments, railways, buildings, many a times obstructing natural drainage.
12. **Lack of catchment protection**: Deforestation, faulty cultivation practices, less in-situ moisture conservation, more runoff and soil loss, silting of downstream channels.
13. **Lack of monitoring**: Lack of monitoring of the health of the land and the water resources. Problem becomes known when it is too late and the remedy is expensive.
14. **People's participation**: Absence of cooperative or community approaches to tackle the problem and too much dependence on government.

On-farm Water Management

Adoption of proven and efficient water management technologies are essential to maximize crop production and minimizing gap between irrigation potential created and utilized. Application losses due to conveyance, methods and scheduling of irrigation also are effective in this direction. Selection of crop(s) and crop sequence(s) also played an important role for enhancement of water use efficiency through crop cultivation and related agronomic practices. Conservation, distribution and utilization of irrigation water are the basic parameters of on-farm water management. Optimum scheduling of irrigation, suitable method adoption, conjunctive use of rain, surface and ground water for crop cultivation having improved agro-technology adoption and provision of drainage. Application of proper amount of water at proper time increased the water use efficiency and crop yield maximization with given amount of water reducing evaporation and deep percolation. Scheduling of irrigation with limited water availability is a big challenge to the irrigation experts that needs rigorous research. Trist (1980) considered the irrigation system as a complex socio-technical phenomenon that involved interaction between a physical, environmental and technological in the process of adoption. Whyte (1991) stated the participatory action research as more effective and communicative measures in the developmental processes. Korten (1980) described the participatory action research as a method of merging both developmental intervention and research activities as an effort to involve the users in systemic process of change to make the socio-technical system more effective and efficient. Majority of the farmers are reluctant to apply irrigation scheduling technologies as they do not expect a net benefit, out of its application (Annandale *et. al.* 1996). Westhurizen and Annandale, 1996 elaborated their work on-farm participation and encouraged the farmers to adopt irrigation scheduling as cost-benefit approach. They formulated uses friendly format for a wide variety of crops irrigated in South Africa. Cheng Xianjun,1996 reported that on-farm irrigation management and irrigation service thereby improved by involving farmers in water management practices and introducing an acceptable water-saving irrigation schedule for growing crops and increasing agricultural production. On-farm water management with emphasis on active participation of the farm at each and every stage (Belloncle,1988).

A series of on-farm water management research activities were conducted as a collaborative project of Indian Council of Agricultural Research, New Delhi and Ministry of Water Resources, Government of India in some of the selected command area either by package intervention or conducting component technology field trials in the irrigation command. Pramanik and Mallick, (1996) described on-farm participatory field trials brought positive improvements towards better water use efficiency following the recommended methods of water management practices in the farmers field in DVC command. Agarwal and Khanna (1983) also reported efficient soil and water management for enhancing agricultural productivity in Haryana. Dhindwal and Poonia (1994) reported the success story of on-farm water management in Haryana. Rajput(1993) also described that on-farm water management proved much more effective than that on-station field trials conducted at several states.

The possible solutions and recommendations

1. Development of micro-level water resources (tank-well system, small springs, perennial and seasonal streams) with a command of 2-10 ha to reduce dependency on major and medium irrigation projects through R & D projects should be explored;

2. Effective harvesting of excess rainwater during heavy to medium rainfall period as well as artificial recharge of groundwater should fully be exploited wherever possible through participatory research and development along with mass awareness programme;
3. Planned intervention is required to reduce the negative effects of surplus groundwater exploitation by two-fold actions: (1) to control excessive draw down and (2) to prevent water quality deterioration and degradation.
4. The gap between technology generation and adoption in the farmers' field is to be minimized; The methods, systems, scheduling of irrigation based on different approaches, its transformation in farmers' language, water-harvesting structures and water-lifting devices, water measurement should come into farmers' practices.

Agricultural Surface drainage in waterlogged areas

Excess water in the crop root zone soil is injurious to plant growth. Artificial drainage is essential on poorly drained agricultural fields to provide optimum air and salt environments in the root zone. Crop yields are drastically reduced on poorly drained soils, and, in cases of prolonged waterlogging, plants eventually die due to a lack of oxygen in the root zone. Waterlogging in irrigated regions may result in excess soil salinity, i.e., the accumulation of salts in the plant root zone.

Sources of excess soil water that result in high water tables include: high precipitation in humid regions; surplus irrigation water and canal seepage in the irrigated lands; and artesian pressure. Agricultural drainage is the removal of excess water from the soil surface and/or soil profile of cropland, by either gravity or artificial means. In some soils, the natural drainage processes are sufficient for growth and production of agricultural crops, but in many other soils, artificial drainage is needed for efficient agricultural production.

Agricultural drainage improvement can help reduce the year-to-year variability in crop yield, which helps reduce the risks associated with the production of abundant, high quality, affordable food. Improved access of farm equipment to the field provides more time for field activities, can help extend the crop production season, and helps reduce crop damage at harvest.

Agricultural drainage conditions around the world can be grouped according to three agro-climatic zones: the temperate zone, the arid and semi-arid zone and the humid and sub-humid zone.

In the temperate agro-climatic zone the major role of drainage is to prevent waterlogging, i.e., excess water in the root zone of crops from surplus rainfall.

It also serves the purpose of improving the traffic ability of the farms for farm machinery etc. The field drainage system is generally sub-surface.

In arid and semi-arid zone the function of drainage is to prevent irrigation induced waterlogging and salinization of soils. These zones suffer from population pressure and intensive agriculture through irrigation is resorted to which makes drainage provision essential to maintain the sustainability of agricultural lands.

In humid and semi-humid regions the role of drainage is a combination of waterlogging control, salinization prevention and flood protection. Generally open drainage systems are applied as the rainfall intensities are so high that the sub-surface drainage becomes insufficient or exorbitantly inexpensive.

Benefits of drainage

a) It provides better soil environment for plant growth by creating favourable soil aeration conditions

b) It improves the soil structure and in turn increases the soil infiltration

c) High infiltration capacity reduces soil erosion.

d) It maintains desirable soil temperature, which accelerates plant growth and bacterial activity

e) It promotes increased leaching of salts and prevents accumulation of salts in the crop root zone

Problems of ill drainage

a) Limitation of aeration.

b) Accumulation of CO2 and toxic substances like H2S, ferrous sulfide etc. in the crop root zones

c) Reduced water uptake due to reduced activity of roots as a result of oxygen stress

d) Reduced nutrient uptake

e) Development of soil salinity and alkalinity

Surface drainage systems

Safe removal and disposal of excess water primarily from land surface or cropped area by a net work of surface drains or constructed channels and through proper land shaping is known as surface drainage. There are four general types of surface drainage systems used in flat areas having a slope of <2% viz.,

1. Random drain system
2. Parallel field drain system
3. Parallel open ditch system
4. Bedding system

Random drain system

This system is usually adopted in areas where the ground surface is characterized by a series and depression (undulating land surface) and where small depressions are to be drained off

Parallel field drain system

The parallel field ditch system is used in places where the surface is uniform and has few noticeable ridges or depressions.

Bedding system

This system is usually adopted in fields with very little slope, usually 0.5% or less and slowly permeable soils. It is essentially a tillage operation wherein the land is ploughed into a series of parallel beds separated by dead furrows, which run in the direction of greatest slope lateral drains are located perpendicular to slope

Parallel open ditch system

The parallel open ditch system is similar to parallel field drain system in all respects except that the drains are replaced by open ditches which are comparatively deeper and have steeper side slopes than the field drains

Sub-surface drainage systems

The removal and safe disposal of excess water that has already entered the soil profile is considered sub-surface drainage. Though several sub-surface systems are available, the most commonly used and effective ones are Tile drainage and Mole drainage systems.

Tile drainage systems

Tile drains removes excess water from the soil through a continuous line of tiles (pipes) laid at specified depth and grade. The pipes are made of either concrete or burnt clay.

Random system

The random system is used in areas that have scattered wet areas somewhat isolated from each other. Tile lines are laid more or less at random to drain the wet patches.
Herringbone system:

The system is applicable in places where the main or sub-main is located in a narrow depression i.e., in areas that have a concave surface or a narrow depression with the land sloping to it from both directions. The parallel laterals enter the sub-main from both sides. It is less economical, because considerable double drainage occurs where the laterals and mains join.

Gridiron systems

The gridiron and parallel systems are similar to that of herringbone system except that the laterals enter the main or sub-main from only one side.

Double main system

The double main system is a modification of the gridiron system. It may be used where the sub-main is in a broad, flat depression, which frequently is a natural watercourse and sometimes may be wet because of small amounts of seepage water from nearby slopes.

Intercepting system

This system involves the interception of seepage water that flows over the surface of an impervious sub-soil. The tile line is placed approximately at the impervious layer along which the seepage water travels, so that water will be intercepted and wet condition is relieved. The tile line should be located in such a way that there is at least 60 cm of soil cover over the top of the tile.

Mole drainage system

Mole drainage is a semi-permanent method of sub-surface drainage, similar to tile drain in layout and operation. Instead of permanent tiles a continuous circular mole drain (channel) is prepared below the ground surface in the soil profile at desired depth and spacing using a special implement known as mole plough. The depth of the mole drain varies from 4.5 cm to 120 cm depending on the moling equipment and water table. Diameter of the mole varies from 7.5 to 15 cm. The life the mole drain is 10-15 years. It is adapted to a particular type of soil because the soil stability is more important in this type of sub-surface drainage.

Drainage coefficient

It is defined as the depth of water (cm) to be removed in 24 hours period from the entire drainage area. It ranges from 0.6 to 2.5 cm/day and in extreme cases 10 cm/day.

Biological Drainage

Bio-drainage technology helps in reclamation of salt affected waterlogged areas. Benefits include increased cropping intensity, increased nutrient use efficiency, growing arable crops including pulses and oilseed, which otherwise is not possible on waterlogged soils, and increased employment generation. The driving force behind the bio drainage concept is the consumptive water use of plants.

In natural environments the components of the hydrological system, i.e. rainfall, evapo-transpiration, change in soil-water storage and drainage, are in balance. Periods of high rainfall might temporarily result in increased drainage flows, a rise in the groundwater table and/or soil moisture storage, then over a period of about 5-10 years equilibrium is established. Vegetation plays a vital role in the evapo-transpiration and soil-water storage components of this balance.

When natural vegetation is cleared and replaced by crops or tree plantations, the seepage losses to the groundwater table under the new land use system are either higher or lower than under the pre-clearing situation.Plantations of fast growing tree species such as eucalyptus, when grown on previously cleared land, could result in strongly reduced accessions to the groundwater and in the drying-up of wells and springs.

The deep-rooting characteristics of trees make them extremely efficient users of water. The deep-rooted trees can access water tables down to several metres while shallow-rooted grasses and crops have limited access to underlying water tables, Also, in recharge situations with deep water tables, the deep root systems of trees greatly reduce the opportunity for rainfall/irrigation accessions to the water table.

One of the major factors determining the sustainability of plant water use issalt balance. If the salts moving into the rootzone are not either (i) taken up by the vegetation and harvested or (ii) removed from the rootzone by leaching, the vegetation is doomed to succumb to salinity. Where bio drainage results in salt accumulation, engineering assistance is needed to make the system sustainable. Bio-drainage management mechanisms can be classified as follows.

Dry land / rainfed systems

- Recharge control
- Groundwater flow interception
- Discharge enhancement

Irrigated systems

- Water table control
- Channel seepage interception
- Bio drainage cum conventional drainage systems

Biological Drainage

Bio-drainage technology helps in reclamation of salt affected waterlogged areas. Benefits include increased cropping intensity, increased nutrient use efficiency, growing arable crops including pulses and oilseed, which otherwise is not possible on waterlogged soils, and increased employment generation. The driving force behind the bio drainage concept is the consumptive water use of plants.

In natural environments the components of the hydrological system, i.e. rainfall, evapo-transpiration, change in soil-water storage and drainage, are in balance. Periods of high rainfall might temporarily result in increased drainage flows, a rise in the groundwater table and/or soil moisture storage, then over a period of about 5-10 years equilibrium is established. Vegetation plays a vital role in the evapo-transpiration and soil-water storage components of this balance.

When natural vegetation is cleared and replaced by crops or tree plantations, the seepage losses to the groundwater table under the new land use system are either higher or lower than under the pre-clearing situation.Plantations of fast growing tree species such as eucalyptus, when grown on previously cleared land, could result in strongly reduced accessions to the groundwater and in the drying-up of wells and springs.

The deep-rooting characteristics of trees make them extremely efficient users of water. The deep-rooted trees can access water tables down to several metres while shallow-rooted grasses and crops have limited access to underlying water tables, Also, in recharge situations with deep water tables, the deep root systems of trees greatly reduce the opportunity for rainfall/irrigation accessions to the water table.

One of the major factors determining the sustainability of plant water use issalt balance. If the salts moving into the rootzone are not either (i) taken up by the vegetation and harvested or (ii) removed from the rootzone by leaching, the vegetation is doomed to succumb to salinity. Where bio drainage results in salt accumulation, engineering assistance is needed to make the system sustainable. Bio-drainage management mechanisms can be classified as follows.

Dry land / rainfed systems

- Recharge control
- Groundwater flow interception
- Discharge enhancement

Irrigated systems

- Water table control
- Channel seepage interception
- Bio drainage cum conventional drainage systems

Bio-drainage in waterlogged areas for agricultural sustainability

Waterlogged and subsequent soil salinization are the two major causes of degradation of irrigated agricultural lands making it unfit for crop cultivation. Surface drainage is inefficient to combat the problem adequately. Subsurface Drainage Technology (SSDT) is an effective control measure for both. However, this technology is expensive and has an associated problem of safe disposal of the drainage effluents. Biodrainage is considered by some as an effective alternative, in which water and salt tolerant plant species are grown to remove the salt solutions from saline and waterlogged lands in an eco-friendly and economical manner. The saplings of Acacia biflora to determine the plant's salt uptake capacity from the root zone that was made saline by irrigation with water of different salt concentrations for a period of 6 months likewise Euclytusspp can draw substantial amount of groundwater and help land free from inundation. The consequent plant sap salinity was measured and salt balance analysis was carried out to ascertain salt migration from the soil to the plant. The study revealed that the plants irrigated with the water of highest salinity removed 28.7 g of salt in comparison to the plant under control, which removed 8.4 g of salts. Even the highest salt removal by Acacia biflora saplings was negligible as compared to that by conventional subsurface drainage system. To portray bio-drainage as a land reclamation option, confirmation of salt uptake from saline lands having established plantation needs to be carried out.

Keywords: Waterlogged situations, Salinazation and maintenance of water table and crop cultivations, sustainable agriculture; Water, irrigation, waterlogging, land degradation, salinity;

Background

India has got 17.2 million ha of low-lying area out of 38.0 million ha through out world and 75% of low land is situated in eastern India, Bangladesh, Myanmar and Thailand and rice is the predominant crop out of total waterlogged area in the country is more than 12 million hectare including salt affected waterlogged area.

The states under eastern part of the country in Assam, Orissa, West Bengal and Bihar remain inundated for a period during heavy downpour and excess rain in shorter period of time and unfit the land for crop cultivation. About 16.45 million ha of rainfed lowland in our country, wherein primarily rice is grown under several constraints.The flood plains of Brahmaputra and its tributaries in Assam, the Ganges and its tributaries in eastern UP, Bihar and West Bengal and the Mahanadi river system in Orissa constitute about 50% of the total rainfed lowlands in this zone (Zaman, 2012). Generally, three physiographic-hydrologic types of lands are recognized by pluvial, phraetic and fluxial (Mooremann and Breeman, 1978). Since pluvial lands have free drainage, the lowlands are either phraetic or fluxial. In phraetic lands, the water requirement of rice crop is met both by rainwater and groundwater. The water table remains shallow at least during most part of the growing/ cropping season. Water availability in the root zone during the growing cycle is influenced by the nature and distribution of natural rainfall, inherent soil characteristics and its management. In phreatic lands, rice could be grown with less drought risk. Fluxial rice lands receive water fully or partly from surface flow runoff or floodwater from rivers, streams. These lands remain inundated during the cropping season. The duration of flooding preludes the growing of other upland crops in the rainy season. In fluxiallands, floodwater availability depends on the rainfall pattern and the runoff from the catchment. Floodwater is not dependable source of water for growing rice in drought years. In wet years, excessive rains in the catchment area cause widespread flooding, damaging the standing crops.

In this context, research on bio-drainage for mitigation of both low water table as well as salinity problem is pivotal for sustainable agricultural productivity.

Characteristics of Water Logged Area

An area is said to be waterlogged when the water table rises to an extent that soil pores in the root zone of a crop become saturated, resulting in restriction of aeration, oxygen supply and increase in the level of carbon dioxide concentration, lack of decomposition of organic material and gaseous diffusion, nitrogen deficiency, fall of soil temperature and development of salinity and alkalinity. Working Group constituted by MOWR (1991) has suggested waterlogged area where water table is within 2 m of surface; partially waterlogged area where water table is between 2-3 m of surface and safe area where water table is below 3 m of surface. The resource poor farmers do subsistence farming. The whole agrarian society in the saline and / or waterlogged areas in respect of its social, political, economical and environmental aspects remains staggered and isolated from the rest of the farming community living in the non-waterlogged areas.

The malady due to waterlogging in the state is well known. In the Gangetic plains, one of the main reasons of the flood havoc and drainage congestion is the gradual encroachment of the natural flood plains for habitation and agricultural purposes and the same can not be stopped because of the complex socio-economic condition. In addition, all the major rivers flowing through the state raises outside the geographical boundaries, little progress can be achieved in building of storage works due to inter-state / inter-national aspects involved in the process. The main resources available to the state have, therefore, been to restrict its activities to flood protective measures by construction of flood embankments. But the escalating construction and maintenance cost of which is inevitably very high. Besides, the limitation in financial capability of the state also resulted not to undertake any long term measure to prevent the flood and other attendant problems. However, with the available resources, it has been possible to provide reasonable protection from flood and drainage congestion to more than half of the flood prone areas of the state (approximately 2.02 m ha) by undertaking various flood protection works, appropriate drainage and anti-erosion measures.

Importance of bio-drainage

Recent trends and future demographic projections suggest that the need to produce more food, fuel, fodder and fibre will necessitate proper utilization of land and other natural resources related to crop cultivation. To achieve this target the productive potential of our agricultural land and potential of irrigation have to be increased. Water is the most precious natural resource which is essential for growing crops. Crops receive water in the form of rain, but it does not always meet the full plant water requirements and, wherever possible, supplemental water is provided, which is in other words called irrigation. The economic yield of different crops grown under irrigated land is more and productivity is comparatively higher than those cultivated in rainfed areas. Irrigated agriculture, covering about 17 per cent of the total cropped area of the world, contributes 40 per cent of the total food production (INCID 2003). In India, only one-third area under irrigation produces two-third of the food grains.

The World Food Summit (1996) estimated that 60 percent of the extra food required to sustain the world in the future must come from irrigated agriculture. Irrigation of agricultural land has a long and well documented history. Irrigation will, without doubt, play an important role in keeping the future world population supplied with their food, fibre, bioenergy and bio-industrial feedstock needs. Irrigation has negative impacts also on our natural resource base, which often include shallow water tables. Waterlogging and increase in soil and groundwater salinity are associated with a lack of drainage. Irrigation potential has been created for increasing crop production by utilizing surface and sub-surface sources of water which ultimately caused a rise in the groundwater table followed by waterlogging and secondary soil salinization.

Development of irrigation potential generally upset the natural hydrological balance. In dryland agriculture, the introduced plants and crops rarely have the same rooting depth and annual evaporative potential as the natural vegetation they replace. In the case of irrigation, the component of applied water that is not used by the plants further adds to the water entering the water table. The hydrological changes caused by land use modification lead to changes in the salt-balance. Under rainfed conditions this often results in a lateral redistribution of salts in the landscape. Under irrigated conditions, the extra salts imported via irrigation water have to be removed from the root zone to avoid long-term accumulation. In the past, drainage has often been neglected, but it is now widely accepted with the increase of area under salt affected soils and now drainage has been emerged as an essential component to any irrigation system design. Recently, different detrimental impacts of improper drainage on soil and environment have been recognized.

Even though availability of water is essential for better productivity, but the excess water situation in agricultural field is a serious challenge to agricultural productivity in addition to negative ecological implications like green house gas emission from waterlogged areas and marshy lands. The seepage losses from the conveyance and along distribution system, excess water application in the crop field and recharge of ground water table, inadequate drainage etc. are among reasons for waterlogged condition in agricultural fields. The diversion of flood waters through existing canal systems to low lying areas causes large scale inundation of agricultural fields rendering them unproductive for agriculture. Sub-Himalayan plateau, Indo-Gangetic plains, deltaic Orissa are among major areas plagued with such type of problems due to waterlogging and sub-optimal agricultural productivity due to waterlogged situation. In India out of 8.4 m ha waterlogged area, about 2.46 m ha is waterlogged due to over irrigation and rest is due to run-off congestion.

Presently, about one-third of the world's irrigated area facing the threat of water logging, about 60 Mha has already become waterlogged and 20 m ha salt-affected (Heuperman et al. 2002). According to the estimate of the Ministry of Water Resources of the Government of India, in canal command areas of the country 2.46 m ha is waterlogged and 3.30 m ha salt-affected. India has the largest irrigated area in the world with an ultimate irrigation potential of 139.91 Mha. of which 98.84 m ha has been utilized by the end of tenth plan (MoWR, 207). However, unscientific use of water, along with other natural and manmade causes, leads to water logging, soil salinity and consequently, sub optimal agricultural production (Sarangi and Bundela 2011).

Currently at least 20 per cent of the world's irrigated land is salt affected and/ or irrigated with waters containing elevated levels of salts. Several major irrigation schemes have suffered from the problems of salinity and sodicity, reducing their agricultural productivity and sustainability. (Quadir*et.al.* 2008).

Several major irrigation schemes throughout the world have suffered from the problems of salinity (Gupta and Abrol 2000, Cai*et al.* 2003, Sarraf 2004). Introduction of canal irrigation has led to rising water table and consequent water logging and salinity problems (Kumar, 2004). Water logging is a wider problem in non-irrigated areas, where low-lying depressions serve as discharge areas, and on irrigated lands (Heuperman*et al.* 2002) and it caused salinization (Dubey 2012). In India also Salinization/alkalization and waterlogging have rendered a sizeable area of arable lands unproductive (Dwivedi 2006). Water logging is mainly due to seepage from canal, water stagnation on the surface and/or shallow groundwater levels. High rainfall in low land gradient areas also contributes to water logging.

Attempts made by various departments, organizations and agencies so far for prevention of flood and water logging and reclamation of the degraded lands have not been satisfactory in India. The problems of water logging and salinity can be effectively tackled by conventional engineering based subsurface drainage systems provided these are properly designed, installed, maintained and operated. Conventional subsurface drainage systems are of two types, vertical (tube wells) and horizontal (drain pipe) systems. When properly designed, installed and maintained, these systems can be efficient in lowering the water table and preventing water logging and associated salinization; but these are more expensive, cause environmental problems and they require periodic maintenance. Under such a situation, biodrainage is envisioned as a benign and cost effective technology to lower the rising water table so as to take it well below (>1.5 m) the root zone of crop plants (Angrish*et al.* 2008).

As the conventional technique is sub-surface drainage which is relatively expensive and generates harmful drainage effluents, the viable alternative could be biodrainage, which is simplt pumping of excess soil water by deep-rooted plants using bioenergy. In other words, it may be stated that biodrainage is the vertical drainage of soil water through evapotranspiration by vegetation. The term biodrainage is relatively new, although the use of vegetation to dry out soil profiles has been known for a long time. Biodrainage is economically attractive because it requires only an initial investment for planting the vegetation, and when established, the system could produce economic returns by means of fodder, wood or fibre harvested (Heuperman *et al.* 2002, Ram *et al.*, 2007 and Angrish *et al.*, 2008). There is consensus that biodrainage, when properly implemented, can lower the water table and it could solve problems associated with waterlogged areas and canal seepage. In a demonstratin it has been observed that under ideal conditions, a tree canopy may lower water tables by 1–2 m over a time period of 3–5 years (Kapoor 2002). Fast growing tree species may performed as effective Biodrainage System. The deep rooting characteristics of these trees make them extremely efficient users of water in compared to the crop plants. Fast growing tree species like *Eucalyptus*, known for luxurious water consumption under excess soil moisture condition, are suitable for biodrainage. Other suitable species for biodrainage may be *Acacia nilotica, Tamarixaphylla, Casuarinaglauca, Terminaliaarjuna, Prosopisjuliflora, Meliaazedarach, Pongamiapinnata*and *Syzygiumcuminii* and some more plants. These species can be planted in blocks in the form of farm forestry or along the field boundary in the form of agroforestry. In developing countries like India, since farmers have small holdings and cannot afford to put their whole chunk of land under tree plantations, Agroforestry can be a viable and remunerative option, which provides an additional income as tree biomass (timber, fuel wood etc), in addition to regular income from agricultural crop produce and serving the purpose of biodrainage. Many workers have recommended rehabilitation of such salt-affected, waterlogged lands through tree plantations having biodrainage quality (Bala *et al.* 2009, Ram *et al.* 2011, Roy Chowdhury*et al.* 2011).

Thevariouscomponentsofhydrologicalsystemslikerainfall,evapotranspiration by above-ground vegetation, storage of water in soil profile as well as drainage etc. exist in equilibrium in nature. High rainfall scenario results in short-term increase in surface run-off or drainage flow with rise in ground water table and increased storage of water in soil profile. However, above-ground vegetation has an important role in controlling moisture storage in soil profile due to evapotranspiration. The root systems of trees could intercept saturated zone or unsaturated capillary fringe above water table to control shallow water table. These plants are known as **phreatophytes.** The unsaturated top soil zones are intercepted by the trees by root systems of plants mainly following rainfall or irrigation to remove water from soil profile controlling recharge. Reliance on capability of vegetation to reduce water table has been reported promising (Chhabra and Thakur 1998, Naik and Manjunath 1999, Zang*et al* 1999, Stirzaker *et al* 1999). Several plants species are used for this purpose which are having the main physiological feature of such vegetation is profuse transpiration whenever the root system comes in contact with ground water. However compared to shorter vegetations, tall tree species have greater advantages like high aerodynamic roughness resulting into increased water loss through transpiration. This effect is more pronounced when trees are planted in a row to give clothesline effect (Chorley and Barry 1971). Moreover deeper root systems of trees could reach out for soil water more extensively at a greater depth for draining out waterlogged soils. Biodraiange potential also expectedly correlates significantly with parameters like stomatal density and transpiration rate. Very interestingly and paradoxically it has been discovered that parameters like tracheary element length correlates positively with biodrainage potential whereas the width does not. Actually the biodrainage potential of trees is an outcome of complex interaction of the several biotic and abiotic factors.

Benefits of bio-drainage system

- Involving certain categories of plants that habitually draw water from ground water
- Control waterlogging and Salinization of irrigated lands
- Evacuates undesirable and undrainable water with the help of transpiration from plantation trees
- Controlling of depth of groundwater table
- Help establish water and salt balance in the groundwater
- Acts as sub-surface drainage/ shallow tube-well/bio-pump
- Cost-effective and socially acceptable
- Environment friendly

Drainage and agricultural surface drainage in waterlogged areas

Drainage is the artificial removal of water in excess of the quantity required for the crop. Drainage includes removal of excess water of both surface and subsurface in the root zone of crops. Irrigation and drainage go together and are not mutually exclusive. Irrigation aims at supplying optimum quantities of water throughout the crop period, whereas, drainage aims at removing excess quantity of water in a short time. Often, both may be required together to assure sustained and high level production of crops.

Role of drainage

1. Draining the land provides conditions favorable for crop production;
2. The greatest benefit of drainage relates to aeration. Good drainage facilitates the ready diffusion of oxygen to the root zone and escape of carbon dioxide from the root zone into the atmosphere;
3. Several harmful gases also escape from the root zone into the atmosphere;
4. The activity of aerobic organisms which influence the availability of nutrients such as nitrogen andSulphur to plants depends on soil aeration and hence, drainage improves aerobic organisms;
5. Toxicity in acid soils due to excess iron and manganese is decreased by drainage (due to presence of oxygen in the root zone);

6. Drainage permits roots to grow deeper and spread wider thereby increasing the volume of soil from which nutrients can be extracted;
7. The removal of excess water helps in drying of the soil quickly and optimum soil temperature permits timeliness of field operations;
8. The provision of a good drainage system permits the removal of excess salts in the soil or irrigation water and prevents their build up in the upper soil layers.

Surface Drainage System

1. Random Drainage System
2. Parallel Field Drainage System
3. Parallel Open Ditch Drainage System
4. Bedding Drainage System

Sub Surface Drainage System

1. Tile Drainage System
2. Random Drainage System
3. Herringbone System
4. Gridiron System:
    1. Double Main System
    2. Intercepting System
5. Mole Drainage System

Drainage Coefficient

Defined as depth of water (cm) to be removed in 24 hours period from the entire drainage area (0.6 to 2.5 cm/day in extreme cases 10 cm/day)

Biological Drainage

Bio-drainage technology helps in reclamation of salt affected waterlogged areas. Benefits include increased cropping intensity, increased nutrient use efficiency, growing arable crops including pulses and oilseed, which otherwise is not possible on waterlogged soils, and increased employment generation. The driving force behind the bio drainage concept is the consumptive water use of plants.

In natural environments the components of the hydrological system, i.e. rainfall, evapo-transpiration, change in soil-water storage and drainage, are in balance. Periods of high rainfall might temporarily result in increased drainage flows, a rise in the groundwater table and/or soil moisture storage, then over a period of about 5-10 years equilibrium is established. Vegetation plays a vital role in the evapo-transpiration and soil-water storage components of this balance.

When natural vegetation is cleared and replaced by crops or tree plantations, the seepage losses to the groundwater table under the new land use system are either higher or lower than under the pre-clearing situation. Plantations of fast growing tree species such as eucalyptus, when grown on previously cleared land, could result in strongly reduced accessions to the groundwater and in the drying-up of wells and springs.

The deep-rooting characteristics of trees make them extremely efficient users of water. The deep-rooted trees can access water tables down to several metres while shallow-rooted grasses and crops have limited access to underlying water tables, Also, in recharge situations with deep water tables, the deep root systems of trees greatly reduce the opportunity for rainfall / irrigation accessions to the water table.

One of the major factors determining the sustainability of plant water use is salt balance.

If the salts moving into the root-zone are not either (i) taken up by the vegetation and harvested or (ii) removed from the root-zone by leaching, the vegetation is doomed to succumb to salinity. Where bio drainage results in salt accumulation, engineering assistance is needed to make the system sustainable.

Bio-drainage management mechanisms can be classified as follows

Drylands/ Rainfed systems

- Recharge control
- Groundwater flow interception
- Discharge enhancement

Irrigated systems

- Water table control
- Channel seepage interception
- Bio drainage cum conventional drainage systems

References

Angrish, R., Toky, O.P. and Patel, R.K. 2008.Biodrainage: potential and practice. National level training programme, Ministry of Water Resources, held at Directorate of Human Resource Management. CCS Haryana Agricultural University, Hisar.pp 1-99.

Bala, N.G., Singh, N.K. Bohra, and N.K. Limba. 2009. Increasing Productivity of Waterlogged Zone Of Canal Command Area In Indian Desert. (December):6–11. 60th 5th International Executive Council Meeting & Asian Regional Conference, 6-11, December 2009, New Delhi, India, 2009.

Bhadwal, S. and Singh, R. 2002. Carbon sequestration estimates for forestry options under different land-use scenarios in India. Curr. Sci. 83, 1380–1386.

Biodrainage for preventing waterlogging and concomitant wood yields in arid agro-ecosystem in

Cai X, McKinney DC, Rosegrant MW. 2003. Sustainability analysis for irrigation water management in the Aral Sea region. Agricultural Systems 76: 1043–1066.

Chhabra, R. and Thakur, N.P, 1998. Lysimeter study on the use of bio-drainage to control water logging and secondary salinization in (canal) irrigated arid /semiarid environment. Irrigation and Drainage System.12: 265-288.

Chorley, R.J and Barry, R.G.1971. Introduction to physical hydrology.Routledge (Taylor and Francis).ISBN 0416688101, 9780416688108, London.

DubeyKumud. 2012. SWOT analysis for the application of biodrainage technology to phytoremidiate water logged sites. International Journal of Social Forestry. 5(2): 47-59.

Dwivedi, R. S.2006. Study of Salinity and Waterlogging in Uttar Pradesh (India) Using Remote Sensing Data.Land Degrad. Develop. 5(3): 191 - 199.

Gupta R.K. and Abrol I.P. 2000. Salinity Build-up and Changes in the Rice-Wheat System of the Indo-Gangetic Plains.Experimental Agriculture 36: 273–284.

Heuperman, A.F., Kapoor, A.S. and Denecke, H.W. 2002.Biodrainage - Principles, Experiences and Applications.International Programme for Technology and Research in Irrigation and Drainage (IPTRID).Food and Agriculture Organization of the United Nations (FAO) publication, Italy, p. 79.

INCID. 2003. Bio-drainage - Status in India and other countries, Publication of the Indian National Committee on Irrigation and Drainage, New Delhi, India. Marine and Freshwater Research. 52: 41–52.

India 2007. Annual Publication by Ministry of Information and Broad casting, Publication and Information Bureau, GOI, New Delhi.

Kapoor, A. S. 2002. Biodrainage: A Biological Option for Controlling Waterlogging and Salinity, Tata McGraw-Hill, New Delhi, 2002, pp. 1–332.

Kumar, R. 2004. Groundwater Use in North- West India (eds. Abrol, I. P. et al.), Centre for Advancement of Sustainable Agriculture, New Delhi, 2004, pp. 1–26.

Naik, G.R. and Manjunath, G.T.S. 2000.Project (UKP) areas through bio-drainage concept.8th ICID International drainage workshop on Role of Drainage and challenges in 21st century. 31st Jan- 4th Feb. 2000. New Delhi.

North Western India.Journal of Scientific and Industrial Research. 70: 639-644.

Qadir M., A. Tubeileh, J. Akhtar, A. Larbi, P. S. Minhas and M. A. Khan.2008.Productivity enhancement of salt-affected environments through crop diversification.land degradation & development. Land Degrad. Develop. 19: 429–453.

Ram, J., Dagar, J. C., Lal, K., Singh, G., Toky, O. P., Tanwar, V. S., Dar, S. R. and Chauhan, M. K.. 2011. Biodrainage to combat waterlogging, increase farm productivity and sequester carbon in canal command areas of northwest India. Curr. Sci. 100 (11): 1673-1680.

Ram, J., Garg, V.K., Toky, O.P., Minhas, P.S., Tomar, O.S. Dagar, J.C. and Kamra, S.K. 2007. Biodrainage potential of Eucalyptus tereticornisfor reclamation of shallow water Table areas in north-west India.Agroforest. Syst. 69: 147-165.

Roy Chowdhury, S., Kumar, Ashwani, Brahmanand, P.S., Ghosh, S., Mohanty, R.K., Jena, S.K., Sahoo, N. and Panda, G.C. 2011. Application of Biodrainage for Reclamation of Waterlogged Situations in Deltaic Orissa.Research Bulletin no. 53.Directorate of Water Management (Indian Council of Agricultural Research).Chandrasekharpur, Bhubaneswar-751023, India, p. 32.

Sarangi, A. and Bundela, D. S. 2011. Decision Support Systems in Water Resources Management - A Review. Technical Bulletin - 2011.Water Technology Centre.Indian Agricultural Research Institute. New Delhi, p. 50.

Sarraf, M. 2004. Assessing the Costs of Environmental Degradation in the Middle East and North Africa Countries.Environmental Strategy Notes (No. 9), Environment Department, World Bank: Washington, DC.

Stirzaker, R.J. Cook, F.J. Knight, J.. 1999. Where to plant trees on cropland for control of dryland salinity: some approximate solution. J. Agric. Water Management.39 : 115-133.

Toky O.P., Angrish,R., Datta, K.S., Arora, V., Rani, C., Vasudevan, P. and Harris P.J.C. 2011.

Zang, L., Dawes, W.R, Slavich, P.G., Meyer, W.S, Thorburn, P.J. Smith, D.J. and Walker, G.R. 1999. Growth and ground water uptake responses of Lucerne to changes in ground water levels and salinity: lysimeter, isotope and modelling studies. J. Agric. Water Management. 39: 265-282.

Multiple Choice Questions

1. What is the primary purpose of agricultural drainage?
   a) To increase soil salinity
   b) To remove excess water from the soil
   c) To enhance crop yields only in dry regions
   d) To provide irrigation systems

2. What are the three agro-climatic zones classified for agricultural drainage?
   a) Tropical zone, Arctic zone, and Desert zone
   b) Temperate zone, arid and semi-arid zone, and humid and sub-humid zone
   c) Equatorial zone, Coastal zone, and Highland zone
   d) Urban zone, Industrial zone, and Forest zone

3. Which system is NOT part of surface drainage?
   a) Random Drainage System
   b) Tile Drainage System
   c) Bedding Drainage System
   d) Parallel Field Drainage System

4. What is the drainage coefficient?
   a) The rate of salt removal from the root zone
   b) The depth of water to be removed in a 24-hour period
   c) The amount of water absorbed by plants
   d) The total rainfall in a growing season

5. What is a common biological drainage method?
   a) Construction of canals
   b) Planting deep-rooted trees to lower water tables
   c) Surface irrigation with excess water
   d) Implementing crop rotation
6. Which type of drainage is typically used in humid and semi-humid regions?
   a) Open drainage systems
   b) Sub-surface drainage systems
   c) Tile drainage systems
   d) Random drainage systems
7. Why is bio-drainage considered environmentally friendly?
   a) It increases soil erosion
   b) It allows for chemical applications
   c) It uses plants to remove excess water and salt
   d) It requires heavy machinery for installation
8. Excess water in the crop root zone can lead to which of the following?
   a) Enhanced nutrient uptake b) Improved soil aeration
   c) Plant wilting and reduced yields d) Increased microbial activity
9. What is one of the major factors determining the sustainability of plant water use in bio-drainage?
   a) Soil temperature regulation b) Salt balance in the root zone
   c) The depth of crop roots d) The variety of crops grown
10. Which is a negative outcome of poor drainage in agricultural fields?
   a) Increased crop yields
   b) Accumulation of toxic gases in the root zone
   c) Enhanced soil nutrients
   d) Improved water quality

Answer Key

1	b	2	b	3	b	4	b	5	b	6	a	7	c
8	c	9	b	10	b								

15

Water Management of Different Crops

Water management of different crops (rice, wheat, maize, groundnut, sugarcane, mango, banana and tomato) and implementation of on-farm water management:

1. Rice
2. Wheat
3. Maize
4. Groundnut
5. Sunflower
6. Mustards
7. Pulses
8. Sugarcane
9. Cotton
10. Tomato
11. Potato
12. Mango
13. Banana

Introduction

Surface irrigation is the distribution of water using open canals in which water flows by gravity. This type of irrigation, which includes flood irrigation, requires relatively flat fields and deep soils.

A good irrigation will ensure a uniform and homogenous distribution of water in small and regular doses. The status of moisture in the soil also aims at maintaining the relative humidity in a plot at a stable level, to limit the development of weeds and pathogens. Irrigation is also used to avoid the transmission of phyto-sanitary problems or the leaching of pesticides.

The volumes of water to apply depend on the climatic conditions, the frequencies of the applications, the soil type and the irrigation method. The choice must take into account the efficiency of each method, the water availability, the installation and running costs, the maintenance needs, the spatial requirements and whether it's possible to work while the field is irrigated.

Water Management Practices of Different Crops Rice

Total water requirement is 1100-1250

The daily consumptive use of rice varies from 6-10 mm and total water is ranges from 1100 to 1250 mm depending upon the agro climatic situation. Of the total water required for the crop, 3% or 40 mm is used for the nursery, 16% or 200 mm for the land preparation i.e. puddling and 81% or 1000 mm for field irrigation of the crop.

The growth of rice plant in relation to water management can be divided into four periods viz.,. Seedling, vegetative, reproductiv and ripening. Less water is consumed during seedling stage. At the time of transplanting, shallow depth of 2 cm is adequate and maintained upto 7 days and there after 5 cm of submergence is necessary to facilitate development of new roots. The same water level is required for tiller production during the vegetative phase. At the beginning of the maximum tillering stage the entire water in the field can be drained and left as such for one or two days which is termed as mid season drainage. This mid season drainage may improve the respiratory functions of the roots, stimulate vigorous growth of roots and checks the development of non-effective tillers. Any stress during the vegetative phase may affect the root growth and reduce the leaf area.

During flowering phase 5 cm submergence should be maintained because it is a critical stage of water requirement. Stress during this phase will impair all yield components and cause severe reduction in yield. Excess water than 5 cm is also not necessary especially at booting stage which may lead to delay in heading;

Water requirement during ripening phase is less and water is not necessary after yellow ripening. Water can be gradually drained from the field 15-21 days ahead of harvest of crop. Whenever 5 cm submergence is recommended the irrigation management may be done by irrigating to 5 cm submergence at saturation or one or two days after the disappearance of ponded water. This will result in 30% saving of irrigationwater compared to the continuous submergence.

Groundnut

Total water requirement 500-550 mm

Evapotranspiration is low during the first 35 days after sowing and last 35 days before harvest and reaches a peak requirement between peg penetration and pod development stages. After the sowing irrigation the second irrigation can be scheduled 25 days after sowing i.e. 4 or 6 days after first hand hoeing and thereafter irrigation interval of 15 days is maintained upto peak flowering. During the critical stages the interval may be 7 or 10 days depending upon the soil and climate. During maturity period the interval is 15 days.

Finger millet

Total water requirement: 350 mm

Finger millet is a drought tolerant crop. Pre-planting irrigation at 7 or 8 cm is given. Third day after transplantation life irrigation with small quantity of water is sufficient for uniform establishment. Water is then withheld for 10-15 days after the establishment of seedling for healthy and vigorous growth. Subsequently three irrigations are essential at primordial initiation, flowering and grain filling stages.

Sugarcane

Total water requirement: 1800-2200 mm

Formative phase (120 days from planting) is the critical period for water demand. To ensure uniform emergence and optimum number of tillers per unit area lesser quantity of water at more frequencies is preferable. The response for applied water is more duringthis critical phase during which the crop needs higher quantity of water comparing, the other two phases. Water requirement, number of irrigations etc., are higher during this period. As there is no secondary thickening of stem, elongation of stem as sink for storage of sugar it is desirable to maintain optimum level of moisture during grand growth period. Response for water is less in this stage and this will be still less in the ripening stage. During the ripening phase as harvest time approaches soil moisture content should be allowed to decrease gradually so that growth of cane is checked and sucrose content is increased.

Sugarcane

Sugarcane is long duration crop; duration varies from 12 months in N and 15 – 18 months in S.

Early attempts have been made to decide to determine WR of the crop through transpiration and ET losses. Water requirement ranged from 114 cm to 228 cm; Water requirement was considered 125 cm in N to 304 cm in S. Wwith one year and a half variety, it might be 330 cm.

Advantages over drip irrigation

1 Saves water;
2 Enhances plant growth and yield;
3 Saves energy and labour;
4 Most suited for soil having low water holding capacity and undulating terrain;
5 Reduces weed growth;
6 Improves fertilizer application efficiency;
7 Improves quality of produce;
8 Reduces salt concentration in the root zone;

Maize

Total water requirement: 500 – 600 mm

The water requirement of maize is higher but it is very efficient in water use. Growth stages of maize crop are sowing, four leaf stage, knee high, grand growth,tasseling, silking early dough and late dough stages. Crop uniformly requires water in allthese stages. Of this, tasseling, silking and early dough stages are critical periods.

Cotton

Total water requirement: 550 – 600 mm

Cotton is sensitive to soil moisture conditions. Little water is used by plant with early part of the season and more water is lost through evaporation than transpiration. As the plant grows, the use of water increases from 3 mm / day reaching a peak of 10 mm a day when the plant is loaded with flowers and boll. Water used during the emergence andearly plant growth is only 10% of the total requirement. Ample moisture during flowering and boll development stages is essential. In the early stage as well as at the endthe crop requires less water. water requirement remains high till the boll development stage. If excess water is given in the stages other than critical stages it encourages the vegetative growth because it is a indeterminate plant thereby boll setting may be decreased. Irrigation is continued until the first boll of the last flush opens, and then irrigation is stopped.

Cotton

Cotton farming: About 20 million tons of cotton are produced each year in around 90 countries. China, United States, India, Pakistan, Uzbekistan and West Africa account for over 75% of global production.

Cotton has around 59% share in the raw material consumption basket of the Indian textile industry. It plays a major role in sustaining the livelihood of an estimated 5.8 million cotton farmers and about 40-50 million people engaged in related activities, such as cotton processing and trade. India has the largest cotton cultivated area which constitutes about 30% of the global cotton area. The planting period in India takes place from March to September, while harvesting takes place from October to February. Cotton is produced in 3 zones, the Northern zone, comprising the states of Punjab, Haryana and Rajasthan, the Central zone, comprising Maharashtra, Madhya Pradesh and Gujarat and the Southern zone, comprising Andhra Pradesh, Karnataka and Tamil Nadu.

Worldwide about 73% of cotton is produced in irrigated fields and only 27% under rain-fed conditions (WWF 1999). While enhanced irrigation is critical for increase in production of cotton, it also leads to the extraction of significant amounts of surface and ground waters. As reflected in the table above, over the years an increasing amount of surface water has been withdrawn and diverted for the production of cotton. In India, wherever irrigation is being provided to cotton, the irrigation system in general uses traditional technique of flood irrigation, which is low in efficiency. Moreover, the irrigation schedule does not depend on the actual crop water demand, but rather on the availability of water supply. Hence, inefficient irrigation system and unsustainable water use, particularly in the northern region of India, where cotton is grown under irrigated condition, has several environmental implications. Rivers are being diverted and dammed for large-scale irrigation projects. This has significant downstream effects, including water shortages with adverse impacts on wildlife and water availability for human consumption. Therefore, it is essential to integrate sustainable and efficient utilization of water in cotton production.

Sorghum

Total water requirement: 350-500 mm

The critical periods of water requirement are booting, flowering and dough stages. The crop will be irrigated immediately after sowing. Next irritation is given 15 days sowing to encourage development of a strong secondary root system. irrigation prior to heading and ten days after heading are essential for successful crop production.

Pulses

Total water requirement – 200-450 mm

Mostly the pulse are grown under rainfed condition. Some pulse crops like Redgram, Blackgram, Greengram are grown in summer season as irrigated crop which need 3 to 4 irrigation at critical stags like germination, flowering and pod formation.

Mango

Two representative stations namely Dehradun and Pantnagar, lying in two main mango-growing belts, were selected for the estimation of reference evapotranspiration under per-humid and moist sub-humid, agro-climatic regions of Uttarakhand. FAO Penman-Monteith model was used for the estimation of reference evapotranspiration on weekly meteorological data of ten years. The estimated reference evapotranspiration values were further used for the determination of water requirement of different age groups of mango crop under drip irrigation.

The weekly average value of water requirement of mango crop under drip irrigation varied from 2.11 l/day/plant to 139 l/day/plant in per-humid and 2.45 to 197 l/day/plant in sub humid climatic conditions of Uttarakhand.

On an average, it was found that average annual value of water requirement varied from 2.11 to 197 l/day/plant. It was also found that the water requirement of mango crop increases with the increase in the age of mango tree and becomes constant when the full canopy has been developed at the age of about 30 to 35 years of mango trees.

Mango is one of the most important fruits of India. It is the choicest fruit and often known as the king of fruits. It is grown in 39.16 per cent of the total area under fruits and contributes to 23.09 percent of the total fruit production in the country. The state Andhra Pradesh ranks first with respect to area and production. However, the state Karnataka has the highest productivity. In West Bengal, it covers an area of 62,500 ha with the average production of 0.380 mt and productivity of 6.1 mt/ha. Mango is a rich source of vitamin A and has a fairly good content of vitamin C. It also contains a good amount of minerals particularly potassium. Mango fruits are used for preparation of pickle, chatani, amchur, jam, squash, nectar and many other delicious products.

The daily water requirement of mango varies from 28 to 85 l/d/plant. The total annual water requirement of the crop is 6400 m^3/ha for the planting geometry of 5 m x 5 m. Soil moisture stress from end November to end January is recommended for good flowering. Regular irrigation during fruit setting and development period reduces fruits drop considerably.

Banana

Banana is one of the most commonly grown fruit crop of the country. India produces about 26.217 Mmt of Banana from an area of 0.709 Mha with an average productivity of 37.0 mt/ha. Major producing states are Tamil Nadu, Maharashtra, Karnataka, Gujarat, Andhra Pradesh, Assam and Madhya Pradesh. Tamil Nadu has 0.1244 mha under Banana and the total production during 2008-09 is 6.667 Mmt with 53.6 mt/ha productivity. In Gujarat, banana crop is cultivated in 11 districts covering an area of about 60900 ha. Gujarat is ranking 3rd among the states of India with an average productivity of 58.7t/ ha. Unripe fruits are used for making chips, vegetable flour *etc.* Ripened fruits are used for preparing drink, jam, beer, salad *etc.* Banana fruits are having numerous medicinal uses (As per Aryuveda). From psedostem threads are obtained and used for making rope and cloths.

The water requirement of banana crop by surface method of irrigation (25-30 irrigations) is 1500-1800 mm. By drip irrigation method, water requirement is 900-1080 mm. The drip system should be operated on alternate day for a period of 1.05 to 2.25 hrs during winter and 2.5 to 2.75 hrs during summer at a pressure of 1.2 kg/cm^2.

Moisture management

After sowing of the rainfed crops, inter-culture operations like light hoeing is very essential in cotton and cotton-based cropping systems, to break the upper crust of the soil and improve aeration in the soil to improve available water in the soil for proper plant stand and good growth. It is also essential to keep field free from weeds before application of fertilizer to cotton crop.

Under rainfed situation, moisture management is very important and various moisture conservation practices as per plan were followed on different toposequences.

Irrigation calendar based on IW/CPE ratio

Crop	Date of planting	Days of irrigation after sowing							
		1st	2nd	3rd	4th	5th	6th	7th	8th
Wheat	1st Nov.	21	47	77	100				
	15th Nov.	24	53	80	106				
	1st Dec.	27	50	75	94				
Potato	1st Nov.	7	20	33	51	69	84	98	108
	15th Nov.	5	15	29	47	64	79	91	103
	1st Dec.	5	16	34	52	67	79	90	107
Mustard	15th Oct.	23	53	89					
	1st Nov	27	58	93					
Groundnut	1st Feb.	31	53	67	84	100			
	15th Feb.	26	47	66	79	96			
	1st March	23	42	59	72	89			
Maize	1st Nov	14	31	50	75	87	105		
	15th Nov	17	38	62	87	104			
	1st Dec.	20	42	67	87	100			
Cabbage & Cauliflower	1st Sept	43	63						
	15th Nov	28	48	77					
	1st Oct.	13	20	29	39	50	64	78	
Onion	1st Nov.	14	31	50	75	87	105		
	15th Nov.	17	39	62	87	104			
	1st Dec	20	42	67	87	100			

PULSE CROPS

Pigeon pea (22 cm); Green gram (20 cm), Black gram (20 cm), Cowpea (20 cm), French bean (25 cm), Peas, Chick pea (30 cm); Field Pea (20 cm); Lentil (12 cm), Cluster bean (25 cm)

OILSEED CROPS

Groundnut (25 cm), Mustard (20 cm), Sunflower (25 cm)

Rice (120 cm), Wheat (35 cm), Maize (40 cm), Cotton 950 cm)

Vegetable Crops

Tomato (35 cm), Potato (30 cm)

Water requirement of crops

S.No.	Corps	Duration in days	Water requirement (mm)	No. of irrigations
1.	Rice	135	1250	18
2.	Groundnut	105	550	10
3.	Sorghum	100	350	6
4.	Maize	110	500	8
5.	Sugarcane	365	2000	24
6.	Ragi	100	350	6
7.	Cotton	165	550	11
8.	Pulses	65	350	4

Research based findings of different water management practices of different field crops at various agro-climatic zones

Old Alluvial Zone

The water management technologies developed under specific situations of Damodar Valley Corporation (DVC) command are given hereunder.

Transplanted Rice

a) **Irrigation and Nitrogen:** Levels of irrigation during rainy season did not influence the grain yield of rice under normal and uniformly distributed rainfall. However, the highest value of water use efficiency (WUE) was obtained with the irrigation 512 cm, three days after disappearance of ponded water. The crop responded to nitrogen, phosphorous and potassium up to 80, 60 and 120 kg/ha, respectively as revealed from the results of consecutive three years of field experimentation.

b) **Percolation under various tillage operations:** The percolation rates under tractor ploughing showed always least followed by mouldboard and *deshi* plough during crop growth of *aman* paddy. A dry tillage one month before puddling did not have any effect on percolation rates during crop growth period as well as on yield.

c) **Summer Rice (Transplanted):** The continuous submergence of 5±2 cm water gave the highest grain yield of rice with 100 kg N/ha.

However, the WUE increased with lesser number of irrigation and was found maximum when irrigation was given 5 days after disappearance of ponded water.

Potato

Three irrigations, one each at weekly interval before furrow making followed by three irrigations one each at 20 days intervals (IW/PE of 0.9, depth 5 cm) was the best irrigation schedule for potato.

Mustard

The grain yield of mustard was recorded highest when 2-irrigation of 5.0 cm depth each applied at flowering (33 DAS) and pod setting stage (55 DAS) with 80 kg N/ha. However, WAUE decreased with increasing levels of irrigation.

Winter Maize

The grain yield of maize performed better when irrigation was applied at IW/PE of 1.2 with 5 cm irrigation compared to irrigation at 0.6 IW/PE. The crop significantly responded to increasing levels of nitrogen up to 160 kg N/ha. Water used and Water Use Efficiency values were also found to respond accordingly.

Groundnut (Summer)

Crop responded well with increasing the irrigation frequencies up to 0.9 IW/PE ratios. However, the crop required no irrigation at the later half of the crop growth period due to pre-monsoon shower.

Cauliflower

The curd yield of Cauliflower was increase 23% while the crop was irrigated at IW/PE of 1.2 at the irrigation depth of 5 cm each which required 5 irrigations at about 15 days interval.

Onion

Three irrigations at IW/PE of 1.2 ratio (5 cm depth each) found optimum for growing onion after rice. The crop was transplanted in December.

Crop Sequences

Rice-Potato-*Moong* showed the maximum net return while it was minimum with Rice-Mustard-Sesame sequence.

Summer Rice

Intermittent ponding at early stage (15-40 DAT) followed by continuous ponding did not reduce the grain yield of rice significantly and saving of 15 cm irrigation water could easily be done. Though, the continuous submergence of ponded water gave maximum grain yield of summer rice.

Rice-wheat Sequences

Practice of shallow intense puddling of rice followed by normal tillage of 15 cm depth by tractor drawn cultivator gave maximum levels of productivity of wheat crop gave maximum yield up to 31.66 q/ha with a water use efficiency of 0.755 kg/ha/cm.

Wheat

Irrigation at IW/CPE of 1.2 (having irrigation depth of 5 cm) for wheat (cv: UP 262) with 100 kg N/ha gave significantly highest grain yield of 28.75 q/ha.

Papaya under drip irrigation

Drip irrigation at 80% E0 recorded optimum green fruit yield 43.47 t/ha of papaya (cv: Honey Dew) with a saving of irrigation around 50% as compared to surface irrigation.

Brinjal under drip irrigation

Gravity drip irrigation gave 2 to 3 folds saving of irrigation water. Irrigation to the crop at 125%ET crop gave highest yield (22.08 t/ha) (cv: *Muktakeshi)* with a saving 37% of irrigation water over surface irrigation.

Coriander

Four irrigations, one each at an interval of 12-16 days (y 0.03 Mpa) with 60 kg N/ha gave significantly highest yield of 7.94 q/ha and water use efficiency value of 2.52 q/ha/mm.

Tomato

Irrigation at an interval of 10-14 days with potassium application @80 kg K_2O/ha, tomato (cv: *Sadabahar*) gave maximum yield of 313 q/ha with water use efficiency values of 1.22 q/ha/mm. If irrigation applied at 15-20 days interval, potassium to be increased from 80 to 120 kg K_2O/ha.

Cabbage

Irrigation at 20 mm cumulative pan evaporation (CPE) with 120 kg N/ha found best to give highest head yield of cabbage at a tune of 41.92 t/ha with water use efficiency value of 1.8 t/ha/cm.

Chilli

Three irrigations one each at 1.2 IW/CPE (5 cm irrigation depth) was found optimum to get highest yield of green chilli at a tune of 5.87 t/ha with water use efficiency of 44.54 kg/ha/cm.

Sugarcane under drip irrigation

Drip irrigation at 0.8 E0 was found optimum for cane yield (78.77 t/ha) with 225 kg N/ha where the water use efficiency recorded 8.13 t/ha/cm.

Red and Laterite Zone

Maize

The crop was introduced first as an alternative to rainy season rice under upland condition in drought prone laterite tract of West Bengal. There was no need of irrigation due to adequate rainfall during *kharif*. The crop responded nitrogen up to 100 kgN/ha.

Groundnut

The crop sown during rainy season under upland situation responded up to three irrigations at 0.3 atm tension of 15 cm soil depth and significantly responded to phosphorous application up to 90 kg P_2O_5/ha.

Rice (Direct Seeded)

Drilled seeded rice (cv. *Dular)* had no effect of irrigation due to adequate rainfall but the water use efficiency was increased with nitrogen application up to 80 kg N/ha.

Wheat

1. Wheat (cv. *Sonalika*) was found to increase with increasing levels of irrigation. Four irrigations, one each at panicle initiation, flowering and grain filling stages in addition to crown root initiation stages were found optimum. The crop responded to N-application up to 120 kgN/ha.
2. Straw mulch application @ 5 ton/ha increased the grain yield and WUE of wheat over un-mulched wheat. There was about 25% yield increase due to mulching over the grain yield of 13.26 q/ha under un-mulched.

Safflower

Safflower (cv. A-300) was found superior amongst all the varieties test with three irrigations one each at branching, flowering and seed development stages along with 120 kg N, 60 kg P_2O_5 and 40 Kg K_2O/ha.

Indian Rape

Indian rape requires two irrigations one each at branching and flowering along with 80 kg N/ha that gave highest grain yield of Indian rape.

Horsegram (*Kulthi*)

Growing horsegram with straw mulch ((@ 5 ton/ha)) reduced irrigation requirement of horsegram (*kulthi*) and increased grain yield and water use efficiency significantly.

Crop Sequences

Maize-Safflower and Groundnut-Safflower were proved best and remunerative in the region. Grow second crop of safflower, mustard, wheat and kulthi were possible with limited or residual soil moisture.

On-farm water management research

On-farm water management research in the study area revealed several key findings. During the rainy season, rice yields were comparable between traditional farmer practices and experimental plots, likely due to adequate rainfall. However, aus paddy yields were significantly higher in experimental blocks, highlighting the impact of controlled water management. Fertilizer response varied with submergence depth; farmers' practices of applying heavy fertilizer doses with increased water depth proved less effective.

Farmers readily adopted potato cultivation using high levels of fertilizers and irrigation. However, interventions successfully reduced fertilizer and irrigation inputs without compromising yield; in fact, yields often improved. Recommended water management practices for mustard cultivation increased yield by up to 9% and improved water use efficiency by 36%. Furthermore, the introduction of water management practices for summer-grown groundnut led to increased yields and improved water use efficiency.

These results underscore the importance of adopting efficient water management technologies to maximize crop production and bridge the gap between irrigation potential and actual utilization. Minimizing application losses through improved conveyance, irrigation methods, and scheduling is crucial. Crop selection and sequencing, along with appropriate agronomic practices, also play a significant role in enhancing water use efficiency. Effective on-farm water management encompasses conservation, distribution, and utilization of irrigation water, optimum irrigation scheduling, suitable irrigation methods, and conjunctive use of water sources, improved agro-technology adoption, and proper drainage. Applying the right amount of water at the right time maximizes water use efficiency and crop yield while reducing evaporation and deep percolation.

Efficient irrigation scheduling with limited water availability remains a significant challenge for irrigation experts, requiring further research. This aligns with existing research which characterized irrigation systems as complex socio-technical systems necessitating interactions between physical, environmental, and technological factors. Participatory action research, involving users in the process of change, has also proven effective in optimizing these systems. However, many farmers are hesitant to adopt irrigation scheduling technologies unless they perceive a clear net benefit. Studies encouraging on-farm participation and demonstrating the cost-benefit approach to irrigation scheduling have shown promise in promoting adoption. Similarly, involving farmers in water management practices and introducing acceptable water-saving irrigation schedules have led to improve on-farm irrigation management and increased agricultural production.

Numerous on-farm water management research activities, involving collaborative projects between research institutions and governmental bodies, have been conducted across various irrigation commands, deploying both package interventions and component technology field trials. These participatory field trials have demonstrated positive impacts on water use efficiency by adopting recommended water management practices in farmers' fields. Similar successes in efficient soil and water management for enhancing agricultural productivity have been reported, with on-farm water management consistently proving more effective than on-station field trials.

System as a whole approach

A. Rotational water supply

Rotational water supply is one of the techniques in irrigation water distribution management. It aims at equi-distribution of irrigation water irrespective of location of the land in the command area by enforcing irrigation time schedules.

Each 10 ha block is divided into 3 to 4 sub units (irrigation groups) According to the availability of irrigation water, stabilized field channels and group-wise irrigation requirement, time schedules are evolved. The irrigation will be done strictly in accordance with the group-wise time schedules by the block committees. Within the group, the time is to be shared by the farmers themselves.

The possible solutions and recommendations

1 Development of micro-level water resources (tank-well system, small springs, perennial and seasonal streams) with a command of 2-10 ha to reduce dependency on major and medium irrigation projects through R & D projects should be explored;

2 Effective harvesting of excess rainwater during heavy to medium rainfall period as well as artificial recharge of groundwater should fully be exploited wherever possible through participatory research and development along with mass awareness programme;

3 Planned intervention is required to reduce the negative effects of surplus groundwater exploitation by two-fold actions: (1) to control

excessive draw down and (2) to prevent water quality deterioration and degradation.

4 The gap between technology generation and adoption in the farmers' field is to be minimized; The methods, systems, scheduling of irrigation based on different approaches, its transformation in farmers' language, water-harvesting structures and water-lifting devices, water measurement should come into farmers' practices.

Several innovative water management techniques can improve irrigation efficiency for non-rice crops. These include:

- **Paired Row Technique:** This method involves planting crop rows in pairs on either side of a shared furrow, effectively irrigating two rows with a single furrow. Studies on crops like greengram, blackgram, groundnut, and sunflower have shown that this technique can save approximately 20% of irrigation water and increase crop yields by 15%. In Coimbatore district, a paired row system for cotton saved 29% of irrigation water while maintaining similar yields to conventional furrow systems, resulting in a water use efficiency of 31.1 kg/ha-cm.
- **Alternate Furrow Irrigation:** Irrigating only alternate furrows can significantly reduce water usage. For example, this method saves 34.1% of irrigation water in sugarcane compared to irrigating all furrows in sandy clay loam soils of Coimbatore. Similar water savings have been observed in other crops:
    - **Chillies:** Alternate furrow irrigation with 10 t/ha of coir pith application saves 30.8% of irrigation water compared to all furrow irrigation.
    - **Groundnut:** Growing groundnut in ridges and furrows under well-irrigated conditions saves 24-27% of irrigation water compared to check basin irrigation.
    - **Brinjal:** Alternate furrow irrigation saves 24% of water compared to normal farmer practices.
    - **Tomato (PKM1):** Alternate furrow irrigation saves 34% of irrigation water compared to all furrow irrigation and 55% compared to check basin irrigation, without significantly impacting yields.
    - **Turmeric:** An improved irrigation practice using a 5 cm depth of water and 10 tons/ha of coir pith as mulch saves up to 44% of water compared to normal farmer practices (ridges and furrows).
- **Gradual Widening Technique:** In banana cultivation, irrigating at 1.0 IW/CPE ratio (once in 7 days) from 0-7 months and 1.2 IW/CPE ratio

(once in 5 days) from 7-14 months recorded higher mean yield of 32.7 t/ha with an increased fruit yield of 2.1 t/ha compared to basin irrigation at IW/CPE ratio throughout the crop growth period, besides higher WUE and saving of 140 mm of water. By gradual widening of ring basins from 30x60, 45x90 and 60x120 cm for 0-50, 51-100 and 101-150 days respectively followed. By basin method there was a saving of 25.4% of water and increased WUE of 25.9% over check basin.

- **Surge Irrigation:** This relatively new method involves applying water in intermittent surges, which can significantly improve water distribution and reduce losses. Surge irrigation offers several benefits:
  - Easy operation and efficient manipulation of surge flow rates.
  - Deep percolation losses along furrows are minimized (less than 5% compared to over 25% in continuous flow systems).
  - Irrigable area per unit time increases by nearly 1.5 times.
  - High irrigation and water use efficiencies (14 kg/ha/mm of water with maize).
  - Water savings of 40-60% and reduced labor (30-45 man-hours per ha compared to over 60 man-hours per ha in continuous flow systems).

Crop-Specific Irrigation Recommendations

The following are the farmer's practice and improved technology options for the following crops:

- **Sorghum:** Farmers typically use a flat bed system with irrigation based on weather and visual assessment. Improved options include furrow irrigation once in 15-16 days during the first 20 days after sowing, followed by six irrigations at 6-day intervals. Ratoon sorghum benefits from six irrigations at rationing, 4-5 leaf stage, milking, soft dough, and hard dough stages. Surge irrigation is feasible for long furrows (>100 m) on level land.
- **Pearl Millet:** Traditional methods involve beds and channel irrigation. Optimizing irrigation using an IW/CPE ratio of 0.75 at 4 cm depth is recommended.
- **Finger Millet:** Farmers often use flat bed systems with irrigation based on weather and visual assessment. Irrigating with an IW/CPE ratio of 0.75 at 4 cm depth is an improved option.
- **Maize:** The typical farmer's practice is a flat bed system with irrigation based on weather and visual assessment. Irrigating the field at 10-day intervals is a recommended improvement.

- **Pulses (Blackgram, Greengram, Soybean):** Farmers generally use beds and channels with excessive irrigation. Improved practices include irrigating blackgram and greengram at critical stages (sowing, flowering, and pod formation) with a 4 cm depth, with an optimum irrigation interval of 18 days. For soybean, irrigation at 80 mm CPE once every 11-12 days is recommended.
- **Groundnut:** Traditional methods involve beds and channels with irrigation based on weather and visual assessment. Irrigation at sowing, establishment, and 25 days after sowing is recommended, with an optimum irrigation interval of once every 7-9 days.
- **Gingelly (Sesame):** Farmers typically use a flat bed system with copious irrigation. Irrigating at flowering and capsule formation stages is recommended.
- **Sunflower:** A flat bed system is the common farmer's practice. Improved options include irrigation at an IW/CPE ratio of 0.75 with 20:30:20 Kg of NPK / ha, and surge irrigation under long furrow in level lands.
- **Coconut:** Farmers generally use check basins with copious irrigation. Improved options include drip irrigation at 100 liters of water per tree per day. For stress management, palm basins should be opened to a radius of 1.8m with late showers, and mulching can be done. Other practices include husk mulching, application of green manure and FYM, spreading dried coconut leaves and organic residues, adding tank silt to increase water retention, removing lower senescent leaves during drought, and using pitcher irrigation where water is scarce.

Crop-Specific Irrigation Practices

- **Cotton:** Farmers typically utilize beds and channels for irrigation. Improved techniques include sowing seeds in ridges and furrows, irrigating at an IW/CPE ratio of 0.75, mulching with sugarcane trash (5 t/ha), and spraying Folicot, paraffin wax (10g/L), or kaolin (50g/L). Sprinkler and drip irrigation are also viable options.
- **Banana:** Traditional irrigation involves trench and mound methods. Recommended practices include irrigation at an IW/CPE ratio of 0.75-0.9, or the chain basin method, forming basins around suckers connected by channels. Drip irrigation with high-density planting and fertigation are preferred in well-irrigated areas. Furrows can be gradually widened as the crop develops.
- **Acid Lime:** Farmers commonly use basin irrigation. Improved technology includes drip irrigation at 75% of the water supplied through

basins; drip irrigation is generally preferred.

- **Tomato:** Traditional methods use beds and channels. Recommended options include furrow irrigation and drip irrigation, particularly with micro-sprinklers for hybrid varieties. Irrigation at an IW/CPE ratio of 1.00 during fruit formation and ripening is crucial. Sprinkler irrigation (using 1760 m3 of water) has demonstrated higher water use efficiency compared to other irrigation methods.
- **Sugarcane:** Farmers often apply excessive irrigation via ridges and furrows. Improved practices include irrigation at an IW/PE ratio of 0.9, mulching with sugarcane trash (especially in garden land) to reduce evaporation, and foliar application of kaolin (12.5 kg in 750 liters of water per hectare) to reduce transpiration. Removal of old, dried leaves in 5-7 month old sugarcane plants, alternate or skip furrow irrigation, and irrigating based on sheath moisture percentage is also beneficial. In deep trench planting (30cm deep trenches at 80 cm apart), sets are planted in trenches. Drip irrigation with fertigation is highly suitable, and surge irrigation can be implemented in long fields with light-textured soils.

Problems and Solutions for Non-System Tank Irrigation Systems

Detailed studies have revealed several issues in non-system tank irrigation as follows:-

1. **Encroachment, Siltation, and Pollution:** Encroachment, siltation, and soaking of supply channels impair water inflow. Pollution from tanneries and dyeing factories further degrade water quality (Coimbatore, Erode, and Salem Districts).
2. **Disappearing Tank Chains:** Hydrological interlinking of tank chains has almost disappeared. Reviving tanks and improving their connections can restore benefits through appropriate modernization.
3. **Inefficient Water Supply:** Due to various factors, only 50-60% of the water supply is realized. Crop diversification strategies with non-rice crops are recommended.
4. **Rehabilitation Strategies:** Solutions include de-silting to restore original capacity, planting vegetation on the tank foreshore to control silt flow, connecting small tanks to percolation ponds for groundwater recharge, and rehabilitating tank structures and inflow channels.
5. **On-Farm Development:** Strengthened on-farm development structures are needed to ensure equitable water distribution from head to tail of tank command areas, with active farmer participation.

6. **Poor Quality Water Management:**
   - Conjunctive use of relatively fresh surface water and poor-quality groundwater in appropriate proportions is suggested.
   - Growing salt-tolerant crops in saline irrigation areas with proper drainage is recommended.
   - Community bore wells can enhance crop water availability and yields during periods of erratic canal water supply.

References

Bouman, B.A.M., & Tuong, T.P. (2001). "Field water management to save water and increase its productivity in irrigated rice." Irrigation Science, 20(3), 161-169. https://doi.org/10.1007/s002710100008

Eisa, M. A., & El-Shafie, A. (2018). "On-farm water management for banana crops in arid regions." Water Resources Management, 32(7), 2337-2351. https://doi.org/10.1007/s11269-018-1963-2

Hanjra, M.A., & Qureshi, M.E. (2010). "Global water crisis and its impact on food security." Food Security, 2(1), 1-16. https://doi.org/10.1007/s12571-010-0001-5

Kamble, B., & Ghosh, S. (2020). "Water use efficiency in wheat: A review of the current and future perspectives." Agricultural Water Management, 240, 106286. https://doi.org/10.1016/j.agwat.2020.106286

Kaur, R., & Bansal, S. (2018). "Water management practices for rice cultivation in Punjab, India." Journal of Agricultural Sciences, 10(2), 123-130. https://doi.org/10.5539/jas.v10n2p123

Kumar, P., & Kaur, K. (2021). "Influence of irrigation methods on water productivity of mango (Mangifera indica L.) in tropical conditions." Agricultural Water Management, 250, 106853. https://doi.org/10.1016/j.agwat.2021.106853

Patil, G.B., & Wagh, S.B. (2017). "Water management in groundnut: A study on irrigation scheduling techniques." Indian Journal of Agricultural Sciences, 87(3), 377-384. https://doi.org/10.56093/ijas.v87i3.68237

Sharma, S. K., & Gupta, R. K. (2019). "Water management strategies in sugarcane: A case study from Maharashtra, India." Sugar Tech, 21(2), 231-243. https://doi.org/10.1007/s12355-018-0631-8

Singh, D., & Kumar, A. (2019). "On-farm water management practices for sustainable maize production in North India." Journal of Soil and Water Conservation, 74(2), 169-183. https://doi.org/10.2489/jswc.74.2.169

Singh, R., & Prakash, A. (2020). "Irrigation practices and water management in tomato cultivation: A review." International Journal of Chemical Studies, 8(4), 1862-1868. https://doi.org/10.22271/chemi.2020.v8.i4z.9303

Multiple Choice Questions

1. What is the total water requirement for rice cultivation?

   a) 1000-1100 mm b) 1100-1250 mm

   c) 1500-1600 mm d) 2000-2100 mm

2. During which phase should 5 cm of submergence be maintained for the flowering stage of rice?
   a) Seedling b) Vegetative
   c) Reproductive d) Ripening
3. What percentage of total water requirement is used for land preparation in rice cultivation?
   a) 3% b) 16%
   c) 81% d) 40%
4. What is the total water requirement for groundnut?
   a) 250-300 mm b) 400-450 mm
   c) 500-550 mm d) 600-650 mm
5. In finger millet, how long should water be withheld after seedling establishment for healthy growth?
   a) 5-7 days b) 10-15 days
   c) 20-25 days d) 30-35 days
6. How many irrigations are recommended for cotton crop during its growth?
   a) 8 b) 10
   c) 11 d) 15
7. What is the total water requirement for sugarcane?
   a) 1000-1500 mm b) 1800-2200 mm
   c) 2200-2500 mm d) 1500-1800 mm
8. At what stage of maize cultivation is the peak water requirement observed?
   a) Sowing b) Four leaf stage
   c) Tasseling and silking d) Late dough
9. How much water can be saved through proper irrigation management in rice cultivation?
   a) 20% b) 30%
   c) 40% d) 50%
10. What is the average daily water requirement for mango plants under drip irrigation in sub-humid conditions?
   a) 2.11 l/day/plant b) 28 l/day/plant
   c) 60 l/day/plant d) 139 l/day/plant

Answer Key

1	b	2	c	3	b	4	c	5	b	6	c	7	b
8	c	9	b	10	a								

16

Participatory Water Management

Participatory water management, principles of sustainable farming and crop water productivity and strategic water management practices

Participatory irrigation management

Farmers' participation in management of water resources is one of the prime requirements to improve the irrigation efficiency and enhance food production to meet the future requirement. The objectives participatory irrigation management (PIM) are to achieve equity in water distribution, make best use of rain, surface and ground water conjunctively, develop community responsibility to collect water charges and create a sense of ownership of water resources and the irrigation system among the users so as to promote economy in water use and preservation of system, improve service deliveries through better operation and maintenance and achieve optimum utilization of available resources.

Participation of farmers in the realm or irrigation water management implies an active role of beneficiaries in all the facets of irrigation water management and its attendant linkages with the main system of management in agricultural/ agronomic activities. This role is very different from the traditional passive role of farmers to look to the department for irrigation, for water supply and its distribution. The timing of farmers' involvement is crucial. Farmers' participation is most effective, when it takes place from initial stages of project planning, including the stage of project formulation. Such involvement forms part of ideal conditions for a partnership between farmers and government. The present theme "more crops per drop" is feasible provided the farmers are encouraged and supported to play an active role in the operation and management of the irrigation system. However, to operationalize this concept, there is a need to involve all the farmers within the command area, so that collective and collaborative action becomes possible.

Importance of Participatory Water Management

The per-capita availability of cultivable land as well as water is going to be stagnated, wherein irrigation is the only key factor leading to increased crop and water productivity through increasing cropping intensities and incorporating more profitable cropping patterns accessible to the small holders. Indian agriculture has got resilience with introduction of fertilizers and irrigation responsive HYVs during mid-70s'. Wherein, water is considered as major indicator of agricultural growth for livelihood, environmental and socio-economic security of the country. The integrated management of water resources could only be possible through adoption of efficient and optimum use of irrigation water, which could only be ensured by judicious and economic use of irrigation potential whatsoever created to increase crop production. The integrated water resources management also includes the concept of rainwater management that has got an immense important on the way to develop the rainfed farming system. In the areas of medium to heavy rainfall, there is ample scope of tapping excess rain water through suitable water harvesting structures constructed for this purpose for its subsequent uses as irrigation or to apply lifesaving irrigation to the crops which should also come under the purview of water management for crop production. Farmers' participatory based action research programme on adoption of water saving technologies so far generated for increasing crop production could be useful for maintaining sustainable growth and development of agriculture. The on-farm studies in these aspects showed that there were positive impacts of irrigation water on sustainable agricultural development as regards to increased crop productivity, cropping intensity, input use efficiency, farmers' income, employment generation and poverty alleviation from disadvantaged rural sects. The availability of irrigation facilities enabled the farmers to raise at least one extra crop a year with consequent implications for household food security with higher yields of the crops grown along with employment opportunity per unit of arable land.

Irrigation potential is major indicator of agricultural growth that summarizes its strong positive correlation with poverty alleviation. The availability of cultivable land is exhausted, wherein irrigation helps in increasing crop and water productivity through increasing cropping intensities providing more profitable cropping patterns accessible to the small holders. The integrated management of water resources could only be possible through adoption of efficient and optimum use of irrigation water, which could only be ensured by judicious and economic use of irrigation potential whatsoever created to increase crop production. The integrated water resources management also includes the concept of rainwater management that has got an immense important on the way to develop the rainfed farming system. Previously, rainfed agriculture was associated primarily with conservational land treatments to check soil erosion and land degradation. But in the areas of medium to heavy rainfall, there is ample scope of tapping excess rain water through suitable water harvesting structures constructed for this purpose for its subsequent uses as irrigation or to apply lifesaving irrigation to the crops which should also come under the purview of water management for crop production. The basic objective of the Act is to enhance livelihood security in rural areas by providing at least 100 days of guaranteed wage employment in a financial year to every household whose adult members volunteer to do unskilled manual work. This work guarantee can also serve other objectives: generating productive assets, protecting the environment, empowering rural women, reducing rural-urban migration and fostering social equity, among others that were considered under National Rural Employment Guarantee Scheme (NREGS) 2005 of the Government of India.

Objectives of Participatory Water Management

To establish linkage between irrigation water management and poverty;

To assess the role of irrigation management in poverty alleviation, the linkages and pathways through which irrigation contributes in poverty alleviation and the magnitude of anti-poverty impacts of irrigation and to find out key-determinants factors to enhancing anti-poverty impacts of irrigation; and in a word, a number of water management technologies for field crops have been developed over years to enhance agricultural production. The area of crop(s) under irrigation was also increased substantially. The irrigation potential was created at the cost of huge amount of national exchequer. But there is still a large gap between the irrigation potential created and utilized. Likewise the gap between technologies generation and adoption are also being widening at threshold level. That stands the water management technologies developed so far were not also being practiced at the actual users' end. There are gaps in the existing system, which acts as constraints on adoption of water management technologies.

To address the problems and to bridge the gap, the constraints analysis on issues and priorities of water management would be helpful at end users levels on adoption of water-efficient management practices for growing field crop(s). Particularly wherein the necessity of food for ever-growing population will require irrigated agriculture and more than 80% of available water is being used as irrigation.

Water resource management is a multidisciplinary activity, managing irrigation water will need agronomy and crop husbandry, efficient methods and system of irrigation needs soils scientists and engineers. Effective minimization of wastage of water used for irrigation and application of right quantity of water at right time will be the key to successful management of this crucial resource.

More than 98% of the irrigated lands are under the coverage of surface irrigation where more than 50% of water as considered as wastages. The pressurized irrigation (drip and sprinkler) methods have become the tool of water saving technologies to get rid of such situations.

A series of field experiments on field crops were conducted to find out the optimum water and nutritional requirement of different filed crops and was under operation up to 1985. The project then shifted to Research Farm at Memari under Old Alluvial Zone in 1985 to carry out water management research on different field crops to cater the needs of Damodar Valley Corporation (DVC) command and continued to work up to 1995. The project was again shifted and started its functioning from the headquarter of the Viswavidyalaya at Central Research Farm, Gayeshpur to carry out research work on water management of different field crops, including fruits, vegetables and flowers grown under different deep tube well command in the district of Nadia, West Bengal under assured supply of water in New Alluvial Zone since 1995.

Observations

A large numberof site specific and cost effective water management technologies as regards to suitable method and optimum scheduling of irrigation for various field crops, conjunctive use of rain/surface and ground water for crop cultivation and water economy, implication of drainage in monsoon season to restore the soil workability condition for sustainable crop production, some unorthodox measures for applying water through PVC pipes to avoid application and conveyance losses, on-plot conservation of rain water and subsequent uses in crop production programmes were developed in the controlled condition experimental field trials and appeared to be fruitful for achieving higher yields, saving a significant quantity of valuable water, a greater employment opportunity and higher farm family income. With a view to adopt these proven water management technologies in the actual farming condition, a number of interventions or its modified forms as applicable in the specific farming situation were tried in the fields dominated by the resource poor farmers for adoption with a view to exploit the valuable water resources for remunerative crop yields and also to elevate the socio-economic status of the farming community. The technologies are found accepted and adopted by the farmers in their fields are as follows:

The results of the experimentation got relevancy particularly in installing small-scale irrigation system based on farmers' participatory approach for sustainable crop production for maintaining sustainable growth and development of agriculture. Although there is a difference in the objectives of water management where there is abundant availability of water from continuous flow connected with perennial sources or lakes and rivers and that of from restricted or limited supply from collection of excess rain through water harvesting structures. The former is associated with to obtain as much as yield of crop per unit area under water application and later is associated with increasing productivity of crop per unit of water application even under rainwater recycling and its effective management for the purpose of irrigation.

Researchers and planners have studied in depth the impacts of irrigation water on crop productivity, cropping intensity, input use, farmers' income and employment generation and poverty alleviation. The availability of irrigation facilities has enabled the farmers to raise nearly one extra crop a year with consequent implications for household food security with higher yields of the crops grown along with labour employment per unit of land. The opportunities of rural development has thus been assigned with possibilities of water management interventions through adoption of National Rural Employment Guarantee Scheme of The Government of India creating and assuring at least 100 man days for single farmer a year keeping pace with rural development.

On-farm Water Management

Adoption of proven water management technologies ensures to achieve high water productivity in agriculture. Trist (1980) considered the irrigation system as a complex socio-technical phenomenon that involved interaction between physical, environmental and technological parameters in the process of adoption. Whyte (1991) considered the participatory action research as more effective and communicative measures in the developmental processes. Korten (1980) described the participatory action research as a method of merging both developmental intervention and research activities as an effort to involve the users in systemic process of change to make the socio-technical system more effective and efficient. Xianjun (1996) reported that on-farm irrigation management and irrigation service thereby improved by involving farmers in water management practices and introducing an acceptable water-saving irrigation schedule for growing crops and increasing agricultural production. Pramanick and Mallick, (1996) described on-farm participatory field trials brought positive improvements towards better water use efficiency following the recommended methods of water management practices in the farmers field in DVC command in West Bengal. Dhindwal and Poonia (1994) also reported the success story of on-farm water management. Zaman and Gayen (2013) stated the adoption of rainwater harvesting with existing ponds ensured more water resource availability and utilization in coastal saline region of Sagar Island. Under this stated situations, 4 (four) proven water management technologies were considered for demonstration in farmers field for growing crops in Andaman & Nicobar Island to combat water scarcity ensuring effective water resource utilization.

Planning for Proper Water Resource Utilization

So proper planning for effective water utilization is need of the hour. The network of media (radio, television, newspaper and journal-magazine) and information technology have to play vital to disseminate the proper education on water uses, water-saving technologies generated for agricultural purpose including pressurized irrigation system (drip, sprinkler and subsurface). The proven and need-based technologies developed in the field should acts as practical tool and media should act as communication channel forecasting the alarming situations out of water. The planners, researchers and policy makers should come into hand-to-hand to bridge the gap between technology development in the field and adoption in actual field condition. As because more than 90% of such generated technologies, so far, at the cost of heavy national exchequer, have not been implemented by actual users. Posters, brochures, banners, hoardings, exhibitions, seminars and symposiums are to be organized more to create effective awareness campaigning towards more and more peoples' participation and that is the only the key to save the nations from anticipated devastated situations that would come out of water scarcity. The water managers, researchers and policy makers will have a crucial role to play for creating efficient strategies for equitable sharing of water minimizing possible conflicts and unintended consequences.

India is the seventh largest having 329.73 million ha of geographical area and second most populous country having more than a billion population in the world. There is no other substitute than to use every square inch of land and every drop of water effectively and efficiently. Though there was markedly increase in annual food grain production from 51 (1950-51) million tones to 206 million tones (2000-01), still the dearth on hunger is visualized. The demand for food is increasing with ever-increasing population growth rate. So there is no option other than to increase food production. There is very little or no scope of horizontal expansion due to land limitation, the question of vertical expansion is the only way to fit with the situation. Genetically improved crop variety, hybrids, mechanized agriculture, higher rate of fertilizers and chemicals application may give one or two more green revolution, but the question lies how far it can sustain greenness? Adoption of improved crop cultivars, cropping system, increasing cropping intensities, precision farming can hold good to get out of this situation where whole system is to be considered. In this context, participatory research with active partnership of farmers is considered as key approach for sustainable agriculture and food security and that has successfully been executed over a period of time in Indo-Gangetic plains of West Bengal in Eastern Region of India for selection of appropriate rice cultivars, the staple food crop of the region *vis-à-vis* other technology development activities.

Participatory Research on Zero Tillage

Delayed monsoon caused *kharif* rice transplantation is usually severely affected the *rabi*cropping where the time didn't allow land preparation to keep pace with available residual soil moisture. Growing wheat under such rice based crop sequences in lowland eco-system is thus, might be good proposition under zero/minimal tillage operation which not required any further time for land preparation while was fully recharged with moisture and optimal for seed germination. So sowing crops under minimal tillage/zero tilled condition is better option for growing crops like wheat and mustard. Wheat sowing was made possible by zero tilled seed drill machine in low land situation where sufficient residual soil moisture persists for seed germination. After *kharif* rice harvest wheat was sown on 5.12.07 with 'Pantnagar Zero tilled fertilizer Seed Drill 2001 Model'.. Seed rate was high @ 150 kg/ha and complex fertilizer Suphala (15:15:15) @ 150 kg/ha was used to get the ratio of 1:1 of seed and fertilizer. Seed and fertilizer calibration were done before sowing of wheat. Seed cv. UP 262 was sown at the depth of 5-7 cm. Soil moisture content at 0-30 cm depth varied with the field to field ranging from

26.15 to 32.17% during sowing of wheat in typical clay soil. Depth of irrigation for zero tillage (ZT) wheat was 4 cm (3 hours/bigha) against the conventional tillage (CT) 6 cm (4.5 hours / bigha). Three irrigations were applied for ZT and CT at CRI stage, end of maximum tillering and grain filling stage. Nitrogen @ 60 kg / ha was top-dressed in the form of urea in two splits – one at CRI and 2nd at maximum tillering stage. Crop received 90.4 mm rainfall during growing period. Date of 50% flowering was at mid February and wheat harvesting was completed at the end of March.

Results: Zero-tilled (ZT) wheat with three irrigations recorded the highest grain yield of 1900 kg/ha as compared to lowest grain yield of 1200 kg/ha. The mean of 18 farmers' yield was 1536 kg/ha. The conventional tillage (CT) recorded the highest grain yield of 2800 kg/ha and the lowest grain yield 2500 kg/ha with three irrigations. Conventional tillage in clay soil was not promosing due to drainage congestion masde by irrigation applied particularly at CRI stage of wheat which resulted yellowing of leaf at early stage. Water use efficiency of ZT wheat was 8.30 to 11.60 kg/ha/mm with three irrigation CT with three irrigations that recorded the highest WUE of 12.20 kg/ha/mm.

Farmers' acceptance for growing wheat under zero tilled conditions was mainly attributed due to higher benefit cost ratio obtained than that of conventional tillage.

Table 16.1: Performance and water use efficiency of wheat under ZT situation

Item	**Zero Tillage**			**Conventional ***
Farmer	Case 1	Case 2	Case 3	Average
Area covered (ha)	0.13	0.06	0.09	0.13
Plant height (cm)	67.6	71.6	59.6	68.3
Tillers/ m row	66.1	51.2	32.8	45.6
Panicle length (cm)	12.6	10.8	6.9	9.4
Flag leaf length (cm)	19.7	19.7	16.8	17.6
Flag leaf width (cm)	1.86	1.9	1.5	1.7
Grain yield (q/ha)	24.6	18.2	11.6	20.2
No. of irrigation	3	3	2**	3
Water used (mm)	226	214	189	243
WUE (kg/ha/mm)	8.71	8.50	6.16	7.20

*Two passes of harrowing done by Tractor & seeds were broadcasted, ** No irrigation at CRI

Performances of zero-tilled wheat in low land rice ecosystem

After *aman*(winter) rice harvest wheat was sown on 28.11.05 with 'Pantnagar zero tilled fertilizer seed drill 2001. Seed rate was high @ 150 kg/ha and complex fertilizer Suphala (15:15:15) @ 150 kg/ha was used aiming seed : fertilizer ratio of 1:1. Seed and fertilizer calibration were done before sowing of wheat. Seed cv. UP 262 was sown at the depth of 5-7 cm. Soil moisture content at 0-30 cm depth varied from 24.5 to 27.6% during sowing of wheat in typical clay soil. Depth of irrigation for zero tillage (ZT) was 4 cm (3 hours/bigha) against the conventional tillage (CT) 6 cm (4.5 hours / bigha). Three rrigations were applied at CRI, maximum tillering and flowering stages. Nitrogen @ 60 kg / ha was top dressed in the form of Urea in two splits – one at CRI and 2nd at maximum tillering stage. Date of 50% flowering was 10/2/06 wheat harvest was done on 25.3.06. On-farm trials were conducted in three farmers' field in the DTW command.

Results

In general wheat sowing in rice based crop sequence is very much delayed beyond November due to delayed harvesting of the crop. Wheat sowing was possible by zero tilled seed drill machine in low land situation where sufficient residual soil moisture persists for seed germination. Performance of zero tilled wheat was promising in clay soil. Crop growth was much better under zero tillage. Three irrigations under ZT situation of Farmer AjoyGhosh recorded maximum grain yield of 24.6 q/ha which was 21.8% yield increase over conventional tillage with 3 irrigations, whereas, two irrigations reduced the grain yield by 111.8% under ZT situation. Conventional tillage in clay soil was not promising due to drainage congestion made by irrigation at CRI stage which results yellowing of leaf at early stage. Water use efficiency was higher (8.5 to 8.71 kg/ha/mm) under ZT with 3 irrigations over ZT two irrigations or 3 irrigations at CT.

Participatory Water Management Research on Rice during *kharif* season (Drum Seeder)

Drum seeding of wet rice was done during *kharif* season. Land situation was medium upland having good drainage after seeding. Shallow puddling was done for drum seeding in clay soil. One pass of Power tiller followed by a bullock drawn country plough was used for puddling. Seeds were soaked for 24 hours and incubated for sprouting (0.5 cm). Philippines 12-row plastic drum seeder was used. Calibration of seed dropping was done through closing the excess whole by polythene tape. Two farmers participated in the adaptive trial. Crop sequence was Rice-rice. Performances of wet direct seeding (DR) was compared against transplanted rice (TR).

Table 16.2: Crop performances of drum seeding during kharif season

Study items	**Farmer: Rabi Pal**		**F2: ProdyutGhosh**	
	DR	**TR**	**DR**	**TR**
Area (m^2)	22 X 19	22 X 19	39 X 41	39 X 41
Date of sowing	27.7.06	27.7.06	27.706	27.7.06
Variety	Ratna	Ratna	IET 4094	IET 4094
Line spacing (cm)	20 X conti	20 X 15	20 X conti	20 X 15
Seed rate (kg/ha)	40	40	40	40
NPK (kg/ha)	60:30:30	60:30:30	60:30:30	60:30:30
Date of harvesting	2.11.06	8.11.06	2.11.06	8.11.06
Crop performances				
Plant height (cm)	85	81	92	86
Effective tillers/m^2	381	303	346	325
Panicle length (cm)	21.9	23.1	20.5	21.3
Panicle weight (g)	1.65	1.79	1.46	1.56
Grain yield (t/ha)	4.30	3.53	3.90	3.40
Straw yield (t/ha)	6.05	4.70	5.40	4.85
Harvest Index (%)	41.7	42.8	41.9	42.8

Observations

Performance of wet seeded rice in medium land situation of the command was good as compared to transplanted one. Yield increased to the tune of 22.8 and 14.7% in case of variety Ratna and IET 4094. Plant growth and yield parameters also supported the fact. Wet seeded rice matures one week earlier than transplanted one. Farmers are not motivated with the drum seeding due to high rainfall during July and August and there is a problem of plant emergence due to stagnation of water in the field after wet seeding of sprouted seeds. Drum seeding may be advocated at les rainfall area and in upland area.

Demerits

Calibration of seed dropping is essential to maintain the exact seed rate of rice other wise it cause less or more plant population. Rainfall after seeding may cause harm but 2 days after seeding rain will not damage crop emergence. Weed infestation will be more if puddling is not done properly. Tillering was less but the higher population per unit area due to drum seeding compensated it. Disease pest infestation was noted higher in wet seeded crop.

Residual moisture utilization through *paira* cropping in rice with mustard

Aman rice is generally harvested in the middle of November. Brown mustard (rai) and toria (rape seed) were broadcasted over the standing rice 10-15 days before harvest of winter rice. Soils were saturated during seeding.
Complex fertilizer like Suphala @ 150 kg/ha was applied at time of seeding, 10 days ahead of rice harvest. Seed rate was @ 7.5 kg/ha. Irrigations were given at maturity for toria and branching and siliqua maturity for mustard. Mustard and toria were harvested on 5/3/06 and 3/2/06 respectively. Two farmers participated in the adaptive research. Area covered of 0.13 ha each.

Table 16.3: Performances of mustard and toria as paira crop

Item	Mustard	Toria
Farmer	JoydevGhosh	BasudevGhosh
Date of seeding	3//11/05	5/11/05
Seed rate (kg/ha)	7.5	7.5
Irrigation	23/12, 15/1	23/12
Plant height (cm)	96.3	56.0
Branching	5.4	2.6
Siliqua /plant	68.5	21.5
Seed yield (q/ha)	11.5	7.5
Water used (cm)	17.8	12.5
WUE (kg/ha/mm)	6.46	6.0

Observations

*Paira*cropping of mustard and toria was very promising for residual moisture utilization after rice harvest where irrigation water supply was limited. One and two irrigations were sufficient for toria and mustard. Water use efficiency of mustard and *toria*varied from 6.0 to 6.46 kg/ha/mm.

Participatory Research for Evolving High Value Rice Cultivars

The massive impact of plant breeding in productivity increase became evident by the middle of the sixth decade of the last century. Modern plant breeders work in their experimental farms with set conditions. They evolve new varieties that are then tested in several experimental farms over a state or across the country. The best performing lines are then tested by on-farm trials in farmers' fields. In this process the involvement of farmers who are the actual users of the new variety is at minimum. As a result, the new variety most often fails to satisfy the demands of farmers. The problem is more pronounced due to this technology adoption gap in West Bengal representing widely diverse agro-climatic-edaphic and socio-economic regions. In order to circumvent this shortcoming the concept of farmers' participation at the early stage of plant breeding programme is being preferably conducted. In this programme, the farmers select plants or lines from segregating population of crosses raised in research farm and they carry forward their own selections in their respective zones. The major benefit accrued from this technique is the wide diversity of lines each selected to suit the specific requirements of the farmers in a given zone. The usual shortcoming of a plant breeder who often fails to envisage the zone specific adaptive features while selecting from his breeding population(s) is obviated by farmers' participation at early stage. The farming community knows local resources, markets, socio-economic situation and needs. The community also has the indigenous technology knowledge for post harvest value addition in women dominated cottage industry chains. Organizations based breeders generally ignore such micro-economic parameters.

Fifteen farmers from eight districts of West Bengal, covering seven distinct agro-ecological zones participated in this programme. The farmer selected single plants from F_3 population of a rice cross involving *Govindabhog* (scented, small grain, local cultivars) and IET 13544 (*Basmati*) grown in Research Farm of BCKV at Chakdah. The details of the selection protocols were exercised.

The plants selected by farmers revealed widely different preferences. It is evident that the farmers selected plants that, in their opinion, most suited the ecological conditions in which they cultivated rice. The most significant conclusion drawn from this experiment is that it is well nigh impossible for the plant breeder working in a research farm to envisage the widely diverse ecological conditions prevalent in West Bengal and to anticipate the farmers' distinctive preferences in order to accommodate these in his selection criteria. That farmers' participation at the early stage of plant breeding programme can speed up evolution of eco-specific lines has become quite clear in the present study. Four lines of semi-dwarf high yielding small grain aromatic rice have been selected and are being utilized by the farmers themselves.

Participatory Varietals Evaluation for Selection of Summer Rice Cultivars

In West Bengal, there were no specific summer rice varieties and the farmers usually grow early duration long slender recommended winter rice cultivars under all the *boro* rice situations. Depending upon the topography of lands, there is a need for selection of specific summer rice cultivars for this situation. In this context, farmers can choose from a set of established varieties by growing them in their own fields for two or more seasons. The participating farmers themselves are convinced about the merit of their own selected variety or varieties and hence, further extension work on the varieties can be seduced. The study aimed at examining the significance of the participatory system of adoption of rice cultivars under specific agro-ecological situations. The comparative performance of ten rice genotypes, including nine established varieties and one as per farmers choice, were taken for demonstration on 5 (five) farmer's fields in alluvial zone during rainy and summer seasons.

Nine established varieties of rice, IR 36, IR 64, IR 72, IR 68305, Iron 4, Ajaya, IET 9994, PSBRC 4, IET 4786 (as Check) and experimental selection Gontra I (pure line selection from a Vietnamese variety – made by the members of Gontra Farmers' Cooperative Society, Chakdah, Nadia were used for on-farm trial (in farmers' fields) at Gontra, Chakdah under the guidance of the Scientists of the RRS, New Alluvial Zone. The plot size was 5m x3m and row-to-row distance was 20 cm. Single seedling transplantation was done. The experiment was in randomised block design with 3 replications. Standard cultural practices were adopted. The experiment was conducted in three consecutive seasons, *Boro* 1997-98, *Kharif* 1998 and *Boro* 1998-99.

The mean grain yield per plot of these ten entries were IR36: 10.03 kg (6866 kg/ha), IR 64: 8.90 kg (5933 kg/ha), IR 72: 6.20 kg (4133 kg/ha), IR 68305: 9.50 kg (6333 kg/ha), IRON 4: 9.80 kg (6533kg/ha), Ajaya: 10.10 kg (6733 kg/ha), IET9994: 8.80 kg (5866 kg/ha), PSBRC4: 9.86 kg (6533 kg/ha), IET 4786 (Check): 6.50 kg (4333 kg/ha) and GOTRA I: 11.80 kg (7866 kg/ha), respectively. *Gontra I* turned out as the best performer. This line is now in high demand in the state. *Gontra I* is presently in IVT at the national level. *Gontra II* is in the pipeline.

Participatory Community based Seed Production (Seed Village) Approach

Seed is the basic input of agriculture and productivity is largely influenced by timely availability of quality seeds to the farmers. After completion of NSP I, NSP II and NSP III, only 25-30% quality seeds are being supplied by the public and private seed industries. The gap between demand and supply of quality seeds is the major constraint in enhancing productivity. To overcome the problem, it is imperative to produce situation specific quality seeds through active participation of the farmers.

Hence, a demonstrative model for community seed production has been established and is in operation in the village, Gontra, district of Nadia where farmers' Cooperative Society acted as an apex body and scientists of BCKV and State Department of Agriculture are working as technical experts. Selected farmers from different villages under the society are the seed growers. The work was initiated in 1997 *kharif* season with 22 farmer participants and an area of 6 ha with a production target of 21 tonnes. The result was so encouraging that in the year 2002 *Kharif* the number of farmer participants swelled 350 with 88 ha of land and a production of 350 tonnes of seed. In this seed production programme, the farmers' cooperative society played a vital role for processing of seed and distribution of quality seed to the farmers of other villages. In this community approach the whole village is now converted into seed village. The seed production activity has generated more rural employment. In the seed village the farmers are getting more yields, which ultimately reflect, in state yield. The implementation of seed village concept has a great potential in enhancing productivity of rice in the state as well as in improvement of rural employment. This venture has also resulted in increased total food production.

Participatory Technology Development for Eco-system Management

In view of effective management of risk prone low land eco-system prevailing in more than 4 lacs ha in the state, the farmers usually grow low land rice varieties but they have poor harvest due to congestion of drainage, occasional heavy downpour, frequent flush flood and inundation. People's technology for specific lowland eco-system has been developed as integrated farming with rice-fish-vegetables for effective utilization of low land eco-system. Field models were tried for 3-year period from 1998 to 2000 based on participatory approach at a low lying plot of land (1.36 ha) in Banamalipara at Chakdah block of Nadia district (West Bengal, India). The land normally remained fallow submerged under 2.0 to 2.5 m of water from July to December during pre-experiment years. The land was reshaped with earthwork in to a deep pond, a low terrace area and a high bund around. This reshaping presented ingress of runoff water in to the plot during the rainy season. Indian major carps (mixed), *catla, rohu*and*mrigal* were stocked in June @ 6000 fingerlings /ha. The fishes were fed rice bran and cow dung (1:1). The fishes were harvested periodically from August up to December (100-150 g each). The fields dried up gradually in December and the remaining fishes moved to perimeter ditches when high yielding rice of 120 days duration (*Gontra* I) was transplanted in the 3rd week of January following standard practices under irrigation. The dykes were made along with the periphery of rice crop. The *bundhs* along with the excavated pond like structure (embankment of the *bheri*) enclosed an area of 65 decimal (0.27 ha).

The results revealed that the pond produced fish through out the year @ 1.31 t $ha^{-1}year^{-1}$. Summer (*boro*) rice was grown on the low terrace with an average yield of 6.25 t $ha^{-1}year^{-1}$. The vegetables were grown on the *bundhs*through out the year. Hybrid tomato was grown in two successive winter seasons yielding a total of 6.25 t $ha^{-1}year^{-1}$ and cowpea (as vegetable) for the third year producing 7.33 t ha^{-1}. Fish farming in rainy season did not have any adverse effect on the summer rice rather the cost of rice cultivation was reduced up to 20 %. The unproductive piece of land in a risk prone low-lying situation was, thus converted in to a highly productive and sustain ably profitable venture. This participatory exercise is now expanding in the entire zone with farmers' own initiative with reducing institutional intervention.

Participatory Rice Biodiversity Conservation for Commercial Utilization

The participatory role of farmers towards enhancement of agro-diversity has been emphasized for evaluation of value added rice cultivars for commercial utilization keeping in view the apprehension that many landraces and site-specific genetic resources are fast depleting, particularly with ingress of HYVs (which are showing signs of diminishing return), can be real.

The strategy of participatory conservation is important because:

1. The local landraces and crop varieties are potential source of valuable genes that could benefit the farming community and
2. The success of conservation depends to a large extent on the personal motivation of the farmer who may intimately know the genotype.

The present investigation addresses the problem, and provides empirical evidences of a unique farmers' participatory approach in conservation, propagation and utilisation of 120 (one hundred and twenty) folk varieties of rice at Gontra, a village in Nadia district of West Bengal.

Farmers themselves collected the traditional rice land races from different localities of West Bengal, from relatives belonging to different agro-eco regions. The collected genotypes were maintained and grown each year under supervisions of the farmers' cooperative society based on community approach at Gontra village of Nadia district of West Bengal for evaluation, multiplication and cataloguing. Technical help was made available from the university.

Performance of the genotypes under different eco-systems and interpretation of data were successfully reviewed by the farmers for selection of rice landraces and for its commercial utilization.

The genotypes were selected against their high value characteristics like aroma, quality of parboiled rice, puffing, flattening, popping ability and high market prices under alternative uses of rice. Even from the limited gene pool desirable genotypes could be selected and distributed amongst the farmers of different agro-eco regions for further evaluation.

The farmers proved capable in testing the genotype and in finding out to the location-specific cultivars for commercial cultivation.

Six genotypes out of a collection of 120-genotypes were selected for commercial cultivation during the period of three years field experimentation based on farmers' participatory approach.

This community gene bank has been efficiently utilized through farmer's intuitive knowledge for selection of elite genotypes from the limited gene pool. Selection based on farmers' participatory approach may prove effective and advantageous compared to institutional procedures since farmers remain equipped with better back ground information of the folk varieties. The genotypes were collected by the farmers and from the farmers with their involvement at all stages and such gene bank is more useful because farmers around could directly and conveniently examine the genotypes in the field and select genotypes for their specific demand.

The participatory conservation is considered to be quicker and efficient model of bio-diversity conservation and also for commercial utilization.

Participatory Approach in Creating Marketing Facilities

Creating marketing facilities under its own avenues are in charter of demand of the farmers. The Farmers' Cooperative Society alone was not in position to dispose all the co-products developed in the process. In the selection process of high value rice among conserved biodiversity, for alternate use of rice like pooped, puffed and flattened, the initiation was taken to engage rural women in such extending enterprises in SHGs for creating market to meet such demand through them. At present, more than 52 (fifty two) of such SHGs are operating in the locality.

Participatory Based Fertilization through Leaf Colour Chart in Summer Rice

In view to evaluate the efficacy of Leaf Colour Chart developed from IRRI, Philippines to monitor the fertilizer application to summer rice, the method was applied through participatory mode. More than 250-farmers were provided with the leaf colour chart for its application in Chakdah block of Nadia district (West Bengal) as well as on-farm trials at village Gontra was also carried out. Though the results are in observatory stage, the economizing N-fertilizer through LCC has got tremendous prospect be accepted by the farmers themselves.

People's awareness and participation

The delay and failure of water programme in India is the lack of community participation at every level of its implementation. The success of any water programme depends upon the integration of end users and all stake holders right from programme inception. This is the main principle of integrated water resources development and management. Mass awareness programme and people's participation in the adoption of efficient methods of irrigation by the farmers is the need of hour today. The prevailing low water charges in canal or deep tube well system act as a barrier to adopt these interventions. However, in spite of these limiting factors, farmers can still promote the use of drip and sprinkler systems if proper awareness is created among them about the likely benefits in terms of water saving and more irrigation coverage resulting in increased crop production besides arresting the waterlogging and soil salinity etc. Women are very important stakeholders in water resources management. The workload of women is directly influenced by the availability of water either for irrigation or for domestic purposes. Nearly half of agricultural labour force is provided by women and more than 30 per cent of poor households in rural India are headed by women. With the development of irrigated agriculture, which is labour intensive, increases the workload of women. Similarly, they are required to carry water from far off places if it is not available nearby. It is the women who face lot of problems due to lack of sanitation facilities. Women must be encouraged and allowed to take greater participatory role in all the activities of water management. They should be represented in Water Users' Association, NGO and voluntary organizations. In the local scale, the involvement of women self-help groups in water management needs to be promoted in a major way. The 'Water Resources Day' shall be organized compulsorily starting from school to university level in collaboration with the government and NGO as a means of people's and societal participation in 'conservation of water resources and preservation of its quality

Participatory Approach on Adoption of Water Management Technologies

The followings are the objectives of participatory approach:

1. To create a sense of ownership of water resources and irrigation system amongst the users to promote economy in water use and preservation of the system.
2. To improve service deliveries through better operation and maintenance
3. To achieve optimum utilization of available resources through sophisticated deliveries as per crop needs.
4. To achieve equity in water distribution
5. To increase production per unit of water, where water is scarce and to increase production per unit of land where water is adequate
6. To make the best use of natural precipitation and ground water in conjunction with flow irrigation for increasing irrigation & cropping intensity
7. To facilitate the users to have a choice of crops, cropping sequence, timing, period and frequency of water supply depending on soils, climate and other infrastructure facilities (roads, market, cold storages etc.) available to maximize the incomes and returns.
8. To encourage collective and community responsibility of the farmers
9. To create healthy atmosphere between the irrigation agency and the users.

Adoption of Water Management Technologies

Adoption of proven and efficient water management technologies are essential to maximize crop production and minimizing gap between irrigation potential created and utilized. Application losses due to conveyance, methods and scheduling of irrigation also are effective in this direction. Selection of crop(s) and crop sequence (s) also played an important role for enhancement of water use efficiency through crop cultivation and related agronomic practices. Conservation, distribution and utilization of irrigation water are the basic parameters of on-farm water management. Optimum scheduling of irrigation, suitable method adoption, conjunctive use of rain, surface and ground water for crop cultivation having improved agro-technology adoption and provision of drainage. Application of proper amount of water at proper time increased the water use efficiency and crop yield maximization with given amount of water reducing evaporation and deep percolation. Scheduling of irrigation with limited water availability is a big challenge to the irrigation experts that needs rigorous research. Trist (1980) considered the irrigation system as a complex socio-technical phenomenon that involved interaction between a physical, environmental and technological in the process of adoption. Whyte (1991) stated the participatory action research as more effective and communicative measures in the developmental processes. Korten (1980) described the participatory action research as a method of merging both developmental intervention and research activities as an effort to involve the users in systemic process of change to make the socio-technical system more effective and efficient. Westhurizen and Annandale, 1996 elaborated their work on-farm participation and encouraged the farmers to adopt irrigation scheduling as cost-benefit approach. They formulated uses friendly format for a wide variety of crops irrigated in South Africa. Cheng Xianjun,1996 reported that on-farm irrigation management and irrigation service thereby improved by involving farmers in water management practices and introducing an acceptable water-saving irrigation schedule for growing crops and increasing agricultural production. On-farm water management with emphasis on active participation of the farm at each and every stage (Belloncle,1988).

A series of on-farm water management research activities were conducted as a collaborative project of Indian Council of Agricultural Research, New Delhi and Ministry of Water Resources, Government of India in some of the selected command area either by package intervention or conducting component technology field trials in the irrigation command. Pramanik and Mallick, (1996) described on-farm participatory field trials brought positive improvements towards better water use efficiency following the recommended methods of water management practices in the farmers field in DVC command Zaman and Mallick,1987). Agarwal and Khanna (1983) also reported efficient soil and water management for enhancing agricultural productivity in Haryana. Dhindwal and Poonia (1994) reported the success story of on-farm water management in Haryana. Rajput(1993) also described that on-farm water management proved much more effective than that on-station field trials conducted at several states.

Precision irrigation

Precision irrigation demands increasingly accurate information about crops, soil and weather. Much research has shown that crop temperature is a reliable plant water stress indicator and, thus, a promising tool for irrigation water management. This paper briefly describes the status of a research project to develop an unmanned aerial vehicle with a Global Navigation Satellite System (GNSS) receiver and an infrared temperature sensor. Laboratory tests determined the key features for accurate temperature measurements, flight altitude and image stabilization. Field experiments are being conducted to: (1) assess the sensitivity of crop temperatures as water status indicators in sugar beet; and (2) determine the sensitivity of a remote-sensing system to derive temperature-based water stress indices. Two irrigation treatments were established: (1) full, which met crop water needs; and (2) deficit, in which irrigation was withheld over the experimental period. The preliminary results indicate that crop temperature is sensitive to soil water deficit in sugar beets.

Keywords: precision irrigation, infrared thermometer sensor, unmanned aerial vehicle, remote sensing

Strategic Water Management

Zero/Minimal Tillage

Delayed monsoon caused *kharif* rice transplantation is usually severely affected the *rabi* cropping where the time didn't allow land preparation to keep pace with available residual soil moisture. Growing wheat/mustard/lentil/linseed like crops under such rice based crop sequences in lowland eco-system is thus, might be good proposition under zero/minimal tillage operation which not required any further time for land preparation while was fully recharged with moisture and optimal for seed germination. So sowing crops under minimal tillage/zero tilled condition is better option for growing crops like wheat and mustard. Growing of these crops was made possible by zero tilled seed drill machine in low land situation where sufficient residual soil moisture persists for seed germination.

Results: Zero-tilled (ZT) wheat with three irrigations recorded the highest grain yield of 1900 kg/ha as compared to lowest grain yield of 1200 kg/ha. The mean of 18 farmers' yield was 1536 kg/ha. The conventional tillage (CT) recorded the highest grain yield of 2800 kg/ha and the lowest grain yield 2500 kg/ha with three irrigations. Conventional tillage (CT) in clay soil was not permitting crop growing due to drainage congestion made by irrigation. Irrigation applied particularly at CRI stage of wheat which resulted yellowing of leaf at early stage. Water use efficiency of ZT wheat was 8.30 to 11.60 kg/ha/mm with three irrigations CT with three irrigations that recorded the highest WUE of 12.20 kg/ha/mm. Farmers' acceptance for growing wheat under zero tilled conditions was mainly attributed due to higher benefit cost ratio obtained than that of conventional tillage.

Use of Drum Seeders

Drum seeding of wet rice was done during *kharif* season. Land situation was medium upland having good drainage after seeding. Shallow puddling was done for drum seeding in clay soil. One pass of Power tiller followed by a bullock drawn country plough was used for puddling. Seeds were soaked for 24 hours and incubated for sprouting (0.5 cm). Philippines 12-row plastic drum seeder was used. Calibration of seed dropping was done through closing the excess whole by polythene tape. Performances of wet direct seeding (DR) was compared against transplanted rice (TR) and found remunerative and cost effective.

Results: Performance of wet seeded rice in medium land situation of the command was good as compared to transplanted one. Rice yield increased to the tune of 22.8 and 14.7% in case of variety Ratna and IET 4094. Plant growth and yield parameters also supported the fact. Wet seeded rice matures one week earlier than transplanted one. Farmers are motivated to use the drum seeding in less rainfall areas for effective utilization of soil moisture wherein in high rainfall during July and August might cause a problem of plant emergence due to stagnation of water in the field after wet seeding of sprouted seeds. Hence, the drum seeding may be advocated as appropriate technology at less rainfall area as well as in upland conditions for judicious moisture utilization.

People's awareness and participation

The delay and failure of water programme in India is the lack of community participation at every level of its implementation. The success of any water programme depends upon the integration of end users and all stake holders right from programme inception. This is the main principle of integrated water resources development and management. Mass awareness programme and people's participation in the adoption of efficient methods of irrigation by the farmers is the need of hour today. The prevailing low water charges in canal or deep tube well system act as a barrier to adopt these interventions. However, in spite of these limiting factors, farmers can still promote the use of drip and sprinkler systems if proper awareness is created among them about the likely benefits in terms of water saving and more irrigation coverage resulting in increased crop production besides arresting the waterlogging and soil salinity etc. Women are very important stakeholders in water resources management. The workload of women is directly influenced by the availability of water either for irrigation or for domestic purposes. Nearly half of agricultural labour force is provided by women and more than 30 per cent of poor households in rural India are headed by women. With the development of irrigated agriculture, which is labour intensive, increases the workload of women. Similarly, they are required to carry water from far off places if it is not available nearby. It is the women who face lot of problems due to lack of sanitation facilities. Women must be encouraged and allowed to take greater participatory role in all the activities of water management. They should be represented in Water Users' Association, NGO and voluntary organizations. In the local scale, the involvement of women self-help groups in water management needs to be promoted in a major way. The 'Water Resources Day' shall be organized compulsorily starting from school to university level in collaboration with the government and NGO as a means of people's and societal participation in 'conservation of water resources and preservation of its quality'

References

Agarwal, M.C. and Khanna, S.S. 1983. Efficient soil and water management in Haryana. Haryana Agricultural University, Hissar, p 118

Belloncle, G. 1988. Is user participation in hydro-agricultural scheme possible? Integrated Agriculture in Africa. Vol.1 Proc. of the seminar. pp 81-94

Dhindwal, A.S. and Poonia, S.R. 1994. On-farm water management – a success sory. Intensive Agriculture. 32(1-2)37-38

Hazra, C.R., 1989. Crop yields increased dramatically in Tejpura watershed. Indian Farming, 39: 37-39

Korten, D. 1980. Community Organisation and Rural Development: A Learning Process Approach. Administrative Review, 40 (Sept-Octo):No.5

Pramanik, M. and S. Mallick, 1996. Farmers participatory approach for improvement of present status of irrigation water utilisation in DVC canal Command. Proc. of ICID/FAO Workshop, Rome, Water Report 8

Rajput, R.K. 1993. On-farm water management. Directorate of Water Management Research, Government of India. pp 35

Trist, E. 1980. The evaluation of Socio-technical system: A conceptual framework and Action Research Programme. Toronto.

Westhurizen, A.J. and Annandale, J.G. 1996. Encouraging irrigation scheduling: A Cost-benefit Approach. Proc. of ICID/FAO Workshop, Rome, Water Report 8

Whyte, W.F. 1991. Participatory Action Research. Newbury Park. California; London & New Delhi, Sage Publications.

Xianjun, C. 1996. Introduction of water saving irrigation scheduling through improved water delivery: a case study from China. Proc. of ICID/FAO Workshop, Rome, Water Report 8;

Zaman, A. and Mallick, S. 1987. Feasibility of growing second crop in winter after rainy season crop at rainfed condition in semi-arid region of West Bengal. Environment & Ecology, 5(2): 239-243

Zaman, A., 1993. Effect of watershed management program on land and crop productivity in drought prone region of West Bengal. Environment & Ecology, 11(2): 442-445

Multiple Choice Questions

1. What is the primary goal of participatory irrigation management (PIM)?
   a) Reduce water charges
   b) Achieve equity in water distribution
   c) Limit crop production
   d) Increase reliance on government irrigation supply

2. In which phase of project planning is farmers' participation most effective according to the text?

   a) During implementation
   b) At the initial project stages
   c) After project completion
   d) Post-harvest evaluation

3. What percentage of India's irrigation facilities are currently under surface irrigation, according to the text?

   a) 50%
   b) 80%
   c) 90%
   d) 98%

4. Which of the following is mentioned as a benefit of increased irrigation facilities for farmers?

   a) Increased salt levels in soil
   b) Ability to raise an extra crop per year
   c) Higher rates of urban migration
   d) Reduced employment opportunities

5. What does the text suggest about the gaps in the irrigation potential created and utilized?

   a) It is minimal and manageable
   b) There is a significant gap that needs addressing
   c) It is nonexistent
   d) It only exists in urban areas

6. How are pressurized irrigation methods characterized within the text?

   a) Ineffective for necessary water conservation
   b) A tool for water saving technologies
   c) Primarily used in traditional farming
   d) A major contributor to agricultural waste

7. What is regarded as a significant indicator of agricultural growth related to irrigation?

   a) Fertilizer application rates
   b) Cropping intensity
   c) Pesticide use
   d) Livestock numbers

8. Which participatory research method is highlighted for its effectiveness in development processes?

   a) Public surveys
   b) Participatory action research
   c) Experimental farming
   d) Traditional farming practices

9. What is one recommended method for improving water resource awareness among farmers?
   a) Providing free water access
   b) Organizing seminars and campaigns
   c) Increasing water prices
   d) Eliminating all irrigation fees
10. Which crop is mentioned as being effectively grown under zero/minimal tillage according to the text?
   a) Rice
   b) Corn
   c) Wheat
   d) Barley

Answer Key

1	b	2	b	3	d	4	b	5	b	6	b	7	b
8	b	9	b	10	c								

17

Virtual Water in Agriculture

Importance of virtual water in agriculture

Virtual water is with the meaning of the concept when water embedded in goods and services and is exchangeable from one country to a different. The country is under water-scarce; the water that's 'saved' is often used towards other ends. Water-scarce countries (like Israel) discourage the export of relatively water-intensive crops (like orange, water melon) precisely to stop large quantities of water from being exported to different parts of the earth. In recent years, the concept of virtual water on trade has gained weight both within the scientific also as restricted by the political debate. The notion of the concept is ambiguous. It changes between an analytical, descriptive concept and a political induced strategy. As an analytical concept, virtual water trade represents an instrument that permits the identification and assessment of policy options not only within the scientific but also within the political discourse. As a politically induced strategy, the question of virtual water trade are often implemented during a sustainable way, whether the implementation are often managed during a social, economical, and ecological fashion, and that countries the concept offers a meaningful option. the info that underlie the concept of virtual water can readily be wont to construct water satellite accounts and brought into economic models of international trade like the Computable General Equilibrium of Global Trade Analysis Project such a model are frequently wont to study the economic implications of changes within the water system or water policy, also because the water resource implications of economic development and trade liberalization. The data pertaining to virtual water is product of GTAP.

Keywords: Virtual water; foodstuff and food; water content in different food materials; water deficit areas;

Introduction

The fresh water availability per person having shrunk by two thirds over the past 40 years, a new technology needs to be developed. Some eminent are of the opinion that a number of countries may be required to export water in the form of agricultural goods with virtual water embedded. In some cases, actual water may be exported from one country to another due to water scarcity critical issues. The numbers will rest on those 31 countries which were come down from 62 that are projected to still have water in abundance in 2050.

The ideas of virtual water were given in the year of 1990 and were first defined by Prof. Allan in the year of 1993 as the water is embedded in commodities. Basically, water used to produce agricultural and industrial product is called the virtual water of the product. The third World Water Forum held in Kyoto, Japan, and was marked by the popularity of a phrase that had emerged in the 1990s. The emphasis was given to discussions on global food trade during the decade and added a new dimension to the debate on world water management. The highlighted phrase, "Virtual Water", can be traced to Israeli economists. The term "virtual water" was finally coined at a seminar at the School of Oriental and African Studies (SOAS), University of London in the year of 1993. The idea had been described even earlier by Professor J.A. Allan as embedded water. Water is essential for the production of food such as cereals, vegetables, and meat and dairy products. The amount of water consumed or used in the production process of a product is called the "virtual water" contained in the product. The 'virtual' named, because it is not contained anymore in the product but it is transform only. For easy understand, to produce a kilogram of wheat we need about 1,000 liters of water and for one kilo meat production we need about 5,000 liter of water. Now a day rising global population and increasing water scarcity and human requirements for drinking and sanitation are pushing demand to scarcity levels. Daily requirement of drinking water approximately three liters, people consume "virtual water" in the form of food and fiber that require huge quantity of water for their production. As per United Nations World Water Development (UNWWD) Report, people in developed nations have on average 3,000 liter virtual water consumption per day.

The virtual water content of the products is given in Table. This "virtual water trade" is nothing but the virtual water content of the product times the trade volume of that product as mention here for easy understands. The way for the importing country is that it does not have to consume that amount of water in domestically producing the product.

Global Trade Analysis Project

GTAP (the Global Trade Analysis Project) may be a global network of researchers (mostly from universities, international organizations, and economic and climate/resource ministries of governments) who used to conduct quantitative and qualitative analysis of international policy issues, including nationwide trading policy, climate policy, and globalization linkages to gender quality and employment. The consortium produces a uniform global economic database (the GTAP Data Base) which is widely making the most of in the research community to review prospective international policy around these issues.

Per Capita Fresh Water Availability

The fresh water availability per person having shrunk by two thirds over the past 40 years, a new technology needs to be developed. Some eminent are of the opinion that a number of countries may be required to export water in the form of agricultural goods with virtual water embedded. In some cases, actual water may be exported from one country to another due to water scarcity critical issues. The numbers will rest on those 31 countries which were come down from 62 that are projected to still have water in abundance in 2050. The ideas of virtual water were given in the year of 1990 and were first defined by Prof. Allan in the year of 1993 as the water is embedded in commodities. Basically, water used to produce agricultural and industrial product is called the virtual water of the product. The third World Water Forum held in Kyoto, Japan, and was marked by the popularity of a phrase that had emerged in the 1990s. The emphasis was given to discussions on global food trade during the decade and added a new dimension to the debate on world water management. The highlighted phrase, "virtual water", can be traced to Israeli economists as estimated by the GTAP. GTAP (the Global Trade Analysis Project) is a global network of researchers (mostly from universities, international organizations, and economic and climate/resource ministries of governments) who conduct quantitative analysis of international economic policy issues, including trade policy, climate policy, and globalization linkages to inequality and employment. The consortium produces a consistent global economic database (the GTAP Data Base) which is widely used in the research community to study prospective international economic policy around these issues.

Virtual water

The term "virtual water" was finally coined at a seminar at the School of Oriental and African Studies (SOAS), University of London in the year of 1993. The idea had been described even earlier by Professor J.A. Allan in the year 1900 as "embedded water". Water is essential for the production of food such as cereals, vegetables, and meat and dairy products. The amount of water consumed or used in the production process of a product is called the "virtual water" that contained in the product. The 'virtual' named, because it is not contained anymore in the product but it is transform only. In a word, for easy understanding, to produce a kilogram of wheat needs about 1,000 liters of water and for one kilo meat production it needs about 5,000 liter of water. Now a day rising global population and increasing water scarcity and human requirements for drinking and sanitation are pushing demand to scarcity levels. Daily requirement of drinking water approximately three liters, people consume "virtual water" in the form of food and fiber that require huge quantity of water for their production. As per United Nations World Water Development (UNWWD) Report, people in developed nations have on average 3,000 liter virtual water consumption per day. The virtual water content of the products is given in Table 1.

This "virtual water trade" is nothing but the virtual water content of the product times the trade volume of that product as mention here for easy to understand. The way for the importing country is that it does not have to consume that amount of water in domestically producing the product. Professor J.A. Allan in the year early 1990 gave the concept of virtual water who defined virtual water as the water embedded in commodities. Producing goods and services requires water; the water used to produce agricultural or industrial products is called the virtual water of the product as an indicator of water use. Virtual water is an essential tool in calculating the real water use of a country, or known as water footprint, which is equal to the total domestic use, plus the virtual water import, minus the virtual water export of a country.

Hence, virtual water is the water used to create the goods and services that ready for consumption or use. In a word, inherent water content in fruits, vegetables, flowers, foodstuff and food grains, this could be felt but not apparently visible. But, this is the existing amount of water not that was used in the cultivation process at needs when grown. Virtual water as introduced by (Allan, 1998) is "the volume of water that has been used to produce a commodity and that is thus virtually embedded in it". Therefore, virtual water is actually not that water which has been used for its cultivation. But in a word, it could be expressed how much water inherently present which could be quantified by the amount of water required to produce per unit of foodstuff and food grain. Interestingly, it is noticeable that the amount of water needed for animal products is much higher than what is required to produce vegetables or fruits. Livestock needs water to drink and to serve in addition for the water needed for their food. It is like multi-process task, food production (in the plants) then consumption (in the livestock) and at the end production again (their meat after slaughter). In case of agricultural and horticultural crops, the yield potential of such crops is required to be known to quantify the actual water required to produce per unit of such crop. In case of rice, there is a thumb-rule procedure to calculate such virtual water. If the variety has potential of giving yield 6 ton per hectare, it has been estimated virtual water that about 6000 liter of water to produce 1 kg of such rice, similarly, it takes 1,600 liter of water on average to produce 1 kg of wheat.

Table 17.1: Status of virtual water status of different foodstuff

Foodstuff	Virtual water present in liter
1 glass of milk	200
1 cup of tea	35
1 cup of coffee	140
1 orange	50
1 apple	70
1 glass of wine	120
1 potato	25
1 piece of pizza	2400
1 kg of rice	6000
1 kg of wheat	1600

Source: www.sciencemediacentre.co.nz

Classification of virtual water

Virtual water for a product can be divided into three components (i) blue (evaporated rainwater), (ii) green (evaporated ground-/surface water), and (iii) gray (polluted water due to pesticides and fertilizers used) depending on the water used for the production. As an example, one kilogram of cabbage needs 280 liters of water. This amount is parted to 56% (156.8 l) green water, 12% (33.6 l) blue water, and 32% (89.6 l) gray water (Mekonnen and Hoekstra, 2011).

Hence, there are three kinds of elaborative water foot prints (1) green water footprint (volume of rainwater evaporated), (2) blue water footprint (volume of surface or groundwater evaporated) and (3) grey water footprint (volume of polluted water) wherein green water is the water transpired by the plant that comes from rain water stored in soil, blue water is the water in our surface and groundwater reservoirs and grey water is the water that becomes polluted during production, say in agriculture because of the leaching of nutrients and pesticides. Grey water footprint is defined as the volume of freshwater that is required to assimilate the load of pollutants based on natural background concentrations and existing ambient water quality standards.

Water foot print

Water footprint of nation has been defined as the total volume of freshwater that is used to produce the goods and services consumed by population of the country. Since not all goods consumed in one particular country are produced in that country, the water footprint consists of two parts: use of domestic water resources and use of water outside the borders of the country (Hoekstra and Chapagain, 2006). In broader sense, it is obvious that virtual water and water footprint are closely related. The virtual water is estimated by the data on water used at the production site, footprint is about water used at the consumer place. Following a table, from the previous source, contains these amounts for different products.

Hence, water footprint making a link between consumption in one place and impacts on water systems elsewhere. Present day, water managers, policy makers and scientists in the field are taking account of this global average and aggregate number on the basis of:

1. Actual water footprint of certain food crops at the precise production location;
2. Ratio green/blue/grey water footprint; and
3. Local impacts of the water footprint based on local vulnerability and scarcity.

Concept of water use

Water foot print is an indicator of water use that looks at both direct and indirect water use of a consumer or producer. Usually, water use is measured in terms of either water volume consumed (evaporated) and are being polluted per unit of unit as industrial wastages; Water foot print can be calculated for its processing, product manufacturing and consumer's level either individual or in group (municipality, province, state, nation) and farmers producers organization (public/private). Freshwater availability on earth is determined by annual precipitation above land. One part of the precipitation evaporates and the other part runs off to the ocean through aquifers and rivers. Both the evaporative flow and the runoff flow can be made productive for human purposes. The evaporative flow can be used for crop growth or left for maintaining natural ecosystems; the green water footprint measures which part of the total evaporative flow is actually appropriated for human purposes. The runoff flow is the water flowing in aquifers and rivers those can be used for all sorts of purposes, including irrigation, washing, processing and cooling. The blue water footprint measures the volume of groundwater and surface water consumed, i.e. withdrawn and then evaporated. The grey water footprint measures the volume of water flow in aquifers and rivers polluted by humans. In this way, the green, blue and grey water footprint measured different sorts of water appropriation. Wherever it is necessary, one can further classify the water footprint into more specific components. In case of the blue water footprint, it can be considered relevant to distinguish between ground and surface water use. In case of the grey water footprint, it can be considered valuable to distinguish between different sorts of pollution. In fact, preferably, these more specific pieces of information are always underlying the aggregate water footprint figures.

Water foot print depends on the question that one would like to address. The green water footprint measures total evaporation and is meant to feed the debate about the allocation of water to different purposes in a context of limited availability. Information about increased or reduced evaporation is relevant from the perspective of catchment hydrology and potential downstream effects. Research has shown that crops can sometimes result in increased evaporation when compared to natural vegetation (particularly in the period of rapid crop growth), and other times in reduced evaporation (e.g. because of soil deterioration or reduced aboveground biomass). In many cases the differences are not very significant at basin scale. The change in evaporation is interesting from the perspective of catchment hydrology and potential downstream effects, but not for the debate on how limited freshwater resources are allocated over different purposes. The water footprint is designed for the latter debate. The purpose of the green water footprint is to measure human's appropriation of the evaporative flow, just like the blue/grey water footprint aims to measure human's appropriation of the runoff flow. The green water footprint measures the part of the evaporated rainwater that has been appropriated by human being and is therefore not available for nature. The water footprint thus expresses the cost of a crop in terms of its total water use.

National water footprint

Water footprints of nations: water used by people as a function of their consumption pattern that shows the extent of water use in relation to consumption of people. The water footprint of a country is defined as the volume of water needed for the production of the goods and services consumed by the inhabitants of the country. Hoekstra and Chapagain (2006) descried water footprints of nations as water use by people as a function of their consumption pattern; the water footprint showed the extent of water use in relation to consumption of people. The water footprint of a country is defined as the volume of water needed for the production of the goods and services consumed by the inhabitants of the country.

The internal water footprint is the volume of water used from domestic water resources; the external water footprint is the volume of water used in other countries to produce goods and services imported and consumed by the inhabitants of the country. The study calculated that the water footprint for each nation of the world for the period 1997–2001. USA appeared to have an average water footprint of 2480 m^3/capita/year, while China has an average footprint of 700 m^3/capita/year, while India's water footprint is 980 m^3/capita/

year that ranks below the global average of 1,243 cubic meters, its 1.4 billion people collectively contribute to a significant 12% of the world's total water footprint. The global water footprint is 7450 m^3/year, which is 1243 m^3/capita/year in average. Water footprint has calculated that on average one kilogram of beef needs about 15 thousand liters of water, 4% of that used is blue, and most of the total volume of water (98%) refers to the water footprint of the feed for the animals. The water footprint of beef is primarily impacted by how much and what the cattle eat, and where the feed comes from. This is because of their great size, beef cattle eat massive quantities of feed but are inefficient in converting that feed to meat (compared to, say, chickens or pigs). In a word, it could be classified that more feed equated with more water.

The four major direct factors determining the water footprint of a country are:

1. Volume of consumption (related to the gross national income);
2. Consumption pattern (e.g. high versus low meat consumption);
3. Climate (growth conditions); and
4. Agricultural practice (water use efficiency);

Environmental impacts of water footprint

1. Pollution in runoff is called nonpoint source (NPS) pollution, and it has the largest environmental impact on groundwater and surface water. There are many sources and strategies for controlling nutrient-contaminated runoff.
2. Cattle producers can control pollution, minimizing runoff from manure storage and feedlots with proper placement. Runoff also can be stored and treated. Methods of minimizing agricultural runoff include surface water diversions, drip trenches, grass filter buffers, sediment basins, subsurface drainage, and evaporative or shallow holding ponds.
3. Most beef is processed in slaughterhouses, which produce several streams of wastewater, including cleaning water and processing water from the slaughtering line. Treatment of the latter generally is a challenge due to its highly variable organic content, so it is advisable to treat it in-house because its organic load can overwhelm municipal wastewater treatment plants.
4. Dissolved air flotation (DAF) is a first stage of treatment that removes about 80% and 65% of organic load and nitrogen load, respectively. The leftover sludge is generally treated with anaerobic digestion, which uses microorganisms to degrade sludge. Biogas produced by the process can be captured to fuel steam production or to generate electricity, lowering

operating costs. Often, processing plants produce enough energy to satisfy their energy needs with a surplus that can be sold.

5. DAF and anaerobic digestion often are used with other stages such as dehydration, nitrification-denitrification, and final clarification.

Agricultural technologies to reduce arsenic contamination

Rice is the only crop, next to sugarcane (sugarcane requires 7-8 months period of growing season where rice requires about 3-4 months only) but requires highest amount of water to grow. Hence, water requirement of rice is highest amongst the agricultural crops. A huge amount of groundwater extraction for growing rice in summer season not only causes more amounts of virtual water but also water table dwindling and arsenic and heavy metal contamination in groundwater. Under these situations, ready-to-use interim agricultural technologies to reduce arsenic contamination in agro-ecosystems of affected Ganga-Bhagrathi delta basin (1) irrigation with surface water enriched by rainwater harvest (2) use of arsenic-adsorption ponds (3) phyto-remediation (4) arsenic-reducing parboiling and cooking methods (5) cover cropping (6) alternative cropping system and (7) increasing water use efficiency

Virtual water trade

it is also known as trade in embedded or embodied water) refers to the hidden flow of water if food or other commodities are traded from one place to another.

Virtual water import: virtual water import of a country or region is the volume of virtual water associated with the import of goods or services into the country or region. It is the total volume of water required (in the export countries) to produce the import products. Viewed from the perspective of the importing country, this water can be seen as an additional source of water that comes on top of the domestically available water resources

Production site-specific virtual water: the virtual water content of a product is the volume of water used to produce the product, measured at the place where the product was actually produced

Consumption site-specific virtual water: virtual water content of a product can also be defined as the volume of water that would have been required to produce the product in the place where the product is consumed.

Virtual water flow: virtual water flow between two nations or regions is the volume of virtual water that is being transferred from one place to another because of product trade

Virtual water balance: virtual water balance of a country over a certain time period is defined as the net import of virtual water over this period, which is equal to the gross import of virtual water minus the gross export. A positive virtual water balance implies net inflow of virtual water to the country from other countries. A negative balance means net outflow of virtual water.

Virtual water import dependency: Countries with net import of virtual water depend, de facto, on the water resources available in other parts of the world. The virtual water import dependency of a country or region is defined as the ratio of the net virtual water import of the country or region to its total water footprint.

Local water use efficiency: local water use efficiency can be increased by creating awareness, charging prices based on full marginal cost and by stimulating water-saving technology.

Water allocation efficiency: Choices on the allocation of water can be more or less 'efficient', depending on the (economic, social and ecological) value of water in its alternative uses.

Global water use efficiency:: water-scarce country can therefore aim at importing products that require a lot of water in their production (water-intensive products) and exporting products or services that require less water (water-extensive products). This is called import of virtual water (as opposed to import of real water, which is generally too expensive) and will relieve the pressure on the nation's own water resources. Grains and produces are contributors to dietary inorganic arsenic intake throughout a market basket survey, 40 commodities anticipated to transport in a minimum of 90% of dietary inorganic arsenic intake were identified.

Water footprint assessment

There are four-phase processes that quantify the water print (1) maps the green, blue and gray water footprints, (2) assessment of the sustainability, (3) efficiency and equitability of water use and (4) identifies which strategic actions on prioritization basis consequently as to form a footprint sustainable.

There are four major direct factors determining the water footprint of a country are (1) volume of consumption (related to the gross national income); (2) consumption pattern (e.g. high versus low meat consumption); (3) climate (growth conditions); and (4) agricultural practice (water use efficiency).

Importance of water footprint

Freshwater persist as a scarce resource; its annual availability is restricted and demand is growing. The water footprint of humanity has exceeded sustainable levels at several places & is unequally distributed among people. Good information about water footprints of communities and businesses will help to know how the determination achieves more sustainable & equitable use of freshwater. There are many spots surrounded by the world where serious water depletion or pollution takes place: rivers running dry, dropping lake and groundwater levels, and species due to contaminated water. The water footprint helps to point out the link that exists between our daily consumption of products and seeing that a result the problems of water depletion and pollution that exist elsewhere, within the regions where our goods are produced. Nearly every product skin texture a smaller or larger water footprint, which is of interest to both consumers that buy those products and businesses that produce, process, trade, or sell those products in some stage of their supply chain. According to United Nations World Water Project Report (2012), people in developed nations subsist dressed in an average a 3,000-liter virtual water consumption profile every day.

Water footprint network

Water Footprint Network was established in October 2008 by diversity of major global players from business, civil society, multilateral organizations, and academia. Water Footprint Network and its partners strive to develop and apply the Water Footprint to support the transition to sustainable and equitable water use and management globally.

Water footprint network aims to come back to broadly shared global standards on water footprint accounting. An attempt to prevent what went on surrounded by the case of the carbon footprint: a mess of a variety of definitions approaches & methods in categorize that it's difficult to properly assess claims during this field. A remarkable set of worldwide standards on the water footprint will make efforts of companies & communities that strive for a reduction of their water footprint more transparent. The Network has two work programmes: a technical and a policy work programme. Within each programme there are several projects. Many of the projects are executed by our partners and just run under the umbrella of the water footprint network to construct sure fruitful exchange with other activities within the network. Some projects aim at further research and development, other projects aim at the sensible application of the water footprint in various settings. Additionally, there are activities that purely aim at communication, exchange, and dissemination.

Reduction water footprint by consumers

Consumers can reduce their direct water footprint (home water use) by installing water-saving toilets, applying a water-saving showerhead, closing the faucet during teeth brushing, using less water within the garden and not removing medicines, paints or other pollutants through the sink.

The indirect water footprint of a consumer is usually much larger than the direct one. A consumer has basically two options to scale back his/her indirect water footprint. One option is to substitute a consumer product that features a large water footprint by a special kind of product that features a smaller water footprint. Examples: eat less meat or become vegetarian, drink tea rather than coffee, or maybe better drink plain water. Not wearing cotton but artificial fiber clothes save tons of water. But this approach has limitations because many of us don't easily shift from meat to vegetarian and other people like their coffee and cotton. A second option is to stay to an equivalent consumption pattern but to pick the cotton, beef or coffee that features a relatively low tide footprint or that has its footprint in a neighborhood that doesn't have high water scarcity. This needs, however, that buyers have the right information to form that choice. Since this information is usually not available within the world of today, a vital thing consumers can do now's ask product transparency from businesses and regulation from governments. When information is absent there on the impacts of a meticulous article on the water system, consumers can make conscious choices about what they buy.

Reduction water footprint by corporate

The corporate can reduce its operational water footprint by saving water in its own operations and bringing pollution to zero. Keywords are: reduce, recycle, and treat before disposal. For many businesses, however, the supply-chain water footprint is far larger than the operational footprint. It's therefore crucial that companies to address that also. Achieving improvements within the supply chain could also be harder, because not under direct control but they'll be simpler. Businesses can reduce their supply-chain water footprint by making supply agreements with certain standards with their suppliers or by simply changing to a different supplier. In many cases, it probably means quite something, because the entire business model may have to be transformed so as to include or better control supply chains and to form supply chains fully transparent to consumers. Among the varied alternative or supplementary tools which will help improving transparency are: setting quantitative water-footprint reduction targets, benchmarking, product labeling, certification, and water footprint reporting.

National water footprint accountability

Traditionally countries are formulating national water plans by watching the way to satisfy water users. All the countries nowadays consider options to scale back water demand additionally to options to expand supply, they characteristically don't include the worldwide dimension of water management during this way, they are doing not explicitly consider options to put away lots of water through the import of water-intensive products. Additionally, by looking only at water use in their own country, most governments have a blind spot on the difficulty of sustainability of national consumption. As a matter of fact, many countries have significantly externalized their water footprint without watching whether the imported products are connected with water depletion or pollution in producing countries. Governments can and will engage with consumers and businesses to figure towards sustainable consumer products. National water footprint accounting should be a typical component in national water statistics and supply a basis to formulate a national water plan and basin plans that are coherent with national nationwide trading policy and national environmental policy. As a consumer, your water footprint is sustainable when (a) the entire remains below your equal share of the available freshwater resources within the world, & (b) no component of the entire water footprint presses at places where or times when local environmental flow requirements are violated.

Water pricing

Generally, the value of purchased water is way below its real economic cost. Most government subsidies water system on a huge scale by investing in infrastructures like dams, canals, distribution systems, and wastewater treatment. These costs are often not charged to the water users. As a result, there's an insufficient economic incentive for water users to save lots of water. Besides, thanks to the general public character of water, water scarcity is usually not translated into a further component within the price of products and services that are produced with the water, as happens naturally within the case of individual goods. Finally, water users generally don't buy the negative impacts that they cause on downstream people or ecosystems.

Water footprint labeling

In a world where many products are associated with water depletion and pollution, it's very useful to appearance the history of products more transparent. It's good to possess the facts publicly available, therefore the consumer features a choice. Information is often provided on a label or is often made available through the web. This is often most useful for products that always have large effects on water, like products that contain cotton or sugar. For consumers, it might be helpful to integrate a water label in broader labels that include other issues also, like energy and fair trade. Traditionally statistics on water use specialize in measuring 'water withdrawals' and 'direct water use'. The water footprint accounting method takes a way broader perspective. First of all, the water footprint measures both direct and indirect water use, where the latter refers to the water use inside the supply chain of a product. The water footprint thus links final consumers and intermediate businesses and traders to the water use along the entire production chain of a product. This is often relevant because generally the direct water use of a consumer is little if compared to its indirect water use and therefore the operational water use of a business is usually small if compared to the supply-chain water use. Consequently the picture of the particular water dependency of a consumer and business can change radically. The water footprint method further differs in that it's at water consumption (as against withdrawal), where consumption refers to the part of the water withdrawal that essentially gets lost through evaporation, i.e. the part of the water withdrawal that doesn't return to the system from which it had been withdrawn. Besides, the water footprint goes beyond watching blue water use only (i.e. use of ground and surface water). It also includes a green water footprint component (use of rainwater) and a gray water footprint component (polluted water).

Water footprint metaphor

The term "footprint" is usually used as a metaphor to ask the extremely fact that humanity appropriates a big proportion of the available natural resources (land, energy, water). However, a bit reminiscent of the "ecological footprint" and therefore the "carbon footprint", the "water footprint" is quite a metaphor: there's a rigorous accounting framework with well-defined measurable variables and well-established accounting procedures to calculate the water footprints of products, individual consumers, communities, nations or businesses. We discourage people to use the water-footprint concept as a metaphor because of its strength lies in its effectiveness when utilize in the context of strict accounting and measurable reduction targets.

Water footprint and virtual water

The water footprint might be a term that refers to the water wont to make a product. During this context, the 'virtual water content' of a product somewhat than it's 'water footprint'. The water footprint concept, however, features a wider application. The water footprint of a consumer by watching the water footprints of the products and services consumed or about the water footprint of a producer (business, manufacturer, service provider) by surveillance the water footprint of the products and services produced by the producer. Furthermore, the water footprint concept doesn't simply ask a water volume only, like within the case of the term 'virtual water content' of a product. The water footprint may be a multidimensional indicator, not only pertaining to a water volume used, but also making explicit where the water footprint is found , what source of water is in employment , and when the water is employed .

Conclusion

In an era of the rising global population and increasing water scarcity, human requirements for drinking and sanitation are pushing demand to unsustainable levels. One major issue is that, in adding to drinking approximately three liters per day, people won't to consume "virtual water" within the sort of food and fiber that need water for his or her production. With the measure of freshwater available per person having shrunk by two-thirds over the past four decades, a replacement system must be developed. Experts are of the opinion that variety of nations could also be required to export water inside the sort of agricultural goods with virtual water embedded. In extreme situations, actual water could also exist exported from one country to a dissimilar. The burden of a further amount of water will rest on those countries that are projected to still have water in abundance. Rainwater is often utilizing in agricultural production and water in rivers and aquifers are often used for irrigation or industrial or domestic purposes. But during a certain period, one cannot use more water than is out there. A river is often emptied and within the prospect, one cannot take more water from lakes and groundwater reservoirs than the speed with which they're recharged. The water footprint measures the quantity of water available during a certain period that's consumed (i.e. evaporated) or polluted and throughout this way, it provides a measure of the quantity of obtainable water appropriated by humans. The rainwater not used for agricultural production is left to sustain natural vegetation. The bottom and surface water flow not evaporated for human purposes or polluted is left to sustain healthy aquatic ecosystems. However, the water crisis is increasing all over the place the world wherein virtual water as a newly developed concept that's useful for water management and development. This idea can play an important role in solving water problems. This importance is appreciation to the principles of integrated water resource development and management, especially for policy-making and planning processes. Water scarce countries get to understand this idea and apply it consequently as to possess a sustainable life on living planet.

References

Allan, J.A. (1998). Virtual water: A strategic resource, global solutions to regional deficits; https://doi.org/10.1111/j.1745-6584.1998.tb02825.x;

Andrew A. Meharg and Md. Mazibur Rahman (2002). Arsenic contamination of Bangladeshi paddy field soils: implications for rice contribution to arsenic consumption; Environmental Science and Technology Published online doi: 10.1021/es0259842.

Charles F. Harvey, Christopher H. Swartz, A. B. M. Badruzzaman, Nicole Keon-Blute1 Winston Yu, M. Ashraf Ali and Jenn (2002). Arsenic mobility and groundwater extraction in Bangladesh; Science, 298, 1602 – 1606;

Harvey, C. F. Charles , Khandaker N. Ashfaque, Winston Yu, A.B.M. Badruzzaman , M. Ashraf Ali, Peter M. Oates, Holly A. Michael, Rebecca B. Neumann, Roger Beckie, Shafiqul Islam, M. Feroze Ahmed (2006). Groundwater dynamics and arsenic contamination in Bangladesh. Chemical Geology; 235, 3–4: 205-404;

Hoekstra, A.Y. and Chapagain, A. K. (2006) Water footprints of nations: Water use by people as a function of their consumption pattern;

M. M. Mekonnen and A. Y. Hoekstra (2011). The green, blue and grey water footprint of crops and derived crop products; Hydrol. Earth Syst. Sci., 15, 1577–1600, 2011 www.hydrol-earth-syst-sci.net/15/1577/2011/ doi:10.5194/hess-15-1577-2011

Multiple Choice Questions

1. What does the concept of "virtual water" primarily refer to?
   a) Water used directly for irrigation
   b) Water used to produce goods and services
   c) Water stored in aquifers
   d) Water wasted during production
2. Who first defined the term "virtual water" in 1993?
   a) Prof. J.A. Allan
   b) United Nations World Water Development
   c) Mekonnen and Hoekstra
   d) The Global Trade Analysis Project (GTAP)
3. Which country is specifically mentioned as water-scarce and discouraging the export of water-intensive crops?
   a) China
   b) Israel
   c) India
   d) USA

4. What type of water is referred to when discussing "blue water" in the context of virtual water?
   a) Evaporated rainwater
   b) Groundwater and surface water
   c) Polluted water from agriculture
   d) Water used for drinking
5. According to the United Nations World Water Development Report, how much virtual water do people in developed nations consume on average per day?
   a) 1,000 liters
   b) 3,000 liters
   c) 5,000 liters
   d) 7,450 liters
6. Which of the following is NOT a component of the water footprint mentioned in the text?
   a) Green water
   b) Blue water
   c) Gray water
   d) Yellow water
7. What is the estimated amount of water needed to produce 1 kg of beef?
   a) 1,000 liters
   b) 5,000 liters
   c) 15,000 liters
   d) 1,500 liters
8. The virtual water balance of a country is defined as:
   a) Gross export of virtual water minus gross import
   b) Net import of virtual water over a certain time period
   c) Total water consumption of a nation
   d) Total water used for domestic purposes
9. What is the main goal of the Water Footprint Network established in 2008?
   a) To create efficient irrigation systems
   b) To develop and apply the water footprint for sustainable use
   c) To track pollution levels in water bodies
   d) To promote meat consumption worldwide
10. How can consumers reduce their indirect water footprint according to the text?
    a) By drinking only bottled water
    b) By limiting home water use only

c) By choosing products with a lower water footprint

d) By using more water in agriculture

Answer Key

1	b	2	a	3	b	4	b	5	b	6	d	7	c
8	b	9	b	10	c								

18

Advances in Irrigation

Climate-based/smart (digital) irrigation practices and its management, irrigation automation, sensors based irrigation, robotic application in irrigation and artificial intelligence

Climate-smart (digital) irrigation: Aims and Objectives

Climate-smart irrigation (CSI) is a vital part of climate-smart agriculture, aiming to

- Boost the productivity of irrigated farming and farmer incomes while minimizing negative social, political, and environmental consequences.
- Bridge the gap between potential and actual crop yields across different climates.
- Enhance the resilience of irrigated farming and related supply chains to current and future climate change impacts and other risks.
- Adapt irrigated farming to climate change by capitalizing on new opportunities and mitigating adverse effects.
- Reduce greenhouse gas emissions per unit of food, fiber, or fuel produced in irrigated systems, from farm to market.
- Identify and prioritize ways to cut GHG emissions from irrigated systems at various stages, from water sourcing to post-harvest processes.
- Improve the environmental sustainability of irrigated farming and supply chains while safeguarding water access for communities, protecting vulnerable groups' livelihoods, and preserving aquatic ecosystems.
- Improve the sustainability and reducing the GHG emissions of existing crop production systems.

Irrigation is key to stable agricultural production, especially in dry conditions. CSI aims to improve existing and new irrigation systems by incorporating climate-smart principles. While good irrigation infrastructure is essential, it's not enough. Effective CSI also requires supportive policies, strong institutions, good water governance, secure land and water tenure, farmer knowledge, market access, and credit availability.

Climate change creates a chain of risks, impacting water sources, infrastructure, farming & ultimately, socio-economic & environmental factors. Understanding these risks is crucial for effective climate change adaptation and mitigation.

CSI isn't a one-size-fits-all solution. It involves tailoring irrigation infrastructure and practices to local contexts, policies, and institutions. Sound irrigation practices are a cornerstone of CSI.

There is increasing evidence that climate change is impacting regional and seasonal rainfall patterns, and the frequency and severity of extreme weather events. This has major implications for farmers who are already using irrigation, as well as farmers who hope to adopt irrigation as part of their climate change adaptation strategy.

While climate change predictions are subject to uncertainty, the expectation is that annual rainfall will increase in the tropics and higher latitudes but decrease in the already dry arid to semi-arid mid-latitudes, and in the interior of large continents. As a consequence, it is expected that water scarce areas of the world will become drier and hotter.

Climate change is not taking place in the context of pristine watersheds or river basins. Rather, the norm is that climate change is impacting watersheds and river basins that have already experienced centuries of development, land use change and degradation or depletion of environmental resources. In many of these watersheds and basins, consumptive water use is already outstripping sustainable supply.

Irrigation underpins food supplies in many countries that have large and/or rapidly increasing populations, for example China, India, Indonesia and Pakistan. However, in recent years, there are signs of a decline in expansion of irrigated areas for reasons that include the unsustainable consumptive use of water by irrigation. In these countries, the scope for increasing the area equipped for irrigation may be limited, and focus should be on improving water productivity.

The CSI context and focus can also be differentiated and determined by the magnitude of the yield gap between actual and potential crop yields. In countries or regions that are characterized by low-input farming and wide yield gaps, the focus of CSI is likely to be on sustainable intensification of crop production systems. In countries or regions that are characterized by relatively intensified farming and relatively narrow yield gaps, some further yield gap closure may still be possible, but increases in potential yields are needed if production is to be increased substantially. In such cases, the focus of CSI is likely to be on accumulation of knowledge, know-how and lessons learned often over a long period of time.

Evidence informed planning is also fundamental to good irrigation practice. Planners and/or farmers should have a good understanding of the comparative advantages, limitations and suitability of different irrigation methods. They should also be aware that the performance of irrigation schemes is just as much a function of the farmer and irrigation scheduling procedures as of the irrigation methods or equipment that are used. When planning and implementing irrigation schemes it is important to note that farmers, planners, ecologists, politicians and the general public often have different or even contradictory perspectives and views on what constitutes good irrigation practice.

CSI Approach: Prioritizing Existing Irrigation Systems

A Climate-Smart Irrigation (CSI) approach generally begins by focusing on improving the performance, productivity, and profitability of existing irrigated agriculture. This starting point is strategically chosen because increased profitability motivates stakeholders & fosters buy-in. Moreover, well-managed irrigation systems tend to be more resilient to climate change impacts than those that are underperforming. Investing in climate change adaptation or mitigation strategies may be ineffective if farmers lack essential crop and water management skills, or if the irrigation systems themselves are poorly maintained due to issues such as water source overexploitation.

The initial phase of a CSI approach aims to achieve

1. Productive, profitable, sustainable, and non-polluting irrigation systems and practices within a defined area.
2. Competent, empowered, and financially stable farmers.
3. Safeguards that protect environmental flows, ensure equitable water access for marginalized groups, and address gender considerations.

The second phase builds upon these foundations to implement adaptation and mitigation strategies.

Conservation Agriculture: Agronomic Measures for Water Conservation

Agronomic practices play a crucial role in conserving water by minimizing evapotranspiration losses, reducing raindrop splash erosion, controlling topsoil erosion and runoff, enhancing rainwater infiltration through improved soil structure, and ultimately conserving soil moisture for enhanced plant growth and yield. Key techniques include:

a) **Mulching:** Mulching protects the soil surface from erosion by controlling runoff and facilitates rainwater absorption. Mulching also increases water consumption by crops, regulates soil temperature and moisture, lessens evaporation, prevents soil crusting and plant nutrient losses, and suppresses weed growth. Dry soil mulch, created through surface cultivation, minimizes evaporation. Cover mulches, such as black polythene sheets, organic residues, or crop stubble, also act as barriers to evaporation. Mulching is particularly beneficial for post-rainy season (rabi) crops that rely on stored soil moisture.

b) **Anti-evaporants:** Materials that reduce evaporation from the soil surface. Straw, sawdust, paddy husk, and wood bark are commonly used. Hexadecanol, a long-chain alcohol, can also retard evaporation due to its resistance to microbial degradation and long-lasting effects. However, the high cost of soil surface mulching can be a limitation.

c) **Controlling Transpiration Loss:** Approximately 99% of the water absorbed by plants is lost through transpiration via stomata. Strategies to reduce this loss include:

   i) **Windbreaks/Shelterbelts:** Planting windbreaks along field boundaries, perpendicular to prevailing winds, reduces wind velocity and protects crops from hot air. Optimal design features taller plants in the center and shorter plants on the outside. Windbreaks can moderate wind velocity, reduce soil loss, and improve soil moisture retention.

   ii) **Weed Control:** Removing weeds saves water that would otherwise be absorbed by these unwanted plants, improving water use efficiency for the intended crop.

   iii) **Anti-transpirants:** Applying chemical substances like phenyl mercuric acetate (PMA), atrazine, kaolin, lime, cycocel, or mobileaf can reduce transpiration by closing stomata, forming a reflective film on leaves, or cooling leaves by reflecting solar energy.

   iv) **Crop and Variety Selection:** Choosing crops and varieties with low transpiration rates that are well-suited to water-limited environments is essential.

d) **Tillage Practices:** Deep, minimum, zero, and conservation tillage methods can conserve rainwater based on soil, water, and climate conditions. These methods increase water infiltration, improve soil structure and chemistry, prevent topsoil erosion and runoff, and retain plant nutrients. Deep tillage helps conserve monsoon rainfall, while shallow tillage is useful for moisture retention in the rabi season. While conventional tillage can harm soil health, conservation tillage effectively conserves soil moisture, maintains soil quality, reduces erosion, and improves long-term crop productivity.

e) **Land Leveling:** Precise land leveling ensures efficient and uniform irrigation water distribution, rainwater conservation, and effective drainage of excess water during the monsoon season, ultimately leading to higher water use efficiency and crop yields.

f) **Cropping Pattern Selection:** Selecting suitable cropping systems that align with soil, water, and climate conditions is crucial. Crops are categorized as rainfed, limited-irrigation, or fully-irrigated to maximize production and profit per unit of water. Farmers should be encouraged to grow low-water-consumption crops and diversify their cropping programs with the inclusion of high-value crops.

AI Model

An automated irrigation system operates with minimal to no human intervention, primarily requiring surveillance. These systems leverage timers, sensors, computers, and mechanical appliances to automate various irrigation methods, including drip, surface drip, and sub-surface irrigation. Automation streamlines the irrigation process, freeing up farmers to concentrate on other essential tasks.

Automated irrigation systems can be categorized into three types based on their control mechanisms:

- **Real-Time Basis:** These systems utilize electric solenoid valves to divide the field based on maximum flow. These valves are connected to a central controller via electric cables or wireless connections using DC solenoid valves, which require a DC controller. The operation of each valve is programmed with fixed timing based on the crop type and its age. The controller then operates the valves according to these pre-set programs. Some controllers can maintain these fixed timings even during power outages, ensuring consistent valve operation.
- **Volumetric Basis:** Similar to real-time systems, volumetric systems also use solenoid valves and a central controller. However, they incorporate a water meter on the mainline after the header unit. This meter ensures a pre-defined flow to each valve, controlled by the controller. The solenoid valves are turned on and off based on the volume of water applied to the crop, making this a more accurate system compared to real-time operation.
- **Moisture Sensor Based:** These systems aim to maintain the soil around the root zone at field capacity. Electronic moisture sensors are placed at root zone depth and connected directly to the system's controller, ensuring water application aligns with the crop's requirements. Some

systems in this category use hydraulic pressure to operate pop-up spray heads. Low-tech alternatives include clay pots and porous capsule irrigation networks.

Modern automated irrigation systems have revolutionized farming by simplifying tasks and improving efficiency. Manufacturers and suppliers offer customized systems for various agricultural sectors, including horticulture, vineyards, orchards, row crops, greenhouses, and nurseries.

A specific example is an automatic irrigation system using a soil sensor. This system incorporates:

- **Soil Sensor:** Detects soil moisture levels.
- **555 Timer IC with IC Base:** Controls the timing and operation of the system.
- **VTR Switch:** Enables easy activation and deactivation.
- **5V Voltage Regulator:** Ensures stable power supply.
- **Two 9V Batteries:** Provide portable power.
- **Battery Cap:** Secures the batteries.
- **LED:** Indicates system status.
- **Resistor:** Regulates current flow.
- **Relay:** Controls the pump motor based on soil moisture.
- **Pump:** Delivers water to the irrigation system.
- **Pipe:** Distributes water to the plants.

The working principle involves the soil sensor measuring moisture content. The 555 Timer IC processes this data, triggering the relay to activate the pump when the soil is dry. Once the desired moisture level is reached, the relay switches off the pump, conserving water. This automated process reduces manual intervention and ensures efficient irrigation in agriculture, gardens, and educational projects while also conserving water.

Moreover, artificial intelligence (AI) is increasingly being integrated into irrigation systems to predict weather patterns and adjust irrigation schedules accordingly, further optimizing water usage and improving crop yields.

Robotic approach

Another approach involves using a plant irrigation water sprinkler robot. This robot, equipped with a single sprinkler and a water tank, moves throughout the field, automatically spraying water.

Some robotic systems incorporate temperature and conductivity sensors connected to a microcontroller. The microcontroller monitors soil temperature and moisture levels, sending data to a base station via wireless communication (e.g., Bluetooth). The irrigation system can operate in automatic or manual mode. In automatic mode, the microcontroller controls the irrigation motor based on soil moisture content. In manual mode, the user can control the motor via Bluetooth commands. The microcontroller, such as the LPC2148, and Bluetooth module, such as HC-05, operating at 2.4GHz and 3.3V, are selected for their performance, low power consumption, and reliable wireless communication capabilities.

The robot section, the second key component, incorporates a camera and a wireless module. This wireless module allows for remote control of the robot's movement in various directions. The integrated camera enables the monitoring and observation of insects within the field.

This system represents an advancement in robotics technology, utilizing a robot to monitor crop conditions. In an era where regular on-site inspection is often impractical, this system allows for remote verification of crop health from a computer. Furthermore, such systems have potential applications in industrial settings for frequent monitoring and control purposes.

This system successfully provides remote crop condition monitoring through the use of a field-roaming robot, allowing users to observe conditions from a PC. Moreover, the automated irrigation system implemented demonstrates a reduction in water usage for a given amount of fresh biomass production. Beyond the financial benefits of water conservation, the importance of preserving this natural resource justifies the adoption of such irrigation systems.

AI-based irrigation application

Developing an AI-based irrigation application can significantly enhance agricultural efficiency, optimize water usage, and improve crop yield. Below are some key components and features you might consider when designing such an application:

Key Components

1. **Sensor Integration:**
   - **Soil Moisture Sensors:** Measure soil moisture content to determine the water needs of crops.
   - **Weather Sensors:** Capture real-time weather data, including rainfall, temperature, humidity, and wind speed.
   - **Crop Health Sensors:** Use NDVI (Normalized Difference Vegetation Index) sensors or drones to assess crop health.

2. **Data Collection and Analysis:**
   - Collect data from various sensors and weather sources.
   - Use machine learning algorithms to analyze historical data and predict future irrigation needs based on weather forecasts and soil moisture levels.
3. **Irrigation Management:**
   - Develop algorithms to calculate the optimal amount of water required based on soil moisture, crop type, and weather predictions.
   - Implement automation features to control irrigation systems, such as drip or sprinkler systems.
4. **User Interface:**
   - Create a user-friendly mobile or web application for farmers to monitor and control irrigation.
   - Include visual analytics, alerts for low moisture levels, or weather changes, and historical usage data.
5. **Recommendations and Alerts:**
   - Use AI to provide personalized irrigation schedules and recommendations based on specific crops and soil conditions.
   - Send alerts for extreme weather conditions, potential disease outbreaks, or irrigation system failures.
6. **Remote Monitoring:**
   - Allow farmers to monitor their irrigation systems remotely, making it easy to manage water resources efficiently.
   - Use GPS and mapping tools to visualize irrigation coverage and identify areas needing attention.
7. **Sustainability Features:**
   - Promote water conservation practices and recommend eco-friendly alternatives.
   - Offer insights on the environmental impact of irrigation choices.

Implementation Steps

1. **Needs Assessment**
   - Engage with local farmers to understand their needs and challenges regarding irrigation.
2. **Prototype Development**
   - Create a minimum viable product (MVP) that includes basic features like soil moisture sensing and simple irrigation control.

3. **Testing and Feedback**
   - Conduct field trials to test the system's accuracy and efficiency.
   - Gather feedback from users to refine features and improve usability.
4. **Scaling**
   - Expand the application to support more crops, regions, and advanced features like predictive analytics and integration with other farm management software.
5. **Partnerships**
   - Collaborate with agricultural experts, universities, and technology companies to enhance the application's capabilities and efficacy.
6. **Education and Training**
   - Provide training for farmers on how to utilize the AI irrigation system effectively.

Technologies to Consider

- **Machine Learning Frameworks:** TensorFlow, PyTorch, or Scikit-learn for building predictive models.
- **Cloud Platforms:** AWS, Google Cloud, or Azure for data storage and processing.
- **Mobile Development Frameworks:** React Native or Flutter for building cross-platform mobile applications.
- **IoT Platforms:** MQTT or MQTT broker solutions for communication between sensors and the app.

An AI-based irrigation application has the potential to revolutionize agricultural practices by using technology to optimize water usage. By developing a comprehensive solution that integrates AI, IoT, and user-friendly interfaces, you can empower farmers to make informed decisions that lead to more sustainable and efficient agricultural practices.

Sensor-based irrigation applications

Sensor-based irrigation applications use various types of sensors to optimize water management in agricultural settings. These systems aim to improve water efficiency, reduce waste, and increase crop yields by delivering the right amount of water to plants at the right time. Here's an overview of how sensor-based irrigation works, its components, advantages, and popular technologies involved:

Key Components

1. **Soil Moisture Sensors:** These sensors measure the volumetric water content in the soil. They help determine when irrigation is needed based on the moisture levels.
2. **Weather Stations:** Weather sensors collect data on temperature, humidity, rainfall, wind speed, and solar radiation. This information helps predict evaporation rates and irrigation needs.
3. **Flow Meters**: Flow meters monitor the amount of water being used in the irrigation system, allowing for better management and optimization of water resources.
4. **Rain Sensors:** These sensors detect rainfall and can automatically suspend irrigation when sufficient rain has fallen, preventing overwatering.
5. **Remote Monitoring Systems:** Many modern sensor-based irrigation systems come with mobile or web applications that allow farmers to monitor soil moisture levels, weather, and irrigation status in real-time.

How It Works

1. **Data Collection**: Sensors collect real-time data on soil moisture, weather conditions, and crop water requirements.
2. **Data Analysis**: The data is analyzed using algorithms or software that determines the optimal irrigation schedule and amounts.
3. **Automated Irrigation**: Based on the analysis, the irrigation system is automatically activated or adjusted. Some systems allow for manual interventions through an app or web interface.
4. **Reporting**: Many systems provide reports and analytics on water usage, soil health, and crop conditions, enabling farmers to make informed decisions.

Advantages

- **Water Conservation**: By applying water only when necessary, these systems can significantly reduce water consumption.
- **Increased Crop Yields**: Optimized irrigation can enhance plant growth and improve crop health.
- **Cost Efficiency**: Reducing water usage and improving crop yields can lead to lower operational costs and increased profitability for farmers.
- **Sustainability**: Helps promote more sustainable agricultural practices by conserving resources and protecting the environment.

Popular Technologies

- **Drip Irrigation Systems**: Often used in conjunction with soil moisture sensors for highly efficient water delivery directly to the plant roots.
- **Smart Controllers**: Devices that integrate data from various sensors and automatically adjust irrigation schedules and rates.
- **IoT (Internet of Things) Solutions**: These systems connect sensors, controllers, and mobile applications for data monitoring and management.
- **Cloud Computing**: Used for data storage, analysis, and remote access to irrigation systems, enabling farmers to manage operations from any location.

Applications

- **Agricultural Farming**: Enhancing crop growth and yields by optimizing irrigation practices.
- **Landscape and Turf Management**: Efficient watering of recreational fields, parks, and golf courses.
- **Greenhouses**: Automated systems to ensure ideal moisture levels for optimal plant growth.
- **Drought-Prone Areas**: Implementing sensor-based irrigation where water is a limited resource to maximize efficiency.

Sensor-based irrigation applications represent a significant advancement in agricultural technology, making it easier for farmers to manage water resources efficiently and sustainably. As technology continues to evolve, these systems are likely to become even more advanced, contributing to smarter farming practices and sustainable agriculture.

Robotic applications in irrigation

Robotic applications in irrigation have transformed traditional farming practices by enhancing efficiency, precision, and sustainability. Here are some key ways robotics is being utilized in irrigation systems:

1. **Automated Irrigation Systems**: Robotic systems can automate the control of irrigation schedules based on soil moisture levels, weather forecasts, and crop requirements. Sensors placed in the field provide real-time data, allowing for precise water application.
2. **Drones for Aerial Monitoring**: Drones equipped with sensors and cameras can survey large agricultural areas to monitor soil health, moisture levels, and crop conditions. They can identify areas that require irrigation or adjustments in water application.

3. **Robotic Sprinkler Systems**: These systems can move through the fields, applying water precisely where and when it's needed. Equipped with GPS and advanced mapping technology, they can optimize water use by targeting specific areas rather than applying water uniformly.
4. **Soil Moisture Sensors**: Robots can be integrated with sensors that measure soil moisture content at various depths. This data helps in deciding when and how much to irrigate, conserving water and improving crop yields.
5. **Weeding Robots**: While primarily used for weed control, many advanced weeding robots can also assist in irrigation by ensuring that water is applied directly to the crops rather than weeds, further optimizing water resources.
6. **Remote Control and Telemetry**: Some irrigation systems allow for remote monitoring and control through mobile apps or computers. This integration allows farmers to make real-time adjustments, responding to climatic changes or soil conditions promptly.
7. **Data Analysis and Predictive Analytics**: Robotic systems can gather vast amounts of data over time, which can be analyzed to predict irrigation needs based on historical patterns, thus optimizing future water usage.
8. **Smart Drip Irrigation**: Robotics can automate and optimize drip irrigation systems that supply water directly to the plant roots. This method minimizes water loss and enhances the efficiency of water use.
9. **Climate-Adaptive Irrigation**: Robots can be programmed to adapt irrigation schedules based on changing weather patterns, ensuring that crops receive optimal water while reducing waste.
10. **Multi-functional Agricultural Robots**: Some robots are designed to perform multiple agricultural tasks, including planting, harvesting, and irrigation, thus increasing overall efficiency and reducing labor costs.

The integration of robotics into irrigation practices not only helps in maximizing yield and profit but also contributes to sustainable water usage, which is becoming increasingly critical in the face of climate change and water scarcity issues.

References

AngelinBlessy J., Kumar A.2021. Smart irrigation system techniques using artificial intelligence and IoT. Proc. 3rd Int. Conf. Intell. Commun. Technol. Virtual Mob. Networks, ICICV 2021 (2021), pp. 1355-1359, 10.1109/ICICV50876.2021.9388444

Bhattacharya M., Roy A., Pal J.2021. Smart Irrigation System using Internet of Things, Springer Singapore (2021), 10.1007/978-981-15-6198-6_11

Boursianis A.D., Papadopoulou M.S., Gotsis A., Wan S., Sarigiannidis P., Nikolaidis S., Goudos S.K. 2021. Smart irrigation system for precision agriculture - the AREThOU5A IoT platform IEEE Sens. J., 21 (2021), pp. 17539-17547, 10.1109/JSEN.2020.3033526

Jha K., Doshi A., Patel P. 2018. Intelligent irrigation system using artificial intelligence and machine learning: A comprehensive review Int. J. Adv. Res., 6 (2018), pp. 1493-1502, 10.21474/ijar01/7959

Karar M.E., Alotaibi F., Al Rasheed A., Reyad O.2021. A pilot study of smart agricultural irrigation using unmanned aerial vehicles and IoT-based cloud system Inf. Sci. Lett., 10 (2021), pp. 131-140, 10.18576/isl/100115

Loua L.R., Budihardjo M.A., Sudarno S.2021. A review on a machine learning approach of an intelligent irrigation monitoring system with edge computing and the internet of things. IOP Conf. Ser. Earth Environ. Sci., 896 (2021), pp. 0-9, 10.1088/1755-1315/896/1/012029

Munir M.S., Bajwa I.S., Ashraf A., Anwar W., Rashid R.2021. Intelligent and smart irrigation system using edge computing and IoT Complexity, 2021 (2021), 10.1155/2021/6691571

Oruma S.O., Misra S., Fernandez-Sanz L.2021. Agriculture 4.0: An implementation framework for food security attainment in Nigeria's post-Covid-19 era IEEE Access, 9 (2021), pp. 83592-83627, 10.1109/ACCESS.2021.3086453

Raouhi E.M., Lachgar M., Hrimech H., Kartit A.2022. Optimization techniques in deep convolutional neuronal networks applied to olive diseases classification Artif. Intell. Agric., 6 (2022), pp. 77-89, 10.1016/j.aiia.2022.06.001

Rawal S.2019. IOT based smart irrigation system IOT based smart irrigation system (2019), 10.5120/ijca2017913001

Saiz-Rubio V., Rovira-Más F.2020. From smart farming towards agriculture 5.0: A review on crop data management; Agronomy, 10 (2020), 10.3390/agronomy10020207

Street F. 2011. Advancing Institutional Transformation for Minority Women in a Cademia (2011), pp. 20-25 October

Yang X., et al. 2021. A survey on smart agriculture: Development modes, technologies, and security and privacy challenges IEEE/CAA J. Autom. Sin., 8 (2) (2021), pp. 273-302, 10.1109/JAS.2020.1003536

Multiple Choice Questions

1. What is one of the primary aims of Climate-smart Irrigation (CSI)?
   a) To increase crop variety
   b) To enhance the productivity of irrigated cropping systems
   c) To decrease rainfall patterns
   d) To build more irrigation infrastructure
2. How does Climate-smart Irrigation aim to reduce greenhouse gas emissions?
   a) By promoting traditional farming methods
   b) By reducing crop diversity
   c) By prioritizing water conservation methods in irrigation
   d) By increasing the demand for water

3. Which of the following is NOT considered a component of climate-smart irrigation?
   a) Irrigation hardware
   b) Irrigation software
   c) Improved weather predictions
   d) Market access and farmer know-how
4. What factor is crucial for effective climate change adaptation in irrigation practices?
   a) Using the same irrigation practices across all regions
   b) Understanding local vulnerabilities and risks
   c) Focusing solely on hardware improvements
   d) Ignoring socio-economic impacts
5. In the CSI approach, what is recommended as a first phase for improving irrigation systems?
   a) Increasing the size of farm fields
   b) Promoting international irrigation practices
   c) Enhancing the productivity and profitability of existing systems
   d) Introducing restrictions on water usage
6. What role do moisture sensors play in Automation Irrigation Systems?
   a) Measure crop yield
   b) Determine moisture levels to regulate irrigation
   c) Increase soil salinity
   d) Control pest populations
7. Which of the following is a method to increase the sustainability of crop production in CSI?
   a) Relying solely on rainfall for irrigation
   b) Using high water consumption crops
   c) Implementing conservation agriculture techniques
   d) Expanding irrigated areas without limits
8. What is the purpose of using anti-evaporants in water conservation?
   a) To increase soil acidity
   b) To reduce soil moisture losses due to evaporation
   c) To improve crop nutrition
   d) To control pests in the soil

9. Which of the following describes one mode of the robot-assisted irrigation system mentioned in the text?
   a) It requires continuous manual operation
   b) It solely operates on a timer
   c) It can function in automatic and manual modes
   d) It ignores soil moisture conditions
10. Why is the understanding of different perspectives on irrigation practices important?
   a) To establish control over farmers
   b) To streamline policies without input
   c) To reconcile different interests and optimize practices
   d) To promote a single viewpoint in irrigation management

Answer Key

1	b	2	c	3	c	4	b	5	c	6	b	7	c
8	b	9	c	10	c								

Keywords

Soil and Water Properties

- **Aeration:** Essential for root respiration and microbial activity, referring to the exchange of gases in the soil.
- **Available Water Capacity (AWC):** The volume of soil water that plants can readily access, held between field capacity and wilting point.
- **Clay:** A fine soil particle (less than 0.002mm) characterized by high water-holding capacity and fertility but poor drainage.
- **Cohesion:** The attraction between water molecules, contributing to water retention within soil pores.
- **Electrical Conductivity (EC):** A measure of the salt content in water. High EC indicates a significant presence of dissolved salts.
- **Field Capacity:** The soil moisture content remaining after excess gravitational water has drained, representing the upper limit of plant-available water.
- **Hydraulic Conductivity:** The ease with which soil transmits water, influenced by its porosity and texture. This is crucial for infiltration and subsurface water movement.
- **Infiltration:** The downward movement of water from the soil surface into the soil, affected by soil texture, structure, and existing moisture.
- **Infiltration Rate:** The speed at which water enters the soil (e.g., in cm/hr), generally higher in sandy soils and lower in clays.
- **Salinity:** The overall concentration of salts in water or soil, a critical factor for irrigation water quality.
- **Silt:** Soil particles ranging from 0.002mm to 0.05mm, possessing good water-holding capabilities and intermediate properties between sand and clay.
- **Sodicity:** Refers to water or soil with a high proportion of sodium relative to calcium, which can lead to clay dispersion and poor soil structure.

- **Sodium Adsorption Ratio (SAR):** An index used to assess the sodium hazard in irrigation water, calculated as the ratio of sodium to calcium and magnesium concentrations.
- **Water Quality:** The characteristics of irrigation water, including salinity, pH, and the presence of pollutants, which determine its suitability for crop growth.
- **Wilting Point (Permanent Wilting Point):** The soil moisture level below which plants can no longer extract water, typically around -15 bar suction.

Water Resources and Management

- **Agricultural Water:** Water used for crop production and related farm activities, such as irrigation and pesticide application.
- **Brackish Water:** Water with a moderate salt content, falling between freshwater and seawater. It often requires treatment for effective irrigation.
- **Drainage:** The removal of excess water from the soil to prevent waterlogging. It can also refer to a drainage basin, the area that contributes water to a stream.
- **Grey Water:** Lightly used domestic wastewater (from sinks, showers, etc.) that can be reused for irrigation after appropriate treatment.
- **Groundwater:** Water found beneath the Earth's surface in aquifers, often tapped by wells to supplement surface water for irrigation.
- **Hydrological Cycle:** The continuous movement of water on, above, and below the surface of the Earth, involving evaporation, condensation, precipitation, and runoff.
- **Industrial Water:** Water utilized in industrial processes; it is generally not suitable for irrigation.
- **Irrigation:** The artificial application of water to land to support plant growth.
- **Irrigation Water:** Water specifically designated and allocated for irrigating crops, distinct from potable or industrial water.
- **Rainfall:** Natural precipitation, the primary source of water for many regions, often supplemented by irrigation during dry periods.
- **Rainwater Harvesting:** The practice of collecting and storing rainfall or runoff, often in tanks, for later use in irrigation.
- **Reservoir:** Artificial or natural bodies of water (like dams or lakes) used to store surface water for purposes such as supplying irrigation canals.

- **River Water:** Surface water sourced from rivers and streams, typically diverted into canals for irrigation.
- **Surface Water:** Water bodies found on the Earth's surface, such as rivers, lakes, and canals, as opposed to groundwater.
- **Water Resources:** The total amount of available water, encompassing both surface and groundwater, for irrigation and other human needs.
- **Water Table:** The upper level of the saturated zone in the ground, the depth of which influences well yield and irrigation planning.
- **Wastewater:** Used water from domestic or industrial sources that can be reused for irrigation following treatment (e.g., treated sewage or grey water).

Irrigation Methods and Technologies

- **Basin Irrigation:** A surface irrigation technique where land is divided into basins and then flooded. This method demands well-leveled fields.
- **Border Strip Irrigation:** A surface irrigation method employing long, parallel strips (border strips) outlined by bunds to guide water flow, commonly used on level terrain.
- **Drip Irrigation:** A highly efficient micro-irrigation system that delivers water precisely, drop by drop, to each plant through emitters.
- **Flood Irrigation:** A surface irrigation method where the entire field is inundated with water, often used for crops like paddy rice.
- **Furrow Irrigation:** A surface irrigation technique that uses small, parallel channels (furrows) between crop rows to convey water across the field.
- **Gravity Irrigation:** Any irrigation system that relies on gravity to distribute water, in contrast to pressurized methods.
- **Micro-irrigation:** Low-pressure irrigation systems, primarily drip and micro-sprinkler types, designed to conserve water through targeted application.
- **Raingun Irrigation:** A high-pressure sprinkler system, often referred to as a "motorized cannon," capable of throwing water over long distances, simulating artificial rainfall.
- **Surface Irrigation:** A broad category encompassing all gravity-driven methods that apply water directly onto the ground surface, including basins, furrows, and border strips.

Crop Water Use and Irrigation Performance

- **Class A Pan Evaporimeter:** A standardized device used to measure reference evaporation, which aids in irrigation scheduling.
- **Consumptive Use:** The total amount of water consumed by plants through evapotranspiration (the combined process of evaporation from soil and transpiration from plants). This indicates the crop's water requirements.
- **Conjunctive Use:** A water management strategy that involves the combined utilization of surface water and groundwater to maximize water availability for irrigation.
- **Conveyance Efficiency / Water Conveyance Efficiency:** The ratio of water delivered to the field compared to the total amount withdrawn from the source, accounting for losses during transport.
- **Crop Coefficient (Kc):** A factor used to adjust reference evapotranspiration (ETo) to reflect the specific water needs of a particular crop ($ET\_crop = ETo \times Kc$).
- **Crop Factor (Kp):** A coefficient similar to Kc, but based on Class A pan evaporation (Epan) rather than reference evapotranspiration.
- **Distribution Uniformity / Uniformity Coefficient:** A measure indicating how evenly irrigation water is applied across a field, significantly impacting overall efficiency.
- **Effective Rainfall:** The portion of rainfall that is beneficial for crop water needs, excluding losses from runoff and deep percolation.
- **Evaporation:** The process by which water transforms into vapor and moves from the soil or open water surfaces into the atmosphere.
- **Evapotranspiration (ET):** The combined loss of water from the soil surface through evaporation and from plants through transpiration. It is a key indicator of crop water requirements.
- **Fertigation:** The application of soluble fertilizers through an irrigation system.
- **Irrigation Efficiency:** The ratio of water beneficially used by crops to the total amount of water withdrawn or delivered. Systems like sprinklers and drip irrigation generally achieve higher efficiencies.
- **Irrigation Scheduling:** The process of planning the timing and amount of irrigation based on factors like crop growth stage, soil moisture levels, and climatic conditions.

- **Rainfall:** Natural precipitation; it's the primary source of water and often supplemented by irrigation during dry seasons.
- **Root Zone:** The layer of soil occupied by crop roots, from which water is absorbed. The depth of this zone varies depending on the specific crop.
- **Runoff / Surface Runoff:** Water that flows off the land surface during irrigation or rainfall, representing a water loss unless it is captured.
- **Transpiration:** The process by which plants release water vapor into the atmosphere, primarily through their leaves. This is a component of a crop's total evapotranspiration.
- **Virtual Water:** The amount of water that is indirectly used in the production of goods and services, such as the volume of water needed to produce a unit of a crop.
- **Water Balance:** An accounting of water inputs (rainfall, irrigation) against outputs (evapotranspiration, runoff, deep percolation) in an agricultural system.
- **Water Requirement (Crop):** The total amount of water a crop needs, typically measured as its evapotranspiration over the growing period.
- **Water Use Efficiency (WUE):** A measure of how effectively water is used to produce crop yield or biomass, calculated as yield per unit of water consumed. It serves as an indicator of irrigation performance.

Water Quality Parameters

- **Residual Sodium Carbonate (RSC):** A measure of the carbonate and bicarbonate hazard in irrigation water, which can influence the buildup of sodium in the soil.
- **Saline Water:** Irrigation water containing high levels of dissolved salts (typically $EC > 3$ dS/m), which can be detrimental to many crops.
- **Sodium Adsorption Ratio (SAR):** An index used to assess the sodium hazard in irrigation water, calculated as the ratio of Na to Ca+Mg concentrations. This helps in determining water suitability for irrigation.

Index

Infiltration 17, 49, 52, 58, 63, 68, 69, 71, 72, 73, 74, 79, 83, 92, 124, 140, 163, 165, 166, 167, 171, 172, 174, 176, 179, 183, 191, 214, 244, 245, 246, 249, 268, 274, 275, 281, 282, 293, 375, 376, 389